Fatty Acids: A Human Health Perspective

Fatty Acids: A Human Health Perspective

Editor: Victoria Cook

www.callistoreference.com

Callisto Reference,
118-35 Queens Blvd., Suite 400,
Forest Hills, NY 11375, USA

Visit us on the World Wide Web at:
www.callistoreference.com

ISBN: 978-1-64116-769-7 (Hardback)

Cataloging-in-Publication Data

Fatty acids : a human health perspective / Victoria Cook.
    p. cm.
Includes bibliographical references and index.
ISBN 978-1-64116-769-7
1. Fatty acids. 2. Carboxylic acids. 3. Biochemistry. I. Cook, Victoria.
QP752.F35 F38 2023
612.397--dc23

Table of Contents

Preface

This book was inspired by the evolution of our times; to answer the curiosity of inquisitive minds. Many developments have occurred across the globe in the recent past which has transformed the progress in the field.

Fatty acids are carboxylic acids that are structural components of fats, oils, and all other categories of lipids, except steroids. These acids are the building blocks of fat in human bodies. Fatty acids are classified as saturated or unsaturated. Some common fatty acids found in natural fats include lauric acid, palmitic acid, stearic acid, and linoleic acid. Fatty acids play a prominent role in lipid metabolism, human diet, and human physiology. Omega-3 fatty acids, also called Omega-3 oils, ω-3 fatty acids or n-3 fatty acids, play an important role in body as they are the precursor of several metabolites that help in biochemical signaling, and in controlling cardiovascular diseases. The three forms of omega-3 fatty acids which are involved in human physiology include α-linolenic acid (ALA), eicosapentaenoic acid (EPA), and docosahexaenoic acid (DHA). The book presents researches and studies performed by experts across the globe on various topics related to the effects of fatty acids on human health. It is appropriate for students seeking detailed information in this area of study as well as for experts.

This book was developed from a mere concept to drafts to chapters and finally compiled together as a complete text to benefit the readers across all nations. To ensure the quality of the content we instilled two significant steps in our procedure. The first was to appoint an editorial team that would verify the data and statistics provided in the book and also select the most appropriate and valuable contributions from the plentiful contributions we received from authors worldwide. The next step was to appoint an expert of the topic as the Editor-in-Chief, who would head the project and finally make the necessary amendments and modifications to make the text reader-friendly. I was then commissioned to examine all the material to present the topics in the most comprehensible and productive format.

I would like to take this opportunity to thank all the contributing authors who were supportive enough to contribute their time and knowledge to this project. I also wish to convey my regards to my family who have been extremely supportive during the entire project.

Editor

Nutritional Indices for Assessing Fatty Acids: A Mini-Review

Jiapeng Chen [1] and Hongbing Liu [1,2,*]

[1] Key Laboratory of Marine Drugs, Chinese Ministry of Education, School of Medicine and Pharmacy, Ocean University of China, Qingdao 266003, China; 21180831069@stu.ouc.edu.cn

[2] Laboratory for Marine Drugs and Bioproducts, Pilot National Laboratory for Marine Science and Technology (Qingdao), Qingdao 266237, China

* Correspondence: liuhongb@ouc.edu.cn.

Abstract: Dietary fats are generally fatty acids that may play positive or negative roles in the prevention and treatment of diseases. In nature, fatty acids occur in the form of mixtures of saturated fatty acid (SFA), monounsaturated fatty acid (MUFA), and polyunsaturated fatty acid (PUFA), so their nutritional and/or medicinal values must be determined. Herein, we do not consider the classic indices, such as $\sum$SFA, $\sum$MUFA, $\sum$PUFA, $\sum$n-6 PUFA, $\sum$n-3 PUFA, and n-6 PUFA/n-3 PUFA; instead, we summarize and review the definitions, implications, and applications of indices used in recent years, including the PUFA/SFA, index of atherogenicity (IA), the index of thrombogenicity (IT), the hypocholesterolemic/hypercholesterolemic ratio (HH), the health-promoting index (HPI), the unsaturation index (UI), the sum of eicosapentaenoic acid and docosahexaenoic acid (EPA + DHA), fish lipid quality/flesh lipid quality (FLQ), the linoleic acid/α-linolenic acid (LA/ALA) ratio, and *trans* fatty acid (TFA). Of these nutritional indices, IA and IT are the most commonly used to assess the composition of fatty acids as they outline significant implications and provide clear evidence. EPA + DHA is commonly used to assess the nutritional quality of marine animal products. All indices have their advantages and disadvantages; hence, a rational choice of which to use is critical.

Keywords: fatty acids; nutritional indices; human health

1. Introduction

Fatty acids (FAs) are organic acids with at least one carboxyl (–C(=O)OH, –COOH, or –CO$_2$H) group and a long carbon chain whose links can be double bonds, as in unsaturated fatty acids, or single bonds, as in saturated fatty acids. FAs are generally derived from triglycerides and phospholipids, and are the main components of dietary fats. Most naturally occurring FAs have an unbranched chain of an even number (4–28) of carbon atoms. According to the number of double bonds, the FA catalogue includes saturated fatty acids (SFAs), monounsaturated fatty acid (MUFA), and polyunsaturated fatty acid (PUFA).

FAs are distributed to cells where they serve as fuel for muscular contraction and general metabolism. As biological compounds, FAs play critical roles in human metabolism, health, and disease. Epidemiological studies and clinical trials showed that fatty acids are associated with cardiovascular diseases [1–5], neurological diseases [6–9], non-alcoholic fatty liver disease [10–13], allergic diseases [14–16], and so on. Evidence from metabolomics experiments indicates that they participate in the metabolic pathways of related diseases [8,17–26]. For example, the free FA profile was found to be altered in both leukemia and pre-leukemic diseases, particularly C14:0, C16:0, and C18:0 [26].

FAs play positive or negative roles in the prevention and treatment of diseases. For example, SFAs may increase the risk of developing multiple sclerosis (MS) as well as disease progression, whereas PUFAs may have beneficial effects in MS patients [7]. As another example, some essential FA metabolites may exert health effects such as anti-inflammatory and neuroprotection effects, but they can also produce negative effects such as inflammation, necrosis promoters, and atherosclerosis. In general, FAs are obtained from various dietary sources that possess characteristic FA composition and consequently influence health outcome. From this perspective, the FA composition should be assessed to determine their nutritional and/or medicinal value, especially in fatty-acid-rich foods, food supplements, and herb-based medicines.

In this mini-review, we collated the literature related to fatty acid profile analysis that was published in recent decades since 2000 to understand the implications and applications of various nutritional indices. We did not consider the classic indices such as $\sum$SFA, $\sum$MUFA, $\sum$PUFA, $\sum$n-6 PUFA, $\sum$n-3 PUFA, and n-6 PUFA/n-3 PUFA. The present review may help researchers to evaluate the nutritional value of fatty acids and to explore their potential usage in disease prevention and treatment. It may also help newcomers to the field of fatty acid profile analysis to quickly and accurately select appropriate indices.

2. Nutritional Indices

In this review, we screened articles and summarized the nutritional indices. The results are shown in Table 1.

Table 1. Summary of nutritional indices.

No.	Index	Full Name	Calculation Formula	Application
1	PUFA/SFA	Polyunsaturated fatty acid/saturated fatty acid ratio	$\sum$PUFA/$\sum$SFA	Seaweeds [27–29], crops [30,31], plant oil [32,33], shellfish [34], fish [34–40], meat [41–53], and dairy products [54–57]
2	IA	Index of atherogenicity	[C12:0 + (4 × C14:0) + C16:0]/$\sum$UFA	Seaweeds [27–29,58,59], crops [30,31,60,61], plant oil [33,62], shellfish [63], shrimp [64], fish [36–39,65–73], meat [41–43,48–50,52,53,74–77], and dairy products [54–56,78–89]
3	IT	Index of thrombogenicity	(C14:0 + C16:0 + C18:0)/[(0.5 ×$\sum$MUFA) + (0.5 ×$\sum$n-6 PUFA) + (3 ×$\sum$n-3 PUFA) + (n-3/n-6)]	Seaweeds [27–29,58,59], crops [30,31,60,61], plant oil [62], shellfish [63], shrimp [64], fish [36–39,65–68,70–73], meat [43,48,50,52,53,75,77], and dairy products [54,55,78,80,86–89]
4	HH	Hypocholesterolemic /hypercholesterolemic ratio	(cis-C18:1 + $\sum$PUFA)/(C12:0 + C14:0 + C16:0)	Seaweeds [90], plant oil [62], shellfish [34], fish [34,36–39,72], meat [46–48,52,77,91], and dairy products [54,55,78,86,87]
5	HPI	Health-promoting index	$\sum$UFA/[C12:0+(4 × C14:0) + C16:0]	Milk [92–94] and cheese [57,94,95]
6	UI	Unsaturation index	1 × (% monoenoics) + 2 × (% dienoics) + 3 × (% trienoics) + 4 × (% tetraenoics) + 5 × (% pentaenoics) + 6 × (% hexaenoics)	Seaweeds [27–29,59,96–98], crops [61,99,100], meat [44,101], and milk [102]
7	EPA + DHA	Sum of eicosapentaenoic acid and docosahexaenoic acid	C22:6 n-3 + C20:5 n-3	Shellfish [34] and fish [34,36,37,40,68,103–106]
8	FLQ	Fish lipid quality/flesh Lipid quality	100 × (C22:6 n-3 + C20:5 n-3)/$\sum$FA	Fish [65,66,73,107,108]
9	LA/ALA	Linoleic acid /α-linolenic acid ratio	C18:2 n-6/C18:3 n-3	Lamb [43] and milk [55,109]
10	TFA	*Trans* fatty acid	$\sum$TFA	Seaweeds [90], plant oil [32,33,110], fish [35], lamb [45], and milk [78]

2.1. Polyunsaturated Fatty Acid/Saturated Fatty Acid (PUFA/SFA)

PUFA/SFA is an index normally used to assess the impact of diet on cardiovascular health (CVH). It hypothesizes that all PUFAs in the diet can depress low-density lipoprotein cholesterol (LDL-C) and lower levels of serum cholesterol, whereas all SFAs contribute to high levels of serum cholesterol. Thus, the higher this ratio, the more positive the effect.

All that was missing was MUFA. According to Dietschy's study on dietary FAs and their regulation of plasma LDL-C concentrations in 1998, C18:1 n-9 *cis* (oleic acid), the most common MUFA in dietary food increases the activity of low-density lipoprotein receptors (LDLRs) and decreases the cholesterol concentration in serum [111]. Not all molecular species of SFAs contribute equally to serum cholesterol. C12:0, C14:0, and C16:0 can increase the cholesterol concentration in serum by inhibiting the activity of

LDLRs; C4:0, C6:0, C8:0, and C10:0 were rapidly oxidized to acetyl-CoA in the liver and could not affect the activity of LDLRs, and C18:0 appeared to be biologically neutral and have no effect on circulating LDL-C levels [111].

Notably, not all of the main classes of PUFA positively affect the prevention of cardiovascular disease (CVD). Short-term supplementation with docosahexaenoic acid (DHA)-rich fish oil may modulate the activity of peroxisome proliferators-activated receptor-gamma (PPAR-γ) to protect the cardiovascular system from the unhealthy effects of atherosclerotic lesions [4]. Recent clinical trials support the view that supplementation with eicosapentaenoic acid (EPA) can reduce plasma triglyceride (TG) levels and activate anti-inflammatory, anti-thrombotic, and other mechanisms to prevent atherosclerosis (AS) [2]. However, a narrative review that collated the available data showed that dietary intake of linoleic acid (LA, C18:2 n-6) is inversely correlated with CVD; however, further research is needed to clarify the underlying mechanisms [3].

PUFA/SFA is the most commonly used index for evaluating the nutritional value of dietary foods such as of seaweed (0.42–2.12, except for *Gracilaria changii*), meat (0.11–2.042), fish (0.50–1.62), shellfish (0.20–2.10), and dietary products (0.02–0.175). Chan and Matanjun determined the FA profiles of red seaweed *Gracilaria changii* a mangrove area of Malaysia, and used PUFA/SFA to assess the nutritional quality, finding a value of 6.96 ± 0.98 [28]. PUFA/SFA of chicken is in the range of 0.308 to 2.042 for different dietary treatments [48]. Fernandes et al. compared the FA profile of four species of Brazilian fish, and they used the PUFA/SFA as one of the nutritional quality indices, reporting values between 1.09 to 1.47 [36]. Detailed information about the literature related to PUFA/SFA is shown in Table 2.

Table 2. Application of PUFA/SFA in fatty acid evaluation *.

Materials		PUFA/SFA Value	Reference
Red seaweed	*Amphiora anceps*	0.42	[27]
	Kappaphycus alvarezii	0.57	[27]
	Gelidiella acerosa	0.84	[27]
	Gelidium micropterum	0.30	[27]
	Gracilaria changii	6.96 ± 0.98	[28]
	Gracilaria corticata	2.12	[27]
	Gracilaria dura	1.89	[27]
	Gracilaria debilis	1.17	[27]
	Gracilaria fergusonii	0.58	[27]
	Gracilaria salicornia	0.14	[27]
	Laurencia cruciata	0.79	[27]
	Sarconema filiforme	1.71	[27]
Brown seaweed	*Cystoseira indica*	1.17	[27]
	Padina tetrastromatica	0.85	[27]
	Sargassum fusiforme	0.67 ± 0.31	[29]
	Sargassum horneri	0.56 ± 0.06	[29]
	Sargassum pallidum	0.20 ± 0.09	[29]
	Sargassum swartzii	1.15	[27]
	Sargassum tenerrimum	1.18	[27]
	Sargassum thunbergii	0.39 ± 0.05	[29]
	Spatoglossum asperum	1.38	[27]
Green seaweed	*Caulerpa racemosa*	0.44	[27]
	Caulerpa scalpeliformis	0.88	[27]
	Caulerpa veravalnensis	0.73	[27]
	Ulva fasciata	0.42	[27]
	Ulva reticulata	0.23	[27]
	Ulva rigida	0.33	[27]
Crops	*Cyamopsis tetragonolobaL.*	1.71	[31]
	Lupinus albus	1.53–1.97	[30]
Plant oil	Palm stearin	0.13	[32]
	Sunflower oil	4.75–4.94	[32,33]

Table 2. *Cont.*

Materials		PUFA/SFA Value	Reference
Shellfish	*Cancer edwardsi*	2.10	[34]
	Cervimunida johni	1.81	[34]
	Concholepas concholepas	1.16	[34]
	Heterocarpus reedi	1.47	[34]
	Loxechinus albus	0.20	[34]
	Mesodesma donacium	1.34	[34]
	Pleuroncodes monodon	1.68	[34]
	Pyura chilensis	1.31	[34]
	Venus antiqua	1.06	[34]
Fish	*Carassius gibelio*	1.62–1.70	[35]
	Cilus gilberti	1.15	[34]
	Genypterus chilensis	1.60	[34]
	Hemiramphus brasiliensis	1.09	[36]
	Hyporhamphus unifasciatus	1.11	[36]
	Kutum roach	1.02–1.79	[37]
	Lagocephalus guentheri	1.3	[38]
	Merluccius gayi	1.52	[34]
	Opisthonema oglinum	1.47	[36]
	Orechromis niloticus	0.51–0.56	[39]
	Pinguipes chilensis	0.80	[34]
	Scomber japonicus	0.92	[34]
	Scomberomorus cavalla	1.18	[36]
	Seriola lalandi	0.92	[34]
	Seriolella violacea	0.95	[34]
	Trachinotus carolinus	0.5–1.1	[40]
	Trachurus murphyi	0.95	[34]
Meat	Chicken (Caribro Vishal)	0.308–2.042	[50]
	Chicken (purchased from a hatchery and poultry farm)	0.926–0.945	[48]
	Pig (DanBred × PIC terminal line)	0.46–0.48	[49]
	Pig (Pietrain × (Duroc × Landrace))	0.85–1.29	[44]
	Lamb (Barbarine lamb)	0.13–0.37	[43,45]
	Steer (Blonded Aquitaine steer)	0.29–0.58	[42]
	Calve (75% Charolais breeds)	0.13–0.34	[51]
	Cattle (Nellore cattle)	0.11–0.20	[46]
	Yak (*Phoephagus grunniens*)	0.37–0.55	[41]
	Foal (Galician Mountain × Hispano-Bretón)	0.44–1.06	[52]
	Spanish dry-cured ham	0.19–0.30	[47]
	Bologna sausages	0.27–1.17	[53]
Dairy products	Cheese of Comisana ewe	0.086–0.173	[57]
	Milk of Chios sheep	0.06–0.08	[54]
	Milk of Karagouniko sheep	0.06–0.09	[54]
	Milk of Turcana dairy ewe	0.106–0.175	[55]
	Milk of Friesian × Jersey cow	0.02–0.04	[56]

PUFA/SFA: polyunsaturated fatty acid/saturated fatty acid ratio; * literature from 2000 until April/2020.

2.2. Index of Atherogenicity (IA)

The index of atherogenicity (IA) was developed by Ulbritcht and Southgate in 1991, and characterizes the atherogenic potential of FA [112]. As the PUFA/SFA ratio is too general and unsuitable for assessing the atherogenicity of foods, Ulbritcht and Southgate proposed a new index, IA, based on PUFA/SFA considering the available evidence, and then checked whether the resulting values were in accordance. The formula for calculating IA is:

$$IA = [C12:0 + (4 \times C14:0) + C16:0]/\Sigma UFA \tag{1}$$

The IA indicates the relationship between the sum of SFAs and the sum of unsaturated fatty acids (UFAs). The main classes of SFAs, which include C12:0, C14:0, and C16:0, with the exception of C18:0, are considered pro-atherogenic (they favor the adhesion of lipids to cells of the circulatory and immunological systems) [67,68,113]. UFAs are considered to be anti-atherogenic as they inhibit the accumulation of plaque and reduce the levels of phospholipids, cholesterol, and esterified fatty acids [68,113]. Therefore, the consumption of foods or products with a lower IA can reduce the levels of total cholesterol and LDL-C in human blood plasma [85].

Although the IA is more reasonable than the simple PUFA/SFA ratio for assessing the degree of atherogenicity, there are still some imperfections in the proposed IA formula, which were pointed out by Ulbritcht and Southgate. First, stearic acid (C18:0) should appear in the denominator if sufficient evidence shows that it can reduce the level of LDL-C in human blood plasma in the future. Second, not all PUFAs should be weighed equally. Third, the impact of *trans* fatty acids was not considered due to conflicting evidence [112].

The IA has been used widely for evaluating seaweeds, crops, meat, fish, dairy products, etc. Nantapo et al. analyzed the fatty acid composition of milk during different stages of lactation and found that milk with lower IA is important, and the IA ranges from 4.08 to 5.13 in different stages of lactation [56]. Akintola investigated the techniques of smoking and sun drying to understand the nutritional quality of southern pink shrimp (*Penaeus notialis*) using the IA as an index, and reported values of 0.71 to 0.82 [64].

Detailed information about the literature related to the IA is shown in Table 3. For seaweeds, the species may be the main factor influencing the IA value, which ranges from 0.03 to 3.58. The value ranges from 0.084 to 0.55 for crops, 0.21 to 1.41 for fish, and 0.165 to 1.32 for meat. For dairy products, the value ranges from 1.42 to 5.13. For ruminants, dietary treatment is the main factor influencing the IA.

Table 3. Application of IA in fatty acid evaluation *.

Materials		IA Value	Reference
Red seaweed	*Amphiora anceps*	1.52	[27]
	Ceramium virgatum	0.37 ± 0.027	[59]
	Corallina officinalis	0.48 ± 0.039	[59]
	Gelidiella acerosa	0.80	[27]
	Gelidium micropterum	1.61	[27]
	Gracilaria changii	0.03 ± 0.003	[28]
	Gracilaria corticata	0.38	[27]
	Gracilaria debilis	0.69	[27]
	Gracilaria dura	0.45	[27]
	Gracilaria fergusonii	1.34	[27]
	Gracilaria salicornia	2.87	[27]
	Hymenena sp.	3.58	[59]
	Kappaphycus alvarezii	0.77	[27]
	Laurencia cruciata	0.84	[27]
	Lomentaria clavellosa	3.06 ± 0.611	[59]
	Polysiphonia sp.	1.35 ± 0.206	[59]
	Sarconema filiforme	0.49	[27]
Brown seaweed	*Cystoseira indica*	0.66	[27]
	Dictyota dichotoma	0.29 ± 0.041	[59]
	Laminaria ochroleuca	1.18–1.57	[58]
	Leathesia difformis	0.48 ± 0.021	[59]
	Myriogloea major	0.21 ± 0.019	[59]
	Padina tetrastromatica	0.81	[27]
	Sargassum fusiforme	0.94 ± 0.28	[29]
	Sargassum horneri	1.06 ± 0.06	[29]
	Sargassum pallidum	1.99 ± 0.45	[29]
	Sargassum swartzii	0.61	[27]
	Sargassum tenerrimum	0.66	[27]
	Sargassum thunbergii	1.16 ± 0.10	[29]
	Spatoglossum asperum	0.53	[27]
	Undaria pinnatifida	0.17–0.35	[59]

6

Table 3. *Cont.*

Materials		IA Value	Reference
Green seaweed	*Caulerpa racemosa*	1.61	[27]
	Caulerpa scalpeliformis	0.86	[27]
	Caulerpa veravalnensis	1.17	[27]
	Cladophora falklandica	0.50 ± 0.062	[59]
	Codium decorticatum	0.22 ± 0.002	[59]
	Codium fragile	0.29 ± 0.020	[59]
	Codium vermilara	0.40 ± 0.086	[59]
	Ulva fasciata	1.37	[27]
	Ulva reticulata	1.54	[27]
	Ulva rigida	1.22	[27]
	Ulva sp.1	0.20 ± 0.055	[59]
	Ulva sp.2	0.08 ± 0.004	[59]
Crops	Cumin (*Cuminum cyminum*)	0.46–0.53	[61] [a]
	Guar seed (*Cyamopsis tetragonoloba*)	0.22	[31]
	White lupine (*Lupinus albus*)	0.084–0.107	[30]
	Scabiosa stellata	0.55	[60]
Plant oil	Camelina oil (*Camelina sativa*)	0.05–0.07	[62]
	Sunflower oil	0.09–0.11	[33]
Shellfish	*Chlamys farreri*	0.31–0.37	[63]
	Patinopecten yessoensis	0.29–0.35	[63]
Shrimp	*Penaeus notialis*	0.71–0.82	[64]
Fish	*Abramis brama*	0.37–0.42	[65,66]
	Clupea harengus	0.70 ± 0.10	[66]
	Cynoscion parvipinnis	1.07–1.16	[67]
	Cyprinus carpio	0.36 ± 0.03	[66]
	Dicentrarchus labrax	0.40–0.42	[68]
	Esox lucius	0.43	[65]
	Hemiramphus brasiliensis	0.26	[36]
	Hyporhamphus unifasciatus	0.26	[36]
	Kutum roach	0.58–1.41	[37]
	Lagocephalus guentheri	0.43	[38]
	Leuciscus idus	0.36 ± 0.02	[66]
	Limousin steers	0.70–1.14	[69]
	Micropterus salmoides	0.29–0.68	[70]
	Mugil cephalus	0.91–1.22	[71]
	Oncorhynchus mykiss	0.33 ± 0.01	[66]
	Opisthonema oglinum	0.60	[36]
	Oreochromis niloticus	0.55–0.60	[39]
	Perca fluviatilis	0.37–0.44	[65,66]
	Platichthys flesus	0.41 ± 0.03	[66]
	Rutilus rutilus	0.40	[65]
	Salmo trutta	0.64–0.72	[72]
	Scomberomorus cavalla	0.48	[36]
	Sparus aurata	0.21–0.29	[73]
Meat	Chicken (Caribro Vishal)	0.165–0.634	[50]
	Chicken (purchased from a hatchery and poultry farm)	0.372–0.390	[48]
	Rabbit (Curcuma longa)	0.55–0.69	[75]
	Pig (DanBred × PIC terminal line)	0.27–0.31	[49]
	Lamb (Barbarine lamb)	0.49–0.52	[43]
	Lamb (Gentile di Puglia × Sopravissana)	0.99–1.32	[76]
	Lamb (Ile de France × Pagliarola)	0.71–1.06	[76]
	Lamb (Iranian fat-tailed breed)	0.53–0.77	[74]
	Heifer (Limousin heifer)	0.50–0.57	[77]
	Steer (Blonded Aquitaine steer)	0.51–0.63	[42]
	Yak (Phoephagus grunniens)	0.37–0.43	[41]
	Foal (Galician Mountain × Hispano-Bretón)	0.59–0.62	[52]
	Bologna sausages	0.33–0.60	[53]

Table 3. *Cont.*

Materials		IA Value	Reference
Dairy products	Cheese of Churra ewe	1.61–3.61	[79]
	Cheese of Holstein cow	2.38–3.72	[88]
	Cheese of Italian Friesian and Italian Red Pied cattle (Caciocavallo cheese)	2.43–2.94	[84]
	Curd of cow (Middle Rhodopes)	1.94–5.02	[78]
	Milk of Anglo-Nubian goat	1.89–2.48	[81]
	Milk of goat (market of Sardinia)	2.27–2.91	[89]
	Milk of Nubian goat	1.91–2.32	[82]
	Milk of Saanen goat	2.77 ± 0.08	[85]
	Milk of Swedish Landrace goat	2.47 ± 0.07	[85]
	Milk of Chios sheep	2.00–2.72	[54]
	Milk of Karagouniko sheep	1.76–2.57	[54]
	Milk of Churra ewe	1.71–3.39	[79]
	Milk of Lacaune ewe	1.94–2.53	[80]
	Milk of Turcana dairy ewe	1.42–1.95	[55]
	Milk of cow (Middle Rhodopes)	1.88–4.18	[78]
	Milk of Friesian × Jersey cow	4.08–5.13	[56]
	Milk of Holstein cow	1.83–2.63	[88]
	Milk of Holstein–Friesian cow	1.60–3.79	[83,109] [a]
	Milk of indigenous Indian cow	1.37	[109] [a]
	Milk of Jersey cow	2.4823–3.4360	[87]
	Milk of Sahiwal cow	2.01	[109] [a]
	Milk of Sahiwal × Holstein–Friesian cow	3.14	[109] [a]
	Milk of Italian Friesian and Italian Red Pied cattle	2.49–2.99	[84]
	Yogurt of cow milk (market of Faisalabad)	1.48–2.74	[86]
	Yogurt of sheep milk (market of Faisalabad)	1.42–2.31	[86]

IA: index of atherogenicity; * literature from 2000 until April/2020; [a] recalculated according to the original data in the reference.

2.3. Index of Thrombogenicity (IT)

The index of thrombogenicity (IT) was developed by Ulbritcht and Southgate [112] together with IA in 1991. The formula is:

$$IT = (C14:0 + C16:0 + C18:0)/[(0.5 \times \Sigma MUFA) + (0.5 \times \Sigma n-6\ PUFA) + (3 \times \Sigma n-3\ PUFA) + (n-3 / n-6)] \quad (2)$$

The IT characterizes the thrombogenic potential of FAs, indicating the tendency to form clots in blood vessels and provides the contribution of different FAs, which denotes the relationship between the pro-thrombogenic FAs (C12:0, C14:0, and C16:0) and the anti-thrombogenic FAs (MUFAs and the n-3 and n-6 families) [112]. Therefore, the consumption of foods or products with a lower IT is beneficial for CVH. The IT has been used in many fatty acid composition studies to assess the degree of thrombogenicity. As with the IA formula, the proposed IT formula should be modified as our understanding of MUFA and *trans* fatty acids increases.

The IT has been used in many FA composition studies to assess the degree of thrombogenicity. Chen et al. conducted comparative studies on the fatty acid profiles of four different Chinese medicinal *Sargassum* seaweeds, where the IT was used as one of the nutritional indices to evaluate the potential effects of four *Sargassum* on CVH. The results showed that the IT was between 0.46 and 1.60 [29]. Calabrò et al. compared the fatty acid profile of three cultivars of *Lupinus albus* (Lutteur, Lublanca, and Multitalia) and the IT was used due to the correlation between fatty acids and human health [30].

Detailed information of the literature related to the IT is provided in Table 4. For seaweeds, the value ranges from 0.04 to 2.94 with the exception of *Gracilaria salicornia*, which had an IT value of 5.75 [27]. The ranges of IT values for crops, fish, meat, and dairy products are 0.139–0.56, 0.14–0.87, 0.288–1.694, and 0.39–5.04, respectively.

In brief, both the IA and the IT can be used to assess the potential effects of FA composition on CVH. A FA composition with a lower IA and IT has a better nutritional quality, and its consumption may reduce the risk of coronary heart disease (CHD), but no organization has yet provided the recommended values for the IA and IT. As our comprehensive understanding of the function of FA molecular species deepens, the accuracies of the IA and IT formulas are expected to increase, which might be modified by taking advantage of the massive amount of available data and advanced computer technology.

Table 4. Application of IT in fatty acid evaluation *.

Materials		IA Value	Reference
Red seaweed	*Amphiora anceps*	2.07	[27]
	Ceramium virgatum	0.12 ± 0.005	[59]
	Corallina officinalis	0.28 ± 0.045	[59]
	Gelidiella acerosa	0.52	[27]
	Gelidium micropterum	1.83	[27]
	Gracilaria changii	0.04 ± 0.01	[28]
	Gracilaria corticata	0.63	[27]
	Gracilaria debilis	1.25	[27]
	Gracilaria dura	0.88	[27]
	Gracilaria fergusonii	2.66	[27]
	Gracilaria salicornia	5.75	[27]
	Hymenena sp	2.66	[59]
	Kappaphycus alvarezii	1.17	[27]
	Laurencia cruciata	0.71	[27]
	Lomentaria clavellosa	2.94 ± 1.000	[59]
	Polysiphonia sp	0.61 ± 0.114	[59]
	Sarconema filiforme	0.55	[27]
Brown seaweed	*Cystoseira indica*	0.87	[27]
	Dictyota dichotoma	0.09 ± 0.013	[59]
	Laminaria ochroleuca	1.06–1.89	[58]
	Leathesia difformis	0.14 ± 0.006	[59]
	Myriogloea major	0.09 ± 0.006	[59]
	Padina tetrastromatica	1.20	[27]
	Sargassum fusiforme	0.46 ± 0.21	[29]
	Sargassum horneri	0.65 ± 0.07	[29]
	Sargassum pallidum	1.60 ± 0.56	[29]
	Sargassum swartzii	0.75	[27]
	Sargassum tenerrimum	0.90	[27]
	Sargassum thunbergii	0.76 ± 0.14	[29]
	Spatoglossum asperum	0.50	[27]
	Undaria pinnatifida	0.08–0.26	[59]
Green seaweed	*Caulerpa racemosa*	1.50	[27]
	Caulerpa scalpeliformis	1.38	[27]
	Caulerpa veravalnensis	1.28	[27]
	Cladophora falklandica	0.16 ± 0.048	[59]
	Codium decorticatum	0.12 ± 0.002	[59]
	Codium fragile	0.14 ± 0.013	[59]
	Codium vermilara	0.30 ± 0.080	[59]
	Ulva fasciata	1.56	[27]
	Ulva reticulata	2.90	[27]
	Ulva rigida	1.78	[27]
	Ulva sp.1	0.09 ± 0.028	[59]
	Ulva sp.2	0.04 ± 0.002	[59]
Crops	Cumin (*Cuminum cyminum*)	0.46–0.56	[61] [a]
	Guar seed (*Cyamopsis tetragonoloba*)	0.53	[31]
	Scabiosa stellata	0.23	[60]
	White lupine (*Lupinus albus*)	0.139–0.180	[30]
Plant oil	Camelina oil (*Camelina sativa*)	0.1	[62]
Shellfish	*Chlamys farreri*	0.13–0.17	[63]
	Patinopecten yessoensis	0.09–0.15	[63]
Shrimp	*Penaeus notialis*	0.21–0.30	[64]

Table 4. *Cont.*

Materials		IA Value	Reference
Fish	*Abramis brama*	0.23–0.24	[65,66]
	Clupea harengus	0.26 ± 0.04	[66]
	Cynoscion parvipinnis	0.18–0.29	[67]
	Cyprinus carpio	0.31 ± 0.03	[66]
	Dicentrarchus labrax	0.191–0.63	[68]
	Esox lucius	0.18	[65]
	Hemiramphus brasiliensis	0.21	[36]
	Hyporhamphus unifasciatus	0.44	[36]
	Kutum roach	0.16–0.24	[37]
	Lagocephalus guentheri	0.29	[38]
	Leuciscus idus	0.22 ± 0.05	[66]
	Micropterus salmoides	0.31–0.53	[70]
	Mugil cephalus	0.43–0.58	[71]
	Oncorhynchus mykiss	0.16 ± 0.01	[66]
	Opisthonema oglinum	0.20	[36]
	Oreochromis niloticus	0.82–0.87	[39]
	Perca fluviatilis	0.20–0.21	[65,66]
	Platichthys flesus	0.22 ± 0.02	[66]
	Rutilus rutilus	0.21	[65]
	Salmo trutta	0.21–0.30	[72]
	Scomberomorus cavalla	0.24	[36]
	Sparus aurata	0.14–0.19	[73]
Meat	Chicken (purchased from a hatchery and poultry farm)	0.755–0.784	[48]
	Chicken (Caribro Vishal)	0.288–1.694	[50]
	Rabbit (Curcuma longa)	0.83–1.12	[75]
	Lamb (Barbarine lamb)	1.1–1.15	[43]
	Heifer (Limousin heifer)	1.10–1.34	[77]
	Foal (Galician Mountain × Hispano-Bretón)	0.44–0.80	[52]
	Bologna sausages	0.39–1.55	[53]
Dairy products	Cheese of Holstein cow	3.22–5.04	[88]
	Curd of cow (Middle Rhodopes)	2.02–4.35	[78]
	Milk of goat (market of Sardinia)	2.70–3.20	[89]
	Milk of Chios sheep	1.24–1.46	[54]
	Milk of Karagouniko sheep	1.00–1.47	[54]
	Milk of Lacaune ewe	2.20–2.72	[80]
	Milk of Turcana dairy ewe	1.22–1.76	[55]
	Milk of Holstein cow	2.23–2.90	[88]
	Milk of Jersey cow	3.9813–4.6558	[87]
	Milk of cow (Middle Rhodopes)	2.05–4.03	[78]
	Yogurt of cow milk (market of Faisalabad)	0.39–1.84	[86]
	Yogurt of sheep milk (market of Faisalabad)	0.65–1.68	[86]

IT: index of thrombogenicity; * literature from 2000 until April/2020; [a] recalculated according to the original data in the reference.

2.4. Hypocholesterolemic/Hypercholesterolemic (HH) Ratio

The hypocholesterolemic/hypercholesterolemic (HH) ratio is an index used in the FA profile of lamb meat first proposed by Santos-Silva et al. in 2002 [91]. Due to the high proportion of SFA, the PUFA/SFA is normally low in lambs, so Santos-Silva et al. developed the HH as a new index to assess the effect of FA composition on cholesterol.

Basic on research about dietary FA and the regulation of plasma LDL-C [111], the HH characterizes the relationship between hypocholesterolemic fatty acid (*cis*-C18:1 and PUFA) and hypercholesterolemic FA. Because there was no C12:0 detected in the lambs, Santos-Silva et al. concluded that the formula only includes C14:0 and C16:0 in hypercholesterolemic FA. Later, Mierliță optimized the formula by adding the C12:0 in hypercholesterolemic FA during the studies of sheep milk [55]. The formula is:

$$HH = (cis - C18:1 + \Sigma PUFA)/(C12:0 + C14:0 + C16:0) \qquad (3)$$

Compared with the PUFA/SFA ratio, the HH ratio may more accurately reflect the effect of the FA composition on CVD. The HH ratio has certain limitations. Similar to the IA and IT, the HH might include more kinds of fatty acids such as other molecular species of MUFA and different weights can be assigned to different molecular FA species.

The HH was first used in research on ruminants [46,77,91], which was subsequently extended to dairy products [54,55,78,86,87], marine products [34,36–39,72,90], and other fields [47,48,52,62]. Paiva et al. selected four Azorean macroalgae and used the HH as one of the indices to evaluate their nutritional and health promoting aspects, and found that the HH value ranges from 1.26 to 2.09 [90]. Ratusz et al. analyzed the FA content in 29 cold-pressed camelina (*Camelina sativa*) oils using the HH as a nutritional quality index. A relatively high HH was reported, ranging from 11.7 to 14.7, with a low IA and IT contributing to a decrease in the incidence of CHD [62].

Detailed information about the literature related to the HH is shown in Table 5. For shellfish, the HH value ranges from 1.73 to 4.75, except for *Loxechinus albus*. It is possible that the main food source of *Loxechinus albus* is algae, leading to a high proportion of SFA, so its HH is only 0.21, lower than in other species [34]. For fish, the value ranges from 1.54 to 4.83, with the exception of *Opisthonema oglinum,* which has an HH value of 0.87 [36]. For meat and dairy products, the ranges are 1.27–2.786, 0.32–1.29, respectively.

Table 5. Application of HH in fatty acid evaluation *.

Materials		HH Value	Reference
Red seaweed	*Gelidium microdon*	4.22	[90]
	Pterocladiella capillacea	2.09	[90]
Brown seaweed	*Ulva compressa*	1.90	[90]
	Ulva rigida	1.26	[90]
Plant oil	Camelina oil (*Camelina sativa*)	11.2–15.0	[62]
Shellfish	*Cancer edwardsi*	4.75	[34]
	Cervimunida johni	3.48	[34]
	Concholepas	2.52	[34]
	Heterocarpus reedi	2.91	[34]
	Loxechinus albus	0.21	[34]
	Mesodesma donacium	2.15	[34]
	Pleuroncodes monodon	3.68	[34]
	Pyura chilensis	1.73	[34]
	Venus antiqua	1.90	[34]
Fish	*Cilus gilberti*	1.86	[34]
	Genypterus chilensis	2.93	[34]
	Hemiramphus brasiliensis	2.46	[36]
	Hyporhamphus unifasciatus	2.43	[36]
	Kutum roach	2.04–4.83	[37]
	Lagocephalus guentheri	2.68	[38]
	Merluccius gayi	2.23	[34]
	Opisthonema oglinum	0.87	[36]
	Oreochromis niloticus	1.56–1.63	[39]
	Pinguipes chilensis	1.54	[34]
	Salmo trutta	1.88–2.16	[72]
	Scomber japonicus	2.00	[34]
	Scomberomorus cavalla	1.56	[36]
	Seriola lalandi	2.14	[34]
	Seriolella violacea	2.10	[34]
	Trachurus murphyi	1.73	[34]
Meat	Chicken (purchased from a hatchery and poultry farm)	2.658–2.786	[48]
	Lamb (Merino Branco)	1.92	[91]
	Lamb (Ile de France × Merino Branco)	2.01	[91]
	Cattle (Nellore cattle)	1.56–2.08	[46]
	Heifer (Limousin heifer)	1.27–1.87	[77]
	Foal (Galician Mountain × Hispano-Bretón)	1.76–1.98	[52]
	Spanish dry-cured ham	2.0–2.67	[47]

Table 5. *Cont.*

Materials		HH Value	Reference
Dairy products	Curd of cow (Middle Rhodopes)	0.32–0.74	[78]
	Milk of Chios sheep	0.50–0.61	[54]
	Milk of Karagouniko sheep	0.50–0.68	[54]
	Milk of Turcana dairy ewe	0.88–1.29	[55]
	Milk of cow (Middle Rhodopes)	0.34–0.75	[78]
	Milk of Jersey cow	0.4067–0.5732	[87]
	Yogurt of cow milk (market of Faisalabad)	0.54–1.12	[86]
	Yogurt of sheep milk (market of Faisalabad)	0.82–1.29	[86]

HH: hypocholesterolemic /hypercholesterolemic ratio; * literature from 2000 until April/2020.

2.5. Health-Promoting Index (HPI)

The health-promoting index (HPI) was proposed by Chen et al. in 2004 to assess the nutritional value of dietary fat [94], which focuses on the effect of FA composition on CVD. The formula is:

$$HPI = \Sigma UFA/[C12:0 + (4 \times C14:0) + C16:0]. \tag{4}$$

The HPI is the inverse of the IA. It is currently mainly used in research on dairy products such as milk [92–94] and cheese [57,94,95]. Detailed information about the literature related to the HPI is provided in Table 6. Its values range from 0.16 to 0.68. Dairy products with a high HPI value are assumed to be more beneficial to human health. The HPI has the same shortcoming as the IA, and it requires reliable evidence to optimize the relevant coefficients.

Table 6. Application of HPI in fatty acid evaluation *.

Materials		HPI Value	Reference
Dairy products	Butter of Holstein cow	0.37–0.66	[93,94]
	Cheese of Red Syrian goat	0.37–0.68	[95]
	Cheese of Comisana ewe	0.42–0.50	[57]
	Cheese (Cheddar cheese) of Holstein cow	0.29–0.46	[94]
	Cheese (Provolone Cheese) of Holstein cow	0.38–0.63	[94]
	Cream of Holstein cow	0.31–0.62	[94]
	Milk of ewe (Comisana breed)	0.16–0.28	[92]
	Yogurt of Holstein cow	0.30–0.62	[94]

HPI: health-promoting index; * literature from 2000 until April/2020.

2.6. Unsaturation Index (UI)

The UI indicates the degree of unsaturation in lipids and is calculated as the sum of the percentage of each unsaturated FA multiplied by the number of double bonds within that FA [114]. The calculation formula is:

$$UI = 1 \times (\% \text{ monoenoics}) + 2 \times (\% \text{ dienoics}) + 3 \times (\% \text{ trienoics}) + 4 \times (\% \text{ tetraenoics})$$
$$+ 5 \times (\% \text{ pentaenoics}) + 6 \times (\% \text{ hexaenoics}) \tag{5}$$

Unlike ΣUFA and ΣPUFA, different unsaturated FAs have different weights in the UI. This index indicates the impact of highly unsaturated FA and does not ignore the impact of FAs that have a low degree of unsaturation. In general, the UI more comprehensively reflects the proportion of FA with different degrees of unsaturation in the total FA composition of a species.

The UI is commonly used to determine the composition of macroalgal FA. It can be used as a standard for judging the content of high-quality PUFA, in which macroalgae may be used as alternative sources of high-quality PUFA instead of fish or fish oil [98]. Colombo et al. used the UI to compare macroalgae in cold water with those in warm water, with a high UI value indicating a high degree of total unsaturation. Their results suggested that the fatty acids with a high degree of unsaturation in a membrane lipid can maintain fluidity at relatively low temperature [96].

Detailed information about the literature related to the UI is listed in Table 7. The UI value of seaweeds varies widely from 45 to 368.68, and may be closely related to their species. There is no rule at present. The disadvantage of the UI is that it only focuses on the degree of unsaturation of FAs and does not distinguish between n-6 and n-3 FA. The fatty acids in the n-6 and n-3 series have different physiological effects on the human body.

Table 7. Application of UI in fatty acid evaluation *.

Materials		UI Value	Reference
Red seaweed	*Ahnfeltia plicata*	250 ± 1.01	[98]
	Amphiora anceps	98.01, 97.5	[27,98]
	Callophylis sp	117	[96]
	Ceramium virgatum	284 ± 7	[59]
	Corallina officinalis	202 ± 19	[59]
	Gelidiella acerosa	191.02	[27]
	Gelidium micropterum	98.80	[27]
	Gloiopeltis furcata	54	[96]
	Gracilaria changii	368.68 ± 20.01	[28]
	Gracilaria corticata	257.07	[27]
	Gracilaria debilis	204.85, 205 ± 3.07	[27,98]
	Gracilaria dura	249.10, 249 ± 3.66	[27,98]
	Gracilaria fergusonii	134.75, 135 ± 1.14	[27,98]
	Gracilaria salicornia	50.631	[27]
	Grateloupia indica	286 ± 5.91	[98]
	Grateloupia wattii	181 ± 3.77	[98]
	Hymenena sp	45	[59]
	Hypnea esperi	93.6 ± 4.63	[98]
	Hypnea musciformis	91.3 ± 4.11	[98]
	Kappaphycus alvarezii	140.94, 141 ± 4.05	[27,98]
	Laurencia cruciata	172.95, 173 ± 5.64	[27,98]
	Laurencia papillosa	213 ± 4.89	[98]
	Lomentaria clavellosa	76 ± 12	[59]
	Polysiphonia sp.	143 ± 15	[59]
	Sarconema filiforme	245.54, 246 ± 1.27	[27,98]
	Soliera robusta	77	[96]
Brown seaweed	*Cystoseira indica*	195.44, 195 ± 4.21	[27,98]
	Dictyota dichotoma	321 ± 10	[59]
	Leathesia difformis	272 ± 6	[59]
	Myriogloea major	266 ± 7	[59]
	Padina tetrastromatica	154.49, 155 ± 5.50	[27,98]
	Sargassum fusiforme	125.65 ± 32.25	[29]
	Sargassum horneri	116.16 ± 5.77	[29]
	Sargassum pallidum	62.27 ± 15.05	[29]
	Sargassum swartzii	182.02	[27]
	Sargassum tenerrimum	187.05, 187 ± 4.47	[27,98]
	Sargassum thunbergii	89.87 ± 7.44	[29]
	Spatoglossum asperum	202.83, 203 ± 3.06	[27,98]
	Stoechospermum marginatum	176 ± 3.56	[98]
	Undaria pinnatifida	260-318	[59]
Green seaweed	*Caulerpa racemosa*	106.70, 107 ± 5.67	[27,98]
	Caulerpa scalpeliformis	121.67	[27]
	Caulerpa veravalnensis	141.87, 142 ± 2.96	[27,98]
	Cladophora falklandica	215 ± 7	[59]
	Codium decorticatum	219 ± 2	[59]
	Codium fragile	179 ± 11	[59]
	Codium vermilara	135 ± 16	[59]
	Ulva fasciata	102.92, 103 ± 2.83	[27,98]
	Ulva lactuca	87.5 ± 5.76	[98]
	Ulva linza	124 ± 4.23	[98]
	Ulva reticulata	70.87, 70.9 ± 5.33	[27,98]
	Ulva rigida	93.96, 93.8 ± 5.31	[27,98]
	Ulva tubulosa	99.6 ± 3.23	[98]
	Ulva sp.	76.3 ± 5.40	[98]
	Ulva sp.1	209 ± 20	[59]
	Ulva sp.2	288 ± 10	[59]
Crops	Cumin (*Cuminum cyminum*)	125.21–133.10	[61]
	Soybean (*Glycine max*)	148–155	[99]
Meat	Pig (Pietrain × (Duroc × Landrace))	111–124	[44]
	Dry-cured ham (Landrace × Large White (25% Pietrain) pig)	73 ± 6	[101]
Dairy products	Milk of (New Zealand × California) white rabbit	86–120	[102]

UI: unsaturation index; * literature from 2000 until April/2020.

2.7. Sum of Eicosapentaenoic Acid and Docosahexaenoic Acid (EPA + DHA)

EPA and DHA are n-3 long-chain PUFAs that play essential roles in biological processes in the human body. They can reduce the risk of CVD, hypertension, and inflammation. DHA is a critical component of the retina and the neuronal system and is involved in visual functioning and cognitive functioning in humans [115,116]. The American Heart Association summarized the preventive effect of n-3 PUFA from seafood on CVD in the 2015–2020 Dietary Guidelines for Americans [1].

EPA and DHA can be synthesized from α-linolenic acid in the human body, but exogenous supplementation is still needed when insufficient. α-linolenic acid (ALA; C18:3 n-3) can be converted to EPA and DHA by desaturase and elongase, respectively. EPA and DHA can be supplemented by ingesting ALA. Burdge et al. [115] studied the capacity of humans to convert ALA to EPA and DHA. In a carbon isotope labeling experiment, six young male subjects orally received ^{13}C-ALA as a part of their habitual diet. The results indicated that the subjects had a limited capacity to convert ALA to EPA, and ^{13}C-labeling of DHA was not detected [115]. Brenna et al. summarized related studies and reached a similar conclusion [117]. Although the conversion of ALA to EPA and DHA was observed in tracer studies in all age groups, regardless of whether the study participant was male or female, the efficiency of directly supplementing with EPA to increase the level of EPA was found to be 15-fold that of supplementing with high levels of ALA. The conversion rate of ALA to DHA in infants is only 1%, and is even lower in adults [117]. Therefore, the rate of conversion of ALA to EPA and DHA that is required for health is far from sufficient; direct intake of EPA and DHA is more effective.

EPA + DHA is an index that is recognized worldwide. Recommendations for EPA + DHA intake can be found in various dietary guidelines. According to the Food and Agriculture Organization of the United Nations (UN FAO), the recommended amount is 0.250–2 g/day. Due to the low EPA and DHA contents in terrestrial plants and animals, this index is mostly used to evaluate the nutritional value of seafood and seafood products, particularly fish, which makes it an important nutritional index for seafood. Rincón-Cervera et al. studied the fatty acid composition of fish and shellfish captured in the South Pacific, and the results showed that EPA + DHA ranged between 115.15 and 1370.67 mg/100 g in all studied fish species and between 63.61 and 522.68 mg/100 g in all studied shellfish species [34]. Detailed information about the literature related to EPA + DHA is shown in Table 8. The species of fish and shellfish as well as their nutrition intake are key factors influencing the EPA + DHA value.

2.8. Fish Lipid Quality/Flesh Lipid Quality (FLQ)

FLQ was originally use for fish lipid quality [107,108] or flesh lipid quality [65,66,73]. The purpose of FLQ is similar to that of the EPA + DHA index, but it calculates the sum of EPA and DHA as a percentage of total fatty acids. The formula is:

$$FLQ = 100 \times (C22:6\,n-3 + C20:5\,n-3)/\Sigma FA \qquad (6)$$

FLQ is more suitable for marine products given their higher proportions of EPA and DHA. This index may be considered a supplement to EPA + DHA since the absolute quantity for EPA and DHA is more important. Until now, FLQ has only been used to assess the quality of lipids in fish. Senso et al. examined the fatty acid profile of the fillet of farmed sea bream (*Sparus aurata*) harvested in different seasons using FLQ as the lipid quality index. FLQ was lowest in April [73]. Detailed information about the literature related to FLQ is provided in Table 9. The value ranges from 13.01 to 36.37 for closely related species.

Table 8. Application of EPA + DHA in fatty acid evaluation *.

Materials		EPA + DHA Value	Reference
Shellfish	Cancer edwardsi	205.62 ± 6.19 mg/100 g	[34]
	Cervimunida johni	162.90 ± 2.83 mg/100 g	[34]
	Concholepas concholepas	63.61 ± 0.42 mg/100 g	[34]
	Heterocarpus reedi	186.98 ± 3.88 mg/100 g	[34]
	Loxechinus albus	208.55 ± 10.28 mg/100 g	[34]
	Mesodesma donacium	216.96 ± 9.76 mg/100 g	[34]
	Pleuroncodes monodon	189.83 ± 3.74 mg/100 g	[34]
	Pyura chilensis	522.68 ± 28.02 mg/100 g	[34]
	Venus antiqua	214.34 ± 7.52 mg/100 g	[34]
Fish	Cilus gilberti	294.57 ± 8.76 mg/100 g	[34]
	Dicentrarchus labrax	270–480 mg/100 g	[68]
	Genypterus chilensis	115.15 ± 6.16 mg/100 g	[34]
	Kutum roach	96–250 mg/100 g	[61] [a]
	Merluccius gayi gayi	309.38 ± 6.81 mg/100 g	[34]
	Pinguipes chilensis	507.60 ± 25.32 mg/100 g	[34]
	Scomber japonicus	1370.67 ± 55.79 mg/100 g	[34]
	Seriola lalandi	915.76 ± 19.68 mg/100 g	[34]
	Seriolella violacea	304.04 ± 14.15 mg/100 g	[34]
	Trachinotus carolinus	621–941 mg/100 g	[40]
	Trachurus murphyi	786.90 ± 11.44 mg/100 g	[34]
	Epinephelus coioides	19.9–25.4%	[103]
	Hemiramphus brasiliensis	16.71% ± 0.07%	[36]
	Hyporhamphus unifasciatus	15.53% ± 0.07%	[36]
	Megalobrama amblycephala	5.52–7.36%	[106]
	Opisthonema oglinum	40.86% ± 0.07%	[36]
	Salmo salar	11.80–11.81%	[104]
	Scomberomorus cavalla	35.06% ± 0.07%	[36]
	Sparidentex hasta	45.8–230.4 mg/g lipid	[105]

EPA + DHA: sum of eicosapentaenoic acid and docosahexaenoic acid; * literature from 2000 until April/2020; [a] recalculated according to the original data in the reference.

Table 9. Application of FLQ in fatty acid evaluation *.

Materials		FLQ Value	Reference
Fish	Abramis brama	24.46–30.14	[65,66]
	Clupea harengus	13.01 ± 0.77	[66]
	Cyprinus carpio	13.99 ± 2.15	[66]
	Esox Lucius	36.37	[65]
	Leuciscus idus	24.32 ± 2.47	[66]
	Oncorhynchus mykiss	17.97 ± 2.46	[66]
	Perca fluviatilis	30.14–33.22	[65,66]
	Platichthys flesus	20.25 ± 2.30	[66]
	Rutilus	28.41	[65]
	Sparus aurata	19.35–31.27	[73]

FLQ: fish lipid quality/flesh Lipid quality; * literature from 2000 until April/2020.

2.9. The Linoleic Acid/α-Linolenic Acid (LA/ALA) Ratio

The linoleic acid (LA, C18:2 n-6)/α-linolenic acid (ALA, C18:3 n-3) ratio was developed for guiding infant formula. LA and ALA compete for the same desaturase and elongase enzymes, which they use to synthesize long-chain unsaturated fatty acids. Due to the low conversion rate of ALA, reducing the LA/ALA ratio only provides a modest improvement in the levels of some n-3 long-chain PUFAs; however, the balance may be the most important factor when long-chain PUFAs are not present in infant formulas.

The Definitions & Nutrient Composition section of the Guidelines for Infant Formula published by Food Standards Australia New Zealand (FSANZ) sets the minimum and maximum proportions of LA and ALA, and specifies an LA/ALA ratio within 5:1–15:1.

The LA/ALA ratio has a higher reference value when judging the nutritional value of baby food and infant formula. Tissues of adults have a lower rate of synthesis of n-3 long-chain PUFAs than those of infants, so the LA/ALA ratio in the diet does not have too much of an impact on adults. In the literature we reviewed, the LA/ALA ratio was used in research on ruminants and dairy products as well [43,55,109]. Majdoub-Mathlouthi et al. compared the meat fatty acid composition of Barbarine lambs raised on rangelands and those reared indoors. The results showed that the grazing lambs had lower LA/ALA [43]. Sharma et al. compared the fatty acid profile of indigenous Indian cow milk with exotic and crossbred counterparts. LA/ALA was used to reflect the quality of milk [109]. LA/ALA in indigenous cattle was found to be lower than others, providing scientific data for the superiority of indigenous cow milk [109]. Detailed information of the literature related to LA/ALA is listed in Table 10. Turcana dairy ewe milk has a low LA/ALA value due to the high content of ALA given the inclusion of hemp seed in the diet [55].

Table 10. Application of LA/ALA in fatty acid evaluation *.

Materials		LA/ALA Value	Reference
Meat	Lamb (Barbarine lamb)	6.78–10.05	[43]
Dairy products	Milk of Turcana dairy ewe	0.98–1.36	[55]
	Milk of Sahiwal cow	3.313 ± 0.262	[109]
	Milk of Holstein–Friesian cow	3.446 ± 0.196	[109]
	Milk of Sahiwal × Holstein–Friesian cow	3.065 ± 0.093	[109]
	Milk of indigenous Indian cow	2.464 ± 0.147	[109]

LA/ALA: linoleic acid/α-linolenic acid ratio; * literature from 2000 until April/2020.

2.10. Trans Fatty Acid (TFA)

Most unsaturated FA in the human diet have a *cis* configuration. However, *trans* fatty acid (TFA) is present in the human diet as well. According to the Food and Drug Administration (FDA), TFA is defined as the sum of all unsaturated fatty acids that contain one or more isolated (i.e., non-conjugated) double bond(s) in a *trans* configuration [118,119]. The European Food Safety Authority (EFSA) gives a different definition of TFA, which are also present as either *trans*-MUFA or *trans*-PUFA. *Trans*-PUFAs have at least one *trans* double bond and may therefore also have double bonds in the *cis* configuration. Conjugated fatty acid (CLA) is separated from TFA as an independent section by the EFSA. CLAs may have health benefits that are different from those of TFAs, such as anti-cancer [120,121] and anti-atherosclerosis [122] activities, so it is appropriate to exclude CLA from the definition of TFA.

According to the EFSA, TFAs may originate from various sources, including of bacterial conversion of unsaturated fatty acids in the rumen of ruminants, industrial hydrogenation (used to produce semi-liquid and solid fats; can be used to produce margarine, shortening, biscuits, etc.), deodorization of unsaturated vegetable oils (or occasionally fish oils) with a high content of polyunsaturated fatty acids (a necessary step of refining), and heating and frying oil at excessively high temperatures (>220 °C) [123].

TFA does not play a positive role in any vital functions. On the contrary, the intake of TFA may harm human health. Evidence suggests that ruminant-derived TFA has similar adverse effects on blood lipids and lipoproteins as TFA from industrial sources. Sufficient evidence is still needed to reveal whether a difference exists between equivalent amounts of ruminant and industrially produced TFA in terms of risk of CHD [123]. *Trans*-MUFA is the most common TFA in the human diet. A few clinical trials with normotensive subjects proved that *trans*-MUFA from hydrogenated oil has no effect on systolic or diastolic blood pressure [124]. Prospective cohort studies showed that a consistent relationship exists between higher TFA intake and increased risk of CHD. Conversely, a daily intake of 3.6 g of TFA from milk fat for five weeks did not affect blood pressure or isobaric arterial elasticity [125].

According to population nutrient intake goals from the World Health Organization (WHO)/FAO, the intake of TFA should constitute less than 1% of total energy. For pregnancy and lactation, the lowest possible intake of industrially-produced TFAs is required. According to the EFSA, TFA in the diet is provided by several sources that contain essential FAs and other nutrients [124], Therefore, the EFSA panel concluded that the intake of TFA should be sufficiently reduced within a nutritionally adequate diet to lower the intake of TFA while ensuring the nutrient intake [124]. The 2015–2020 Dietary Guidelines for Americans emphasize that individuals should reduce their intake of *trans* fatty acid to as low as possible by limiting their consumption of foods that contain synthetic sources of *trans* fats. There is no need to eliminate meat and dairy products that contain small quantities of natural TFA from the diet. In the United Kingdom, the recommended intake of TFA is less than 2% of total daily energy or 5 g/day.

The TFA index is currently used in seaweed [90], lamb [45], milk [78], fish [35], and plant oil [32,33,110]. Skałecki et al. compared the fatty acid profiles of Prussian Carp fish (*Carassius gibelio*) fillets with and without skin; the share of TFA was the same in both types [35]. Mishra and Sharma monitored the changes occurring in rice bran oil and its blend with sunflower oil during repeated frying cycles of potato chips with different moisture contents (0.5% and 64.77%) [110]. The results showed that blended oil was better when used to fry dried potato chips, as TFA was the lowest after deep fat frying (increased from 1.15% to 1.80%) [110]. Detailed information of the literature related to TFA is listed in Table 11.

Table 11. Application of TFA in fatty acid evaluation *.

Materials		TFA Value	Reference
Red seaweed	*Gelidium microdon*	1.34% ± 0.20%	[90]
	Pterocladiella capillacea	1.47% ± 0.09%	[90]
Brown seaweed	*Ulva compressa*	7.35% ± 0.63%	[90]
	Ulva rigida	4.89% ± 0.26%	[90]
Plant oil	Palm stearin	0.6%	[32]
	Rice bran oil	1.27–2.91%	[110]
	Sunflower oil	0.2%, 0.84–1.71%	[32,33]
Fish	*Carassius gibelio*	1.06% ± 0.06%, 10.58–37.15 mg/100 g	[35]
Meat	Lamb (Barbarine lamb)	2.23–2.83%	[45]
Dairy products	Curd of cow (Middle Rhodopes)	340–1090 mg/100 g	[78]
	Milk of cow (Middle Rhodopes)	110–210 mg/100 g	[78]

TFA: *Trans* fatty acid; * literature from 2000 until April/2020.

3. Conclusions

In this review, we summarized 10 FA indices that have been commonly used in the literature to characterized FA composition. Among them, PUFA/SFA, IA, IT, HH, HPI, and UI are the most frequently used indices and are widely used to evaluate a variety of research materials, mainly related to CVH. PUFA/SFA is a basic index that simply considers $\sum$PUFA and $\sum$SFA. IA, IT, HH, HPI, and UI were derived based on revising PUFA/SFA, which consider the contribution of different molecular species of SFA, as well as MUFA. However, all of these six indices do not reflect the influence of different molecular species of PUFA. For instance, n-3 PUFA and n-6 PUFA exhibit different effects on CVH. EPA + DHA and FLQ are used in the analysis of fish or shellfish, which are rich in n-3 PUFA. The LA/ALA ratio is an important index for baby food and infant formula. TFA is an indicator of food safety because it has a negative effect on many vital functions. Due to the lack of systematic integration of clinical evidence and literature data related to FA, suggesting ideas and proposals for the update of indices is difficult. Besides, CVH is the main assessment of FA indices used at present. As FA functions continue to be revealed, more indices that can be used for other diseases are expected.

With the present review, we aimed to help researchers evaluate the nutritional value of FAs and to explore their potential usage in disease prevention and treatment, and to help newcomers to the field of FA analysis to quickly and accurately select appropriate indices. The human body is complex, so a reasonable selection of indices can help researchers to more comprehensively evaluate the research materials. The purpose of using an index is only to assess the potential nutritional and/or medicinal value of the research materials; they should not be considered gold standards. The indices should not be used indiscriminately, and the results obtained with the indices should be interpreted with caution. After a reasonable assessment using the indices, a more systematic and complex research process should be used to reach a conclusion about the nutritional effect of the research object on the human body. We recommend that researchers apply these indices to help compare several research objects to select one or more objects of interest for further in-depth research.

Author Contributions: Conceptualization, H.L. and J.C.; writing—original draft preparation, J.C.; writing—review and editing, H.L.; funding acquisition, H.L. All authors have read and agreed to the published version of the manuscript.

Abbreviations

ALA	α-linolenic acid
AS	Atherosclerosis
CHD	Coronary heart disease
CLA	Conjugated fatty acids
CVD	Cardiovascular disease
CVH	Cardiovascular health
DHA	Docosahexaenoic acid
EFSA	European Food Safety Authority
EPA	Eicosapentaenoic acid
FA	Fatty acid
FAO	Food and Agriculture Organization
FDA	Food and Drug Administration
FLQ	Fish lipid quality/flesh lipid quality
FSANZ	Food Standards Australia New Zealand
HH	Hypocholesterolemic/hypercholesterolemic ratio
HPI	Health-promoting index
IA	Index of atherogenicity
IT	Index of thrombogenicity
LA	Linoleic acid
LDL-C	Low-density lipoprotein cholesterol
LDLR	Low-density lipoprotein receptors
MS	Multiple sclerosis
MUFA	Monounsaturated fatty acid
PPAR-γ	Peroxisome proliferators-activated receptor-gamma
PUFA	Polyunsaturated fatty acid
SFA	Saturated fatty acid
TFA	*Trans* fatty acid
TG	Triglycerides
UFA	Unsaturated fatty acid
UI	Unsaturation index
WHO	World Health Organization

References

1. Rimm, E.B.; Appel, L.J.; Chiuve, S.E.; Djoussé, L.; Engler, M.B.; Kris-Etherton, P.M.; Mozaffarian, D.; Siscovick, D.S.; Lichtenstein, A.H. Seafood long-chain n-3 polyunsaturated fatty acids and cardiovascular disease: A science advisory from the American Heart Association. *Circulation* **2018**, *138*, 35–47. [CrossRef] [PubMed]
2. Wu, H.; Xu, L.; Ballantyne, C.M. Dietary and pharmacological fatty acids and cardiovascular health. *J. Clin. Endocrinol. Metab.* **2020**, *105*, 1030–1045. [CrossRef] [PubMed]
3. Marangoni, F.; Agostoni, C.; Borghi, C.; Catapano, A.L.; Cena, H.; Ghiselli, A.; La Vecchia, C.; Lercker, G.; Manzato, E.; Pirillo, A. Dietary linoleic acid and human health: Focus on cardiovascular and cardiometabolic effects. *Atherosclerosis* **2020**, *292*, 90–98. [CrossRef] [PubMed]
4. Naeini, Z.; Toupchian, O.; Vatannejad, A.; Sotoudeh, G.; Teimouri, M.; Ghorbani, M.; Nasli-Esfahani, E.; Koohdani, F. Effects of DHA-enriched fish oil on gene expression levels of p53 and NF-κB and PPAR-γ activity in PBMCs of patients with T2DM: A randomized, double-blind, clinical trial. *Nutr. Metab. Cardiovasc.* **2020**, *30*, 441–447. [CrossRef]
5. Bird, J.K.; Calder, P.C.; Eggersdorfer, M. The role of n-3 long chain polyunsaturated fatty acids in cardiovascular disease prevention, and interactions with statins. *Nutrients* **2018**, *10*, 775. [CrossRef]
6. Tomata, Y.; Larsson, S.C.; Hägg, S. Polyunsaturated fatty acids and risk of Alzheimer's disease: A Mendelian randomization study. *Eur. J. Nutr.* **2020**, *59*, 1763–1766. [CrossRef]
7. Langley, M.R.; Triplet, E.M.; Scarisbrick, I.A. Dietary influence on central nervous system myelin production, injury, and regeneration. *BBA-Mol. Basis Dis.* **2020**, *1866*, 165779. [CrossRef]
8. Zhou, Y.; Tao, X.; Wang, Z.; Feng, L.; Wang, L.; Liu, X.; Pan, R.; Liao, Y.; Chang, Q. Hippocampus metabolic disturbance and autophagy deficiency in olfactory bulbectomized rats and the modulatory effect of fluoxetine. *Int. J. Mol. Sci.* **2019**, *20*, 4282. [CrossRef]
9. Chang, J.P.; Chang, S.; Yang, H.; Chen, H.; Chien, Y.; Yang, B.; Su, H.; Su, K. Omega-3 polyunsaturated fatty acids in cardiovascular diseases comorbid major depressive disorder-results from a randomized controlled trial. *Brain Behav. Immun.* **2020**, *85*, 14–20. [CrossRef]
10. Konstantynowicz-Nowicka, K.; Berk, K.; Chabowski, A.; Kasacka, I.; Bielawiec, P.; Łukaszuk, B.; Harasim-Symbor, E. High-fat feeding in time-dependent manner affects metabolic routes leading to nervonic acid synthesis in NAFLD. *Int. J. Mol. Sci.* **2019**, *20*, 3829. [CrossRef]
11. Chen, L.; Wang, Y.; Xu, Q.; Chen, S. Omega-3 fatty acids as a treatment for non-alcoholic fatty liver disease in children: A systematic review and meta-analysis of randomized controlled trials. *Clin. Nutr.* **2018**, *37*, 516–521. [CrossRef] [PubMed]
12. Chen, Y.; Chen, H.; Huang, B.; Chen, Y.; Chang, C. Polyphenol rich extracts from *Toona sinensis* bark and fruit ameliorate free fatty acid-induced lipogenesis through AMPK and LC3 pathways. *J. Clin. Med.* **2019**, *8*, 1664. [CrossRef] [PubMed]
13. Wu, K.; Zhao, T.; Hogstrand, C.; Xu, Y.; Ling, S.; Chen, G.; Luo, Z. FXR-mediated inhibition of autophagy contributes to FA-induced TG accumulation and accordingly reduces FA-induced lipotoxicity. *Cell Commun. Signal.* **2020**, *18*, 1–16. [CrossRef] [PubMed]
14. Tobias, T.A.; Wood, L.G.; Rastogi, D. Carotenoids, fatty acids and disease burden in obese minority adolescents with asthma. *Clin. Exp. Allergy* **2019**, *49*, 838–846. [CrossRef]
15. Monga, N.; Sethi, G.S.; Kondepudi, K.K.; Naura, A.S. Lipid mediators and asthma: Scope of therapeutics. *Biochem. Pharmacol.* **2020**, in press. [CrossRef]
16. Magnusson, J.; Ekström, S.; Kull, I.; Håkansson, N.; Nilsson, S.; Wickman, M.; Melén, E.; Risérus, U.; Bergström, A. Polyunsaturated fatty acids in plasma at 8 years and subsequent allergic disease. *J. Allergy Clin. Immunol.* **2018**, *142*, 510–516. [CrossRef]
17. Miranda-Gonçalves, V.; Lameirinhas, A.; Henrique, R.; Baltazar, F.; Jerónimo, C. The metabolic landscape of urological cancers: New therapeutic perspectives. *Cancer Lett.* **2020**, *477*, 76–87. [CrossRef]
18. Tripathi, R.K.P. A perspective review on fatty acid amide hydrolase (FAAH) inhibitors as potential therapeutic agents. *Eur. J. Med. Chem.* **2020**, *188*, 111953. [CrossRef]
19. Vadell, A.K.; Bärebring, L.; Hulander, E.; Gjertsson, I.; Lindqvist, H.M.; Winkvist, A. Anti-inflammatory diet in rheumatoid arthritis (ADIRA)-a randomized, controlled crossover trial indicating effects on disease activity. *Am. J. Clin. Nutr.* **2020**, *111*, 1203–1213. [CrossRef]

20. Dos Santos Simon, M.I.S.; Dalle Molle, R.; Silva, F.M.; Rodrigues, T.W.; Feldmann, M.; Forte, G.C.; Marostica, P.J.C. Antioxidant micronutrients and essential fatty acids supplementation on cystic fibrosis outcomes: A systematic review. *J. Acad. Nutr. Diet.* **2020**, *120*, 1016–1033. [CrossRef]

21. Kjølbæk, L.; Benítez-Páez, A.; Del Pulgar, E.M.G.; Brahe, L.K.; Liebisch, G.; Matysik, S.; Rampelli, S.; Vermeiren, J.; Brigidi, P.; Larsen, L.H. Arabinoxylan oligosaccharides and polyunsaturated fatty acid effects on gut microbiota and metabolic markers in overweight individuals with signs of metabolic syndrome: A randomized cross-over trial. *Clin. Nutr.* **2020**, *39*, 67–79. [CrossRef] [PubMed]

22. Song, Y.; Hogstrand, C.; Ling, S.; Chen, G.; Luo, Z. Creb-Pgc1α pathway modulates the interaction between lipid droplets and mitochondria and influences high fat diet-induced changes of lipid metabolism in the liver and isolated hepatocytes of yellow catfish. *J. Nutr. Biochem.* **2020**, *80*, 108364. [CrossRef] [PubMed]

23. Mozaffari, H.; Daneshzad, E.; Larijani, B.; Bellissimo, N.; Azadbakht, L. Dietary intake of fish, n-3 polyunsaturated fatty acids, and risk of inflammatory bowel disease: A systematic review and meta-analysis of observational studies. *Eur. J. Nutr.* **2020**, *59*, 1–17. [CrossRef] [PubMed]

24. Min, S.Y.; Learnard, H.; Kant, S.; Gealikman, O.; Rojas-Rodriguez, R.; DeSouza, T.; Desai, A.; Keaney, J.F.; Corvera, S.; Craige, S.M. Exercise rescues gene pathways involved in vascular expansion and promotes functional angiogenesis in subcutaneous white adipose tissue. *Int. J. Mol. Sci.* **2019**, *20*, 2046. [CrossRef]

25. Syren, M.; Turolo, S.; Marangoni, F.; Milani, G.P.; Edefonti, A.; Montini, G.; Agostoni, C. The polyunsaturated fatty acid balance in kidney health and disease: A review. *Clin. Nutr.* **2018**, *37*, 1829–1839. [CrossRef]

26. Khalid, A.; Siddiqui, A.J.; Huang, J.; Shamsi, T.; Musharraf, S.G. Alteration of serum free fatty acids are indicators for progression of pre-leukaemia diseases to leukaemia. *Sci. Rep.* **2018**, *8*, 1–10. [CrossRef]

27. Kumar, M.; Kumari, P.; Trivedi, N.; Shukla, M.K.; Gupta, V.; Reddy, C.; Jha, B. Minerals, PUFAs and antioxidant properties of some tropical seaweeds from Saurashtra coast of India. *J. Appl. Phycol.* **2011**, *23*, 797–810. [CrossRef]

28. Chan, P.T.; Matanjun, P. Chemical composition and physicochemical properties of tropical red seaweed, *Gracilaria changii*. *Food Chem.* **2017**, *221*, 302–310. [CrossRef]

29. Chen, Z.; Xu, Y.; Liu, T.; Zhang, L.; Liu, H.; Guan, H. Comparative studies on the characteristic fatty acid profiles of four different Chinese medicinal *Sargassum* seaweeds by GC-MS and chemometrics. *Mar. Drugs* **2016**, *14*, 68. [CrossRef]

30. Calabrò, S.; Cutrignelli, M.I.; Lo Presti, V.; Tudisco, R.; Chiofalo, V.; Grossi, M.; Infascelli, F.; Chiofalo, B. Characterization and effect of year of harvest on the nutritional properties of three varieties of white lupine (*Lupinus albus L.*). *J. Sci. Food Agric.* **2015**, *95*, 3127–3136. [CrossRef]

31. Chiofalo, B.; Lo Presti, V.; D'Agata, A.; Rao, R.; Ceravolo, G.; Gresta, F. Qualitative profile of degummed guar (*Cyamopsis tetragonoloba L.*) seeds grown in a Mediterranean area for use as animal feed. *J. Anim. Physiol. Anim. Nutr.* **2018**, *102*, 260–267. [CrossRef] [PubMed]

32. Farajzadeh Alan, D.; Naeli, M.H.; Naderi, M.; Jafari, S.M.; Tavakoli, H.R. Production of *trans*-free fats by chemical interesterified blends of palm stearin and sunflower oil. *Food Sci. Nutr.* **2019**, *7*, 3722–3730. [CrossRef] [PubMed]

33. Filip, S.; Hribar, J.; Vidrih, R. Influence of natural antioxidants on the formation of *trans* fatty acid isomers during heat treatment of sunflower oil. *Eur. J. Lipid Sci. Technol.* **2011**, *113*, 224–230. [CrossRef]

34. Rincón-Cervera, M.Á.; González-Barriga, V.; Romero, J.; Rojas, R.; López-Arana, S. Quantification and distribution of omega-3 fatty acids in south pacific fish and shellfish species. *Foods* **2020**, *9*, 233. [CrossRef] [PubMed]

35. Skałecki, P.; Kaliniak-Dziura, A.; Domaradzki, P.; Florek, M.; Kępka, M. Fatty acid composition and oxidative stability of the lipid fraction of skin-on and skinless fillets of prussian carp (*Carassius gibelio*). *Animals* **2020**, *10*, 778. [CrossRef]

36. Fernandes, C.E.; Da Silva Vasconcelos, M.A.; de Almeida Ribeiro, M.; Sarubbo, L.A.; Andrade, S.A.C.; de Melo Filho, A.B. Nutritional and lipid profiles in marine fish species from Brazil. *Food Chem.* **2014**, *160*, 67–71. [CrossRef]

37. Hosseini, H.; Mahmoudzadeh, M.; Rezaei, M.; Mahmoudzadeh, L.; Khaksar, R.; Khosroshahi, N.K.; Babakhani, A. Effect of different cooking methods on minerals, vitamins and nutritional quality indices of *kutum roach (Rutilus frisii kutum)*. *Food Chem.* **2014**, *148*, 86–91. [CrossRef]

38. Sreelakshmi, K.R.; Rehana, R.; Renjith, R.K.; Sarika, K.; Greeshma, S.S.; Minimol, V.A.; Ashokkumar, K.; Ninan, G. Quality and shelf life assessment of puffer fish (*Lagocephalus guentheri*) fillets during chilled storage. *J. Aquat. Food Prod. Technol.* **2019**, *28*, 25–37. [CrossRef]

39. Tonial, I.B.; Oliveira, D.F.; Coelho, A.R.; Matsushita, M.; Coró, F.A.G.; De Souza, N.E.; Visentainer, J.V. Quantification of essential fatty acids and assessment of the nutritional quality indexes of lipids in tilapia alevins and juvenile tilapia fish (*Oreochromis niloticus*). *J. Food Res.* **2014**, *3*, 105–114. [CrossRef]

40. Rombenso, A.N.; Trushenski, J.T.; Schwarz, M.H. Fish oil replacement in feeds for juvenile Florida Pompano: Composition of alternative lipid influences degree of tissue fatty acid profile distortion. *Aquaculture* **2016**, *458*, 177–186. [CrossRef]

41. Hao, L.; Xiang, Y.; Degen, A.; Huang, Y.; Niu, J.; Sun, L.; Chai, S.; Zhou, J.; Ding, L.; Long, R. Adding heat-treated rapeseed to the diet of yak improves growth performance and tenderness and nutritional quality of the meat. *Anim. Sci. J.* **2019**, *90*, 1177–1184. [CrossRef] [PubMed]

42. Castro, T.; Cabezas, A.; De la Fuente, J.; Isabel, B.; Manso, T.; Jimeno, V. Animal performance and meat characteristics in steers reared in intensive conditions fed with different vegetable oils. *Animal* **2016**, *10*, 520–530. [CrossRef]

43. Majdoub-Mathlouthi, L.; Saïd, B.; Kraiem, K. Carcass traits and meat fatty acid composition of Barbarine lambs reared on rangelands or indoors on hay and concentrate. *Animal* **2015**, *9*, 2065–2071. [CrossRef]

44. Realini, C.E.; Pérez-Juan, M.; Gou, P.; Díaz, I.; Sárraga, C.; Gatellier, P.; García-Regueiro, J.A. Characterization of *Longissimus thoracis*, *Semitendinosus* and *Masseter* muscles and relationships with technological quality in pigs. 2. Composition of muscles. *Meat Sci.* **2013**, *94*, 417–423. [CrossRef] [PubMed]

45. Brogna, D.M.; Nasri, S.; Salem, H.B.; Mele, M.; Serra, A.; Bella, M.; Priolo, A.; Makkar, H.; Vasta, V. Effect of dietary saponins from *Quillaja saponaria L.* on fatty acid composition and cholesterol content in muscle *Longissimus dorsi* of lambs. *Animal* **2011**, *5*, 1124–1130. [CrossRef] [PubMed]

46. Correa, L.B.; Zanetti, M.A.; Del Claro, G.R.; de Melo, M.P.; Rosa, A.F.; Netto, A.S. Effect of supplementation of two sources and two levels of copper on lipid metabolism in Nellore beef cattle. *Meat Sci.* **2012**, *91*, 466–471. [CrossRef] [PubMed]

47. Fernández, M.; Ordóñez, J.A.; Cambero, I.; Santos, C.; Pin, C.; de la Hoz, L. Fatty acid compositions of selected varieties of Spanish dry ham related to their nutritional implications. *Food Chem.* **2007**, *101*, 107–112. [CrossRef]

48. Winiarska-Mieczan, A.; Kwiecień, M.; Kwiatkowska, K.; Baranowska-Wójcik, E.; Szwajgier, D.; Zaricka, E. Fatty acid profile, antioxidative status and dietary value of the breast muscle of broiler chickens receiving glycine-Zn chelates. *Anim. Prod. Sci.* **2020**, *60*, 1095–1102. [CrossRef]

49. Alvarenga, A.; Sousa, R.V.; Parreira, G.G.; Chiarini-Garcia, H.; Almeida, F. Fatty acid profile, oxidative stability of pork lipids and meat quality indicators are not affected by birth weight. *Animal* **2014**, *8*, 660–666. [CrossRef]

50. Mir, N.A.; Tyagi, P.K.; Biswas, A.K.; Tyagi, P.K.; Mandal, A.B.; Kumar, F.; Sharma, D.; Biswas, A.; Verma, A.K. Inclusion of flaxseed, broken rice, and distillers dried grains with solubles (DDGS) in broiler chicken ration alters the fatty acid profile, oxidative stability, and other functional properties of meat. *Eur. J. Lipid Sci. Technol.* **2018**, *120*, 1700470. [CrossRef]

51. Turner, T.; Hessle, A.; Lundström, K.; Pickova, J. Silage-concentrate finishing of bulls versus silage or fresh forage finishing of steers: Effects on fatty acids and meat tenderness. *Acta Agric. Scand.* **2011**, *61*, 103–113. [CrossRef]

52. Lorenzo, J.M.; Crecente, S.; Franco, D.; Sarriés, M.V.; Gómez, M. The effect of livestock production system and concentrate level on carcass traits and meat quality of foals slaughtered at 18 months of age. *Animal* **2014**, *8*, 494–503. [CrossRef] [PubMed]

53. Pires, M.A.; Rodrigues, I.; Barros, J.C.; Carnauba, G.; de Carvalho, F.A.; Trindade, M.A. Partial replacement of pork fat by *Echium* oil in reduced sodium bologna sausages: Technological, nutritional and stability implications. *J. Sci. Food Agric.* **2020**, *100*, 410–420. [CrossRef] [PubMed]

54. Sinanoglou, V.J.; Koutsouli, P.; Fotakis, C.; Sotiropoulou, G.; Cavouras, D.; Bizelis, I. Assessment of lactation stage and breed effect on sheep milk fatty acid profile and lipid quality indices. *Dairy Sci. Technol.* **2015**, *95*, 509–531. [CrossRef]

55. Mierliță, D. Effects of diets containing hemp seeds or hemp cake on fatty acid composition and oxidative stability of sheep milk. *S. Afr. J. Anim. Sci.* **2018**, *48*, 504–515. [CrossRef]

56. Nantapo, C.; Muchenje, V.; Hugo, A. Atherogenicity index and health-related fatty acids in different stages of lactation from Friesian, Jersey and Friesian × Jersey cross cow milk under a pasture-based dairy system. *Food Chem.* **2014**, *146*, 127–133. [CrossRef]

57. Bonanno, A.; Di Grigoli, A.; Mazza, F.; De Pasquale, C.; Giosuè, C.; Vitale, F.; Alabiso, M. Effects of ewes grazing sulla or ryegrass pasture for different daily durations on forage intake, milk production and fatty acid composition of cheese. *Animal* **2016**, *10*, 2074–2082. [CrossRef]

58. Otero, P.; López-Martínez, M.I.; García-Risco, M.R. Application of pressurized liquid extraction (PLE) to obtain bioactive fatty acids and phenols from *Laminaria ochroleuca* collected in Galicia (NW Spain). *J. Pharm. Biomed.* **2019**, *164*, 86–92. [CrossRef]

59. Dellatorre, F.G.; Avaro, M.G.; Commendatore, M.G.; Arce, L.; de Vivar, M.E.D. The macroalgal ensemble of Golfo Nuevo (Patagonia, Argentina) as a potential source of valuable fatty acids for nutritional and nutraceutical purposes. *Algal Res.* **2020**, *45*, 101726. [CrossRef]

60. Rahmouni, N.; Pinto, D.C.; Santos, S.A.; Beghidja, N.; Silva, A.M. Lipophilic composition of *Scabiosa stellata* L.: An underexploited plant from Batna (Algeria). *Chem. Pap.* **2018**, *72*, 753–762. [CrossRef]

61. Pandey, S.; Patel, M.K.; Mishra, A.; Jha, B. Physio-biochemical composition and untargeted metabolomics of cumin (*Cuminum cyminum L.*) make it promising functional food and help in mitigating salinity stress. *PLoS ONE* **2015**, *10*, e0144469. [CrossRef]

62. Ratusz, K.; Symoniuk, E.; Wroniak, M.; Rudzińska, M. Bioactive Compounds, nutritional quality and oxidative stability of cold-pressed Camelina (*Camelina sativa L.*) oils. *Appl. Sci.* **2018**, *8*, 2606. [CrossRef]

63. Wu, Z.X.; Hu, X.P.; Zhou, D.Y.; Tan, Z.F.; Liu, Y.X.; Xie, H.K.; Rakariyatham, K.; Shahidi, F. Seasonal variation of proximate composition and lipid nutritional value of two species of scallops (*Chlamys farreri* and *Patinopecten yessoensis*). *Eur. J. Lipid Sci. Technol.* **2019**, *121*, 1800493. [CrossRef]

64. Akintola, S.L. Effects of smoking and sun-drying on proximate, fatty and amino acids compositions of southern pink shrimp (*Penaeus notialis*). *J. Food Sci. Technol.* **2015**, *52*, 2646–2656. [CrossRef] [PubMed]

65. Łuczyńska, J.; Paszczyk, B. Health risk assessment of heavy metals and lipid quality indexes in freshwater fish from lakes of Warmia and Mazury region, Poland. *Int. J. Environ. Res. Public Health* **2019**, *16*, 3780. [CrossRef]

66. Łuczyńska, J.; Paszczyk, B.; Nowosad, J.; Łuczyński, M.J. Mercury, fatty acids content and lipid quality indexes in muscles of freshwater and marine fish on the polish market. Risk assessment of fish consumption. *Int. J. Environ. Res. Public Health* **2017**, *14*, 1120. [CrossRef]

67. González-Félix, M.L.; Maldonado-Othón, C.A.; Perez-Velazquez, M. Effect of dietary lipid level and replacement of fish oil by soybean oil in compound feeds for the shortfin corvina (*Cynoscion parvipinnis*). *Aquaculture* **2016**, *454*, 217–228. [CrossRef]

68. Monteiro, M.; Matos, E.; Ramos, R.; Campos, I.; Valente, L.M. A blend of land animal fats can replace up to 75% fish oil without affecting growth and nutrient utilization of European seabass. *Aquaculture* **2018**, *487*, 22–31. [CrossRef]

69. Benhissi, H.; García-Rodríguez, A.; de Heredia, I.B. The effects of rapeseed cake intake during the finishing period on the fatty-acid composition of the *longissimus* muscle of Limousin steers and changes in meat colour and lipid oxidation during storage. *Anim. Prod. Sci.* **2020**, *60*, 1103–1110. [CrossRef]

70. Subhadra, B.; Lochmann, R.; Rawles, S.; Chen, R. Effect of dietary lipid source on the growth, tissue composition and hematological parameters of largemouth bass (*Micropterus salmoides*). *Aquaculture* **2006**, *255*, 210–222. [CrossRef]

71. Bouzgarrou, O.; El Mzougui, N.; Sadok, S. Smoking and polyphenols' addition to improve freshwater mullet (*Mugil cephalus*) fillets' quality attributes during refrigerated storage. *Int. J. Food Sci. Technol.* **2016**, *51*, 268–277. [CrossRef]

72. Dal Bosco, A.; Mugnai, C.; Roscini, V.; Castellini, C. Fillet fatty acid composition, estimated indexes of lipid metabolism and oxidative status of wild and farmed brown trout (*Salmo trutta L.*). *Ital. J. Food Sci.* **2013**, *25*, 83–89.

73. Senso, L.; Suárez, M.D.; Ruiz-Cara, T.; García-Gallego, M. On the possible effects of harvesting season and chilled storage on the fatty acid profile of the fillet of farmed gilthead sea bream (*Sparus aurata*). *Food Chem.* **2007**, *101*, 298–307. [CrossRef]

74. Ghafari, H.; Rezaeian, M.; Sharifi, S.D.; Khadem, A.A.; Afzalzadeh, A. Effects of dietary sesame oil on growth performance and fatty acid composition of muscle and tail fat in fattening Chaal lambs. *Anim. Feed. Sci. Technol.* **2016**, *220*, 216–225. [CrossRef]

75. Peiretti, P.G.; Masoero, G.; Meineri, G. Effects of replacing palm oil with maize oil and *Curcuma longa* supplementation on the performance, carcass characteristics, meat quality and fatty acid profile of the perirenal fat and muscle of growing rabbits. *Animal* **2011**, *5*, 795–801. [CrossRef]

76. Salvatori, G.; Pantaleo, L.; Di Cesare, C.; Maiorano, G.; Filetti, F.; Oriani, G. Fatty acid composition and cholesterol content of muscles as related to genotype and vitamin E treatment in crossbred lambs. *Meat Sci.* **2004**, *67*, 45–55. [CrossRef]

77. Wójciak, K.M.; Stasiak, D.M.; Ferysiuk, K.; Solska, E. The influence of sonication on the oxidative stability and nutritional value of organic dry-fermented beef. *Meat Sci.* **2019**, *148*, 113–119. [CrossRef]

78. Ivanova, S.; Angelov, L. Assessment of the content of dietary *trans* fatty acids and biologically active substances in cow's milk and curd. *Generations* **2017**, *4*, 5.

79. Bodas, R.; Manso, T.; Mantecon, A.R.; Juarez, M.; De la Fuente, M.A.; Gómez-Cortés, P. Comparison of the fatty acid profiles in cheeses from ewes fed diets supplemented with different plant oils. *J. Agric. Food Chem.* **2010**, *58*, 10493–10502. [CrossRef]

80. Casamassima, D.; Nardoia, M.; Palazzo, M.; Vizzarri, F.; D Alessandro, A.G.; Corino, C. Effect of dietary extruded linseed, verbascoside and vitamin E supplements on yield and quality of milk in Lacaune ewes. *J. Dairy Res.* **2014**, *81*, 485–493. [CrossRef]

81. Kholif, A.E.; Morsy, T.A.; Abd El Tawab, A.M.; Anele, U.Y.; Galyean, M.L. Effect of supplementing diets of Anglo-Nubian goats with soybean and flaxseed oils on lactational performance. *J. Agric. Food Chem.* **2016**, *64*, 6163–6170. [CrossRef] [PubMed]

82. Kholif, A.E.; Gouda, G.A.; Olafadehan, O.A.; Abdo, M.M. Effects of replacement of Moringa oleifera for berseem clover in the diets of Nubian goats on feed utilisation, and milk yield, composition and fatty acid profile. *Animal* **2018**, *12*, 964–972. [CrossRef] [PubMed]

83. Rutkowska, J.; Białek, M.; Bagnicka, E.; Jarczak, J.; Tambor, K.; Strzałkowska, N.; Jóźwik, A.; Krzyżewski, J.; Adamska, A.; Rutkowska, E. Effects of replacing extracted soybean meal with rapeseed cake in corn grass silage-based diet for dairy cows. *J. Dairy Res.* **2015**, *82*, 161–168. [CrossRef] [PubMed]

84. Esposito, G.; Masucci, F.; Napolitano, F.; Braghieri, A.; Romano, R.; Manzo, N.; Di Francia, A. Fatty acid and sensory profiles of Caciocavallo cheese as affected by management system. *J. Dairy Sci.* **2014**, *97*, 1918–1928. [CrossRef] [PubMed]

85. Yurchenko, S.; Sats, A.; Tatar, V.; Kaart, T.; Mootse, H.; Jõudu, I. Fatty acid profile of milk from Saanen and Swedish Landrace goats. *Food Chem.* **2018**, *254*, 326–332. [CrossRef] [PubMed]

86. Ahmad, N.; Manzoor, M.F.; Shabbir, U.; Ahmed, S.; Ismail, T.; Saeed, F.; Nisa, M.; Anjum, F.M.; Hussain, S. Health lipid indices and physicochemical properties of dual fortified yogurt with extruded flaxseed omega fatty acids and fibers for hypercholesterolemic subjects. *Food Sci. Nutr.* **2019**, *8*, 273–280. [CrossRef]

87. Salles, M.S.; D Abreu, L.F.; Júnior, L.C.R.; César, M.C.; Guimarães, J.G.; Segura, J.G.; Rodrigues, C.; Zanetti, M.A.; Pfrimer, K.; Netto, A.S. Inclusion of sunflower oil in the bovine diet improves milk nutritional profile. *Nutrients* **2019**, *11*, 481. [CrossRef]

88. Vargas-Bello-Pérez, E.; Íñiguez-González, G.; Fehrmann-Cartes, K.; Toro-Mujica, P.; Garnsworthy, P.C. Influence of fish oil alone or in combination with hydrogenated palm oil on sensory characteristics and fatty acid composition of bovine cheese. *Anim. Feed Sci. Technol.* **2015**, *205*, 60–68. [CrossRef]

89. Pittau, D.; Panzalis, R.; Spanu, C.; Scarano, C.; De Santis, E.P. Survey on the fatty acids profile of fluid goat milk. *Ital. J. Food Saf.* **2013**, *2*, 33. [CrossRef]

90. Paiva, L.; Lima, E.; Neto, A.I.; Marcone, M.; Baptista, J. Health-promoting ingredients from four selected Azorean macroalgae. *Food Res. Int.* **2016**, *89*, 432–438. [CrossRef]

91. Santos-Silva, J.; Bessa, R.; Santos-Silva, F. Effect of genotype, feeding system and slaughter weight on the quality of light lambs: II. Fatty acid composition of meat. *Livest. Prod. Sci.* **2002**, *77*, 187–194. [CrossRef]

92. Bonanno, A.; Di Grigoli, A.; Vitale, F.; Alabiso, M.; Giosuè, C.; Mazza, F.; Todaro, M. Legume grain-based supplements in dairy sheep diet: Effects on milk yield, composition and fatty acid profile. *Anim. Prod. Sci.* **2016**, *56*, 130–140. [CrossRef]

93. Bobe, G.; Zimmerman, S.; Hammond, E.G.; Freeman, A.E.; Porter, P.A.; Luhman, C.M.; Beitz, D.C. Butter composition and texture from cows with different milk fatty acid compositions fed fish oil or roasted soybeans. *J. Dairy Sci.* **2007**, *90*, 2596–2603. [CrossRef] [PubMed]

94. Chen, S.; Bobe, G.; Zimmerman, S.; Hammond, E.G.; Luhman, C.M.; Boylston, T.D.; Freeman, A.E.; Beitz, D.C. Physical and sensory properties of dairy products from cows with various milk fatty acid compositions. *J. Agric. Food Chem.* **2004**, *52*, 3422–3428. [CrossRef] [PubMed]

95. Giorgio, D.; Di Trana, A.; Di Napoli, M.A.; Sepe, L.; Cecchini, S.; Rossi, R.; Claps, S. Comparison of cheeses from goats fed 7 forages based on a new health index. *J. Dairy Sci.* **2019**, *102*, 6790–6801. [CrossRef]

96. Colombo, M.L.; Rise, P.; Giavarini, F.; De Angelis, L.; Galli, C.; Bolis, C.L. Marine macroalgae as sources of polyunsaturated fatty acids. *Plant Foods Hum. Nutr.* **2006**, *61*, 64–69. [CrossRef]

97. Poerschmann, J.; Spijkerman, E.; Langer, U. Fatty acid patterns in *Chlamydomonas* sp. as a marker for nutritional regimes and temperature under extremely acidic conditions. *Microb. Ecol.* **2004**, *48*, 78–89. [CrossRef]

98. Kumari, P.; Kumar, M.; Gupta, V.; Reddy, C.; Jha, B. Tropical marine macroalgae as potential sources of nutritionally important PUFAs. *Food Chem.* **2010**, *120*, 749–757. [CrossRef]

99. Ghassemi-Golezani, K.; Farhangi-Abriz, S. Changes in oil accumulation and fatty acid composition of soybean seeds under salt stress in response to salicylic acid and jasmonic acid. *Russ. J. Plant Physiol.* **2018**, *65*, 229–236. [CrossRef]

100. Gomes-Laranjo, J.; Peixoto, F.; Sang, H.W.W.F.; Torres-Pereira, J. Study of the temperature effect in three chestnut (*Castanea sativa* Mill.) cultivars' behaviour. *J. Plant Physiol.* **2006**, *163*, 945–955. [CrossRef]

101. Segura, J.; Escudero, R.; de Ávila, M.R.; Cambero, M.I.; López-Bote, C.J. Effect of fatty acid composition and positional distribution within the triglyceride on selected physical properties of dry-cured ham subcutaneous fat. *Meat Sci.* **2015**, *103*, 90–95. [CrossRef] [PubMed]

102. Rodríguez, M.; García-García, R.M.; Arias-Álvarez, M.; Millán, P.; Febrel, N.; Formoso-Rafferty, N.; López-Tello, J.; Lorenzo, P.L.; Rebollar, P.G. Improvements in the conception rate, milk composition and embryo quality of rabbit does after dietary enrichment with n-3 polyunsaturated fatty acids. *Animal* **2018**, *12*, 2080–2088. [CrossRef] [PubMed]

103. Lin, H.Z.; Liu, Y.J.; He, J.G.; Zheng, W.H.; Tian, L.X. Alternative vegetable lipid sources in diets for grouper, *Epinephelus coioides* (Hamilton): Effects on growth, and muscle and liver fatty acid composition. *Aquac. Res.* **2007**, *38*, 1605–1611. [CrossRef]

104. Larsson, T.; Koppang, E.O.; Espe, M.; Terjesen, B.F.; Krasnov, A.; Moreno, H.M.; Rørvik, K.; Thomassen, M.; Mørkøre, T. Fillet quality and health of Atlantic salmon (*Salmo salar L.*) fed a diet supplemented with glutamate. *Aquaculture* **2014**, *426*, 288–295. [CrossRef]

105. Mozanzadeh, M.T.; Marammazi, J.G.; Yavari, V.; Agh, N.; Mohammadian, T.; Gisbert, E. Dietary n-3 LC-PUFA requirements in silvery-black porgy juveniles (*Sparidentex hasta*). *Aquaculture* **2015**, *448*, 151–161. [CrossRef]

106. Xu, W.; Qian, Y.; Li, X.; Li, J.; Li, P.; Cai, D.; Liu, W. Effects of dietary biotin on growth performance and fatty acids metabolism in blunt snout bream, *Megalobrama amblycephala* fed with different lipid levels diets. *Aquaculture* **2017**, *479*, 790–797. [CrossRef]

107. Abrami, G.; Natiello, F.; Bronzi, P.; McKenzie, D.; Bolis, L.; Agradi, E. A comparison of highly unsaturated fatty acid levels in wild and farmed eels (*Anguilla anguilla*). *Comp. Biochem. Physiol.* **1992**, *101*, 79–81. [CrossRef]

108. Krajnović-Ozretic, M.; Najdek, M.; Ozretić, B. Fatty acids in liver and muscle of farmed and wild sea bass (*Dicentrarchus labrax L.*). *Comp. Biochem. Physiol.* **1994**, *109*, 611–617. [CrossRef]

109. Sharma, R.; Ahlawat, S.; Aggarwal, R.; Dua, A.; Sharma, V.; Tantia, M.S. Comparative milk metabolite profiling for exploring superiority of indigenous Indian cow milk over exotic and crossbred counterparts. *J. Food Sci. Technol.* **2018**, *55*, 4232–4243. [CrossRef]

110. Mishra, R.; Sharma, H.K. Effect of frying conditions on the physico-chemical properties of rice bran oil and its blended oil. *J. Food Sci. Technol.* **2014**, *51*, 1076–1084. [CrossRef]

111. Dietschy, J.M. Dietary fatty acids and the regulation of plasma low density lipoprotein cholesterol concentrations. *J. Nutr.* **1998**, *128*, 444–448. [CrossRef] [PubMed]

112. Ulbricht, T.; Southgate, D. Coronary heart disease: Seven dietary factors. *Lancet* **1991**, *338*, 985–992. [CrossRef]

113. Omri, B.; Chalghoumi, R.; Izzo, L.; Ritieni, A.; Lucarini, M.; Durazzo, A.; Abdouli, H.; Santini, A. Effect of dietary incorporation of linseed alone or together with tomato-red pepper mix on laying hens' egg yolk fatty acids profile and health lipid indexes. *Nutrients* **2019**, *11*, 813. [CrossRef] [PubMed]

114. Logue, J.A.; De Vries, A.L.; Fodor, E.; Cossins, A.R. Lipid compositional correlates of temperature-adaptive interspecific differences in membrane physical structure. *J. Exp. Biol.* **2000**, *203*, 2105–2115. [PubMed]

115. Burdge, G.C.; Jones, A.E.; Wootton, S.A. Eicosapentaenoic and docosapentaenoic acids are the principal products of α-linolenic acid metabolism in young men. *Br. J. Nutr.* **2002**, *88*, 355–363. [CrossRef] [PubMed]

116. Johnson, E.J.; Schaefer, E.J. Potential role of dietary n-3 fatty acids in the prevention of dementia and macular degeneration. *Am. J. Clin. Nutr.* **2006**, *83*, 1494–1498. [CrossRef]

117. Brenna, J.T.; Salem, N., Jr.; Sinclair, A.J.; Cunnane, S.C. α-Linolenic acid supplementation and conversion to n-3 long-chain polyunsaturated fatty acids in humans. *Prostaglandins Leukot. Essent. Fat. Acids* **2009**, *80*, 85–91. [CrossRef]

118. Wayland, M.M. Final determination regarding partially hydrogenated oils. *Fed. Regist.* **2015**, *80*, 116.

119. Food and Drug Administration, H. Final Determination Regarding Partially Hydrogenated Oils. Notification; declaratory order; extension of compliance date. *Fed. Regist.* **2018**, *83*, 23358–233589.

120. Lock, A.L.; Corl, B.A.; Barbano, D.M.; Bauman, D.E.; Ip, C. The anticarcinogenic effect of *trans*-11 18: 1 is dependent on its conversion to *cis*-9, *trans*-11 CLA by Δ9-desaturase in rats. *J. Nutr.* **2004**, *134*, 2698–2704. [CrossRef]

121. Corl, B.A.; Barbano, D.M.; Bauman, D.E.; Ip, C. *cis*-9, *trans*-11 CLA derived endogenously from *trans*-11 18: 1 reduces cancer risk in rats. *J. Nutr.* **2003**, *133*, 2893–2900. [CrossRef] [PubMed]

122. Kritchevsky, D.; Tepper, S.A.; Wright, S.; Czarnecki, S.K.; Wilson, T.A.; Nicolosi, R.J. Conjugated linoleic acid isomer effects in atherosclerosis: Growth and regression of lesions. *Lipids* **2004**, *39*, 611. [CrossRef] [PubMed]

123. European Food Safety Authority (EFSA). Scientific opinion on dietary reference values for fats, including saturated fatty acids, polyunsaturated fatty acids, monounsaturated fatty acids, *trans* fatty acids, and cholesterol. *EFSA J.* **2010**, *8*, 1461.

124. European Food Safety Authority (EFSA). Opinion of the scientific panel on dietetic products, nutrition and allergies on a request from the commission related to the presence of *trans* fatty acids in foods and the effect on human health of the consumption of *trans* fatty acids. *EFSA J.* **2004**, *81*, 1–49.

125. Raff, M.; Tholstrup, T.; Sejrsen, K.; Straarup, E.M.; Wiinberg, N. Diets rich in conjugated linoleic acid and vaccenic acid have no effect on blood pressure and isobaric arterial elasticity in healthy young men. *J. Nutr.* **2006**, *136*, 992–997. [CrossRef] [PubMed]

2

Novel *Vibrio* spp. Strains Producing Omega-3 Fatty Acids Isolated from Coastal Seawater

Mónica Estupiñán [1], **Igor Hernández** [2], **Eduardo Saitua** [2], **M. Elisabete Bilbao** [1], **Iñaki Mendibil** [1], **Jorge Ferrer** [2] and **Laura Alonso-Sáez** [1,*]

[1]   AZTI, Marine Research Division, Txatxarramendi Irla s/n, 48395 Sukarrieta, Spain; mestupinan@azti.es (M.E.); ebilbao@azti.es (M.E.B.); imendibil@azti.es (I.M.)

[2]   AZTI, Food Research Division, Astondo Bidea, Building 609, 48160 Derio, Spain; igor.hernandezo@ehu.eus (I.H.); esaitua@azti.es (E.S.); jferrer@azti.es (J.F.)

*   Correspondence: lalonso@azti.es.

Abstract: Omega-3 long-chain polyunsaturated fatty acids (LC-PUFAs), such as eicosapentaenoic acid (EPA) (20:5n-3) and docosahexaenoic acid (DHA) (22:6n-3), are considered essential for human health. Microorganisms are the primary producers of omega-3 fatty acids in marine ecosystems, representing a sustainable source of these lipids, as an alternative to the fish industry. Some marine bacteria can produce LC-PUFAs de novo via the Polyunsaturated Fatty Acid (*Pfa*) synthase/ Polyketide Synthase (PKS) pathway, which does not require desaturation and elongation of saturated fatty acids. Cultivation-independent surveys have revealed that the diversity of microorganisms harboring a molecular marker of the *pfa* gene cluster (i.e., *pfa*A-KS domain) is high and their potential distribution in marine systems is widespread, from surface seawater to sediments. However, the isolation of PUFA producers from marine waters has been typically restricted to deep or cold environments. Here, we report a phenotypic and genotypic screening for the identification of omega-3 fatty acid producers in free-living bacterial strains isolated from 5, 500, and 1000 m deep coastal seawater from the Bay of Biscay (Spain). We further measured EPA production in pelagic *Vibrio* sp. strains collected at the three different depths. *Vibrio* sp. EPA-producers and non-producers were simultaneously isolated from the same water samples and shared a high percentage of identity in their 16S rRNA genes, supporting the view that the *pfa* gene cluster can be horizontally transferred. Within a cluster of EPA-producers, we found intraspecific variation in the levels of EPA synthesis for isolates harboring different genetic variants of the *pfa*A-KS domain. The maximum production of EPA was found in a *Vibrio* sp. strain isolated from a 1000 m depth (average 4.29% ± 1.07 of total fatty acids at 10 °C, without any optimization of culturing conditions).

Keywords: omega-3 fatty acid; eicosapentaenoic acid (EPA); *Vibrio* sp.; polyunsaturated fatty acid synthase *Pfa*; polyketide synthase (PKS); marine gamma-Proteobacteria

1. Introduction

Omega-3 long-chain polyunsaturated fatty acids (LC-PUFAs), such as eicosapentaenoic acid (EPA) (20:5n-3) and docosahexaenoic acid (DHA) (22:6n-3), have a key physiological role in human metabolism [1]. In mammals, omega-3 LC-PUFAs can only be synthesized from the dietary precursor α-linolenic acid (ALA) [2,3], and thus, they are considered essential nutrients. Omega-3 LC-PUFAs are necessary for normal prenatal and child development and, nowadays, their health benefits are well-recognized as they are associated with the improvement of the immune system, neuromuscular function, and neuropsychiatric disorders, among others [1,4–6]. The Food and Agriculture Organization has recommended a minimum daily intake of 250 mg of omega-3 LC-PUFAs and up to 300 mg in pregnant women due to their critical biological functions in fetal development [7]. The consumer awareness of their health implications has triggered an exponential increase of the global market demand of omega-3 fatty acids with therapeutic objectives [8]. Besides the fish and seafood incorporated in the diet, fish oil supplements are the predominant commercial source of omega-3 fatty acids as some wild fish families (e.g., Scombridae, Clupeidae, and Salmonidae) consume and accumulate omega-3 fatty acids in their tissues [9]. However, due to the severe concerns associated with wild fisheries, such as environmental damage of fish stocks and bioaccumulation of pollutants, new sustainable sources of omega-3 fatty acids are needed to meet the growing demand of food and nutraceutical industries [2].

Microorganisms (bacteria and phytoplankton) are the primary producers of omega-3 PUFAs in marine ecosystems and they transfer these valuable fatty acids to the rest of the food chain. Bacteria also represent a green source of LC-PUFAs, with cheaper downstream purification processes compared with fish oil [9–13]. In the last years, several marine bacterial cultures have been isolated with the ability to produce omega-3 fatty acids and the metabolic pathways behind their synthesis have been identified. Most of the known microbial producers are gamma-Proteobacteria affiliated with the genera *Moritella, Photobacteria, Shewanella, Colwellia, Psychromonas*, and *Vibrio* [14]. In the Flavobacteria class, some isolates affiliated with *Flexibacter* and *Psychroserpens* have also shown the ability to produce omega-3 LC-PUFAs [12]. These isolates have been predominantly retrieved from polar regions and deep ocean habitats characterized by high pressure and/or low temperature [10,12,15] or as part of the gut microbiota of some marine and freshwater fishes [16,17]. A variety of physiological roles of LC-PUFAs in microorganisms have been proposed, including adaptation to psychrophile and piezophile habitats by enhancing membrane fluidity [18–20] and acting as antioxidants [21–24].

While the classical pathway of omega-3 LC-PUFAs' production in phytoplankton is through desaturases and elongases in aerobic conditions [25], in marine bacteria, the synthesis of these secondary metabolites is mainly via the anaerobic polyketide synthase (PKS) pathway, mediated by a polyunsaturated fatty acid (*Pfa*) synthase [26]. The gene cluster involved in the latter pathway was first deciphered in *Shewanella* sp. strain SCRC-2738 by functionally characterizing a 38 Kbp genomic fragment containing the operon responsible for EPA synthesis [27]. This gene cluster is composed of at least five genes located sequentially (*pfa*A-E), functioning as a multi-enzyme for the de novo synthesis of omega-3 LC-PUFAs [26,28]. Interestingly, the same gene cluster is found in eukaryotes and bacteria, suggesting that these genes can be transferred horizontally across both domains [29]. Since the discovery of the *pfa* gene cluster, several strategies to enhance omega-3 production have been developed, such as using genetic engineering techniques for heterologous production of EPA/DHA in recombinant strains, and the optimization of their culturing conditions [12]. However, the number of isolated bacteria with the ability to produce omega-3 LC-PUFAs remains rather low. Yet, cultivation-independent surveys based on a functional marker of the *pfa* gene cluster (the β-ketoacyl synthase/KS domain from the *pfa*A gene) have revealed that the diversity of these genes in marine bacteria is large, and their distribution in oceanic samples is much more widespread than initially thought [30]. These results suggest that the potential to retrieve omega-3 LC-PUFAs producers from seawater has been, so far, underexplored.

Here, we performed a bioprospection of bacterial producers of omega-3 LC-PUFAs in the Bay of Biscay, a region located in the north-east Atlantic Ocean characterized by high fish and shellfish diversity [31]. Both physiological and molecular analyses were conducted in order to identify new bacterial producers of omega-3 LC-PUFAs in coastal seawater and explore the possible role of these fatty acids in their adaptation to varying temperature conditions.

2. Results

2.1. Phenotypic and Phylogenetic Analysis of Marine Isolates

From 217 marine isolates obtained from coastal seawater of the Bay of Biscay, ca. 36% were obtained from shallow waters (5 m depth), while 64% were originated from samples collected at 500 and 1000 m depths. No significant variations in viable and culturable bacteria were detected according to the depth of the samples, with a viability range from 2 to 448 CFU/mL (data not shown). All isolated free-living marine bacteria were Gram-negative and catalase positive. Growth was observed at 4, 10, and 25 °C, and therefore the isolates can be considered as psychrotolerant mesophiles. Five different phenotypes were identified based on the colony morphology when grown at 10 °C. Partial 16S rRNA sequences were successfully amplified in 203 strains. The isolates were affiliated with three genera of gamma-Proteobacteria commonly found in marine seawater: *Alteromonas*, *Pseudoalteromonas*, and *Vibrio* sp [32]. Phylotypes closely related to already described species of each genus were found (>99% identity based on partial 16S rRNA genes), although some isolates affiliated with *Alteromonas* and *Pseudoalteromonas* had <99% identity when compared to existant isolates (e.g., *Alteromonas* sp. 611 and 821 and *Pseudoalteromonas* sp. 318) (Figure S1). The relative abundance of each genus varied with depth (Figure S2). *Pseudoalteromonas* sp. was the predominant phylotype in shallow (5 m) and the deepest (1000 m) water samples, while *Vibrio* and *Alteromonas* sp. were mostly found at the 500 m depth.

2.2. Phenotypic Screening of LC-PUFA Producers by the TTC Test and GC-FID Lipid Analysis

In order to identify potential LC-PUFAs producers, a first phenotype-based screening was performed based on TTC plates (supplemented with 2,3,5-triphenyl-tetrazolium chloride). About 64% positive-red color colonies were found among the purified isolates and were thus considered as potential LC-PUFAs producers. Similar percentages of TTC-positive colonies were detected at different isolation depths (ca. 22%, 25%, and 17% in 5, 500 and 1000 m depths, respectively). Most TTC-positive colonies were affiliated with *Pseudoalteromonas* sp. at 5 and 1000 m depths, and with *Alteromonas* sp. at the 500 m depth (Figure S2). Although isolates from the three genera showed >99% of identity among them (based on partial 16S rRNA genes) (data not shown), TTC phenotypic variations were observed. Based on this initial screening, we selected 20 TTC-positive isolates from the three genera and depth of isolation and two TTC-negative isolates for GC-FID analysis of their fatty acid profiles. The most abundant fatty acid in the lipid profiles within the assayed isolates was 16:1, followed by 16:0 and 18:1 (Figure S3). All TTC-positive isolates that produced significant amounts of omega-3 (>1% total fatty acids—TFA) at 10 °C, specifically EPA (20:5n-3), were affiliated with *Vibrio* sp.

2.3. Molecular Screening of Omega-3 Fatty Acid Producers by the pfaA-KS Functional Marker and GC-MS Lipid Analysis of Vibrio sp. Isolates

A molecular screening based on the presence of the *pfa*A-KS gene domain was performed on 86 TTC-positive isolates. A positive amplification of the *pfa*A-KS region (ca. 400 bp) was found in 63% TTC-positive *Vibrio* sp. isolates whereas none of the TTC-positive *Pseudoalteromonas* and *Alteromonas* sp. were positive for the *pfa*A-KS amplification (Table S1). While the amino acid sequence of the *pfa*A-KS amplified region was identical in all *pfa*A-KS-positive strains (Figure S4), four clusters were identified (I-IV) at the nucleotide level. The same genetic variant of the *pfa*A-KS was shared among strains isolated at different depths (Figure 1). In order to confirm EPA production in *Vibrio* sp. isolates according to this molecular screening, we re-analyzed by GC-MS a total of 11 *pfa*A-KS positive and 5 *pfa*A-KS negative strains, as representatives from the three water depths of isolation (Figure S5, Figure S6). EPA production was confirmed in all *pfa*A-KS-positive strains and absent in negative strains. Interestingly, different *pfa*A-KS genetic variants identified at the nucleotide level produced different levels of EPA when grown at 10 °C (Figure 1). Specifically, the strains affiliated with cluster I produced a lower quantity of EPA (<2% of total fatty acids) as compared to the other clusters (Student's T-test, $p < 0.005$). The most promising candidate for EPA production, *Vibrio* sp. 618, displayed a genetic variant of *pfa*A-KS phylogenetically close to that of *Vibrio splendidus* LGP32, with a reported maximum of 2.94% of eicosenoid fatty acids ($\sum$20:1n-9 + 20:2 + 20:4n-6 + 20:5n-3) at 25 °C in previous studies [33] and *Photobacterium profundum* SS9, a well-known EPA producer in psychrophilic conditions (7% of EPA at 4 °C) [28]. Other *Vibrio* sp. harboring the *pfa*A-KS cluster (*V. splendidus* LGP32, *V. alginolyticus* K08M4, *V. crassostreae* 9CS106, and *V. tapetis* CECT4600) were identified from the NCBI Genbank database (see the materials and methods, Section 4.3) and included in the phylogenetic tree, although their ability to produce omega-3 LC-PUFAs has not been determined yet experimentally.

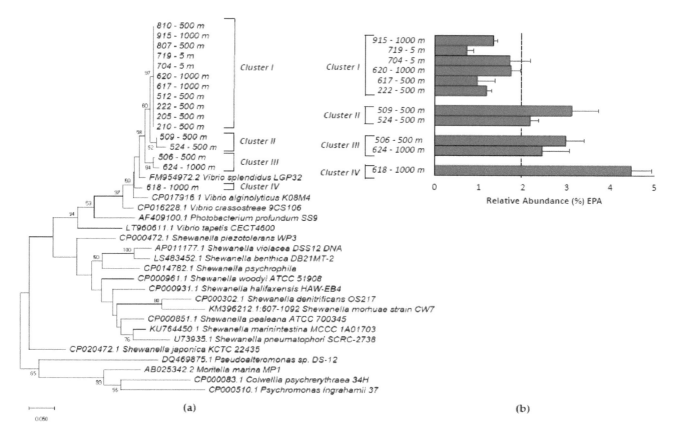

Figure 1. (**a**) Phylogenetic tree of *pfa*A-KS nucleotide sequences. GenBank sequences are identified with the accession numbers and *Vibrio* sp. strains isolated in this study are identified by a number and the depth of isolation in the water column, (**b**) relative abundance of eicosapentaenoic acid (EPA, % of total fatty acids) of representative *Vibrio* sp. isolates in the clusters identified in the phylogenetic tree.

2.4. Phylogenetic Analysis of Vibrio Isolates

In order to conduct a more robust phylogenetic analysis of Vibrio isolates, an analysis of full-length 16S rRNA genes was carried out (Figure 2). Most of the Vibrio sp. isolates were closely related at the phylogenetic level, and two main clades were observed (clade I and clade II, sharing 99.05% identity among them, Figure 2). Both clades included phylotypes retrieved from the three different depths, confirming their presence in the same environments along the water column. Based on 16S rRNA sequence analysis, Vibrio sp. 618 shared a maximum of 99.85% nucleotide identity with sequences deposited in GenBank, with the bioluminescent Vibrio splendidus LGP32 (FM954972.2) being the most closely related. However, it should be noted that bioluminescence was not observed in any of the isolated Vibrio sp. strains under the assayed conditions. Clades I and II contained members harboring the pfaA-KS marker gene (light blue, Figure 2), and some of them were experimentally confirmed EPA producers (dark blue, Figure 2). However, clade I also included Vibrio isolates, which were not positive for the pfaA-KS marker gene. In general, the phylogenetic relationship of most of the Vibrio isolates based on the 16S rRNA gene was not congruent with that observed based on the pfaA-KS marker (Figure 2).

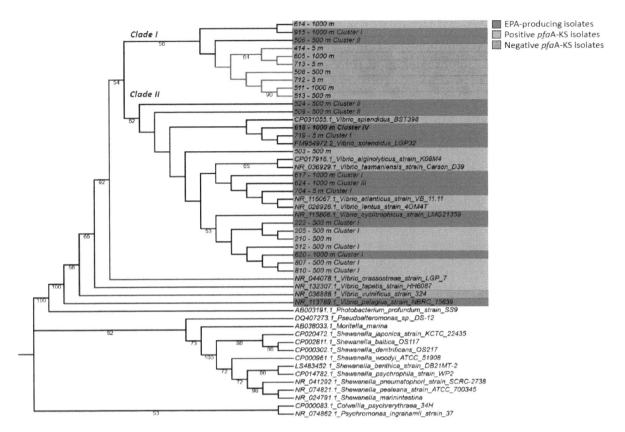

Figure 2. Phylogenetic tree based on 16S rRNA nucleotide sequences. Experimentally demonstrated omega-3 LC-PUFA producers are highlighted in dark blue. Positive *pfa*A-KS isolates are highlighted in light blue and *pfa*A-KS negative isolates in red. Uncolored strains are described gamma-proteobacteria omega-3 producers. GenBank sequences are identified with their correspondence accession numbers. New *Vibrio* sp. isolates are identified by an ID number and depth of isolation (m) in the water column. The best EPA-producer isolate (*Vibrio* sp. 618) is highlighted in bold.

2.5. Genomic Sequence Analysis

We selected the EPA producer *Vibrio* sp. 618, and the non-producer *Vibrio* sp. 414 for whole genome sequencing. Both strains were taxonomically assigned to *Vibrio tasmaniensis* LGP32 (*Vibrio splendidus* LGP32) by the WIMP Nanopore software (Table S2). The presence of the *pfa*/PKS cluster was confirmed in the genome of *Vibrio* sp. 618, while it was apparently absent in *Vibrio* sp. 414. Based on a BLASTn search, the different *pfa* genes were identified in the assembled *Vibrio* sp. 618 genome. The genetic organization of the EPA biosynthesis cluster in *Vibrio* sp. 618 was composed of *pfa*A–E genes and a hypothetical regulatory gene (*epa*R) (Figure 3). This organization has been previously found in other EPA-producing gamma-proteobacteria strains (e.g., *Shewanella* sp.) and has been designated as a type A *pfa* synthase cluster, in which the order of *pfa* genes is the following: *pfa*A [KS-MAT-ACP(4–6)-KR], *pfa*B [AT], *pfa*C [KS-CLF-DH2], *pfa*D [ER], and *pfa*E [PPTase] [17,29].

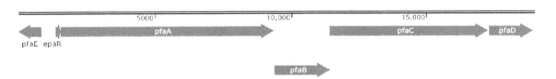

Figure 3. Organization of the genes encoding the EPA biosynthetic polyketide synthase (PKS) cluster present in *Vibrio* sp. 618, as obtained by whole-genome sequencing using the Nanopore MinION platform.

2.6. Effect of temperature on fatty acid profiles of Vibrio sp. EPA-producers and non-producers

We tested the effect of temperature on the fatty acid profiles of an EPA-producer isolate (*Vibrio* sp. 618) and a non-producer isolate (*Vibrio* sp. 414). A clear increment in the percentage of monounsaturated fatty acids (MUFAs 16:1 and 18:1) was observed for both strains under cold temperature, particularly in the case of 16:1 (Figure 4). EPA production (>1% TFA) was observed in *Vibrio* sp. 618 at 10 °C but not at 25 °C (Figure 4 and Figure S7). The average production at 10 °C in the latter strain was 1.82 mg/g of dry weight (4.29% of TFA ± 1.07%) after 24 h and up to 2.34 mg/g (7.33% of TFA) after 120 h.

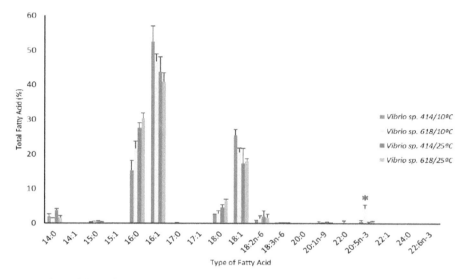

Figure 4. Fatty acid profiles of an EPA producer (*Vibrio* sp. 618) and a non-producer (*Vibrio* sp. 414) grown at 10 and 25 °C. The EPA peak has been highlighted with an asterisk.

The MUFA/SFA ratio was higher in *Vibrio* sp. 414 (non-producer strain) than in *Vibrio* sp. 618 (EPA producer strain) at both temperatures, especially at 10 °C. Accordingly, the unsaturated index values (see the materials and methods for a definition) observed at 10 °C (91% for EPA producer and 82% for the non-producer strain) were significantly higher than the values obtained at 25 °C (64% for the EPA producer and 68% for the non-producer strain, Student's t test, $p < 0.005$, Figure 5). Additionally, we performed a PCR-screening to verify the presence of DES9, a key enzyme involved in MUFA (16:1n-9 and 18:1n-9) synthesis by the desaturation of SFAs (16:0 and 18:0). Both *Vibrio* sp. strains were positive for Δ9-des amplification (Table S1).

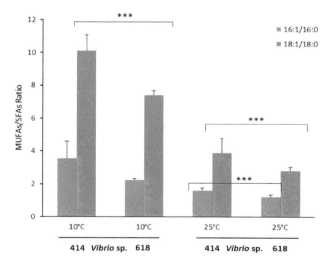

Figure 5. Monounsaturated fatty acids (MUFAs)/saturated fatty acids (SFAs) ratio in *Vibrio* sp. 618 and 414 grown at 10 and 25 °C. (***) indicates statistical significance (Student t-test, $p < 0.005$).

Finally, the growth rate of both strains at 10 and 25 °C was compared (Figure S8). In mesophilic conditions (25 °C), both strains showed a short lag phase, and the maximal growth rates (μmax) were 0.507 and 0.309 h^{-1} for *Vibrio* sp. 618 and 414, respectively. In psychrophilic conditions (10 °C), the non-producer strain showed a substantially longer lag phase (5.8 h) than the EPA producer strain (Figure S8). However, maximal growth rates were similar in both cases (0.179 and 0.151 h^{-1} for *Vibrio* sp. 618 and 414, respectively) and they reached a similar maximal abundance after 45 h of incubation.

3. Discussion

Extreme environments, such as polar waters and the deep ocean, have been the main sources of bacterial producers of omega-3 fatty acids [13,34]. Due to their important role in the fluidity and functioning of cell membranes, LC-PUFAs are presumably key for the adaptation of these microorganisms to cold temperature and high atmospheric pressure conditions [10,19,35,36]. However, a recent survey targeting a molecular marker of the microbial PKS pathway of the synthesis of LC-PUFAs (*pfa*A-KS) in different marine samples has suggested a much more widespread distribution of potential EPA/DHA producers in the ocean [29,30]. Unexpectedly, the maximum relative abundance of potentially EPA-producing bacteria in the latter study was found in surface waters (3 m depth) at 17 °C, in a sample collected off the Scripps Pier (La Jolla, California, CA, USA) [30]. These results suggest that bacterial EPA producers can be potentially isolated from shallow seawater, which would represent a much less challenging source for bioprospection as compared to deep or polar waters.

In this work, we successfully isolated novel pelagic omega-3 fatty acid producers from surface coastal waters (5 m depth), affiliated with *Vibrio* sp. EPA/DHA-producing *Vibrio* sp. had been previously isolated from deep-sea sediments [14,37,38], marine fish larvae, and freshwater fish gut [16,39]. Although it should be noted that the proportion of EPA-producing *Vibrio* sp. was higher in isolates recovered at 500 and 1000 m depths, here we confirm that bacterial producers of omega-3 LC-PUFAs are not restricted to piezophilic and psychrophilic environments [17,40,41].

The plate-based TTC method has been recently proposed for the screening of the production of EPA in Gram-negative bacteria, where positive LC-PUFAs colonies are red colored [42]. In accordance with previous screening studies [17,34,43], we observed that this method causes false positives, as many TTC-positive isolates did not produce EPA/DHA, as confirmed by subsequent GC-FID and GC-MS analyses. In 2016, Dailey and co-workers included for the first time a PCR step for evaluating the presence of the *pfa*A-KS domain in some TTC-positive colonies with previously designed primers [17,30]. The amplification of the *pfa*A-KS domain was experimentally demonstrated in Gram-negative EPA-producing isolates in the latter work, and more recently, in Gram-positive EPA-producing strains [44]. Here, we included a PCR screening step and further measured their lipid content by GC-MS in order to experimentally confirm that *pfa*A-KS positive strains were indeed omega-3 LC-PUFA producers. Based on our results, we strongly recommend the use of *pfa*A-KS PCR screening when analyzing a large number of samples in order to minimize the effort required for GC analysis. This molecular approach has the advantage of being independent of bacterial growth, temperature, and media composition, which are the main sources of variations in the analysis of microbial fatty acid profiles [45–48].

Furthermore, among EPA-producing *Vibrio* sp. strains, we found that different genetic variants of *pfa*A-KS produced different levels of EPA. The isolate *Vibrio* sp. 618, which showed the highest levels of production, harbored a *pfa*A-KS genetic variant similar to that of the isolate *Photobacterium profundum* SS9, which produces a similar quantity of EPA (8.4% of TFA) [49,50], and also belongs to the *Vibrio*naceae family. These results may suggest that the *pfa*A-KS molecular marker might be used as a tool not only for identifying EPA/DHA-producing strains but also for determining genetic variants with different levels of LC-PUFAs production. However, more research should be conducted to confirm this point by increasing the number of isolates tested, and by ensuring a straightforward comparison under similar growth conditions. Additionally, based on the presence of the *pfa* gene cluster in the genomes of some other *Vibrio* isolates (e.g., *V. alginolyticus* strain K08M4, *V. crassostreae* strain LGP7,

V. lentus strain 4OM4T, *V. splendidus* BST398, *V. tapetis* strain HH6087, *V. tasmaniensis* strain Carson D39), we suggest that they are also potential EPA producers, even if there is no experimental evidence in the literature.

From an environmental perspective, it is interesting that *pfa*A-KS-positive/EPA-producing and *pfa*A-KS-negative/EPA-non-producing *Vibrio* sp. strains, in some cases sharing >99% identity at the phylogenetic level, were isolated simultaneously in the same marine samples. The presence of the complete *pfa* gene cluster (*pfa*A–E) was further confirmed by whole genome sequencing in the strain *Vibrio sp.* 618. The same cluster organization has also been annotated in the genome of *Vibrio splendidus* LG32 [51], the closest strain to the *Vibrio* sp. isolates found in this study. Moreover, the *pfa* gene cluster has been identified in three other *Vibrio* genomes (*Vibrio* sp. MED222, *V. splendidus* 12B01, and *V. splendidus* LGP32), while it was absent in other *Vibrio*nales. These results support the view that the *pfa* cluster can be transferred horizontally among *Vibrio* species, as suggested for other taxa [29,30]. Acquiring these genes may provide a selective advantage in terms of responding to changes in environmental conditions, such as decreases in temperature. In order to explore this hypothesis, we tested the effect of temperature on the growth and fatty acid composition of an EPA-producing and a non-producing isolate (*Vibrio* sp. 618 and 414, respectively), which had a similar fatty acid profile under mesophilic conditions (Figure 4). As reported previously for other *Vibrio* isolates [52–54], the major fatty acids of these *Vibrio* strains were 16:1, 16:0, and 18:1. We found that the relative proportion of these fatty acids was temperature dependent for both *Vibrio* sp. 414 and 618. When grown in low temperature conditions (10 °C), a higher proportion of MUFAs was found in *Vibrio* sp. 414 as compared to *Vibrio* sp. 618, while in the latter, EPA production was observed. These results suggest that omega-3 LC-PUFAs are not essential for the adaptation of *Vibrio* sp. to cold environments in accordance with previous observations [14,18]. Specifically, the EPA non-producer *Vibrio* sp. strain 414 adapted to low temperature conditions by increasing the ratio of 16:1 and 18:1 to maintain membrane fluidity as it has been described previously in other Vibrionales, such as *Photobacterium profundum* SS9 [18]. Fatty acid desaturases encoded by the Δ9-*des* gene have been demonstrated to be involved in the production of MUFAs 16:1(n-9) and 18:1(n-9) in aerobic conditions [19] in some gamma-proteobacteria, such as *Psychrobacter urativorans*, *Pseudoalteromonas* sp. MLY15, and the Antarctic strain *Pseudomonas* sp. A3 [18,55,56]. Here, the Δ9-*des* gene was amplified in all *Vibrio* sp. isolates, suggesting its potential role for the adaptation of cell membranes to fluctuating environmental conditions in this genus [57]. Finally, although both *Vibrio* sp. strains were able to grow at 10 °C, the EPA non-producer strain *Vibrio* sp. 414 showed a longer lag phase compared with *Vibrio* sp. 618, which suggests a faster adaptation to low temperature conditions for the EPA-producing strain. In order to verify the role of the omega-3 fatty acids in these strains, more detailed studies should be performed, for example, generating *Vibrio* sp. *pfa* mutants, or by comparing more EPA producer and non-producer isolates.

In summary, in this work, we successfully isolated bacterial producers of EPA from shallow coastal seawater affiliated with *Vibrio* sp., confirming the potential of retrieving novel marine omega-3 producers from the surface ocean. Given the easy access to this environmental niche as compared to the deep ocean or polar environments, this represents an important advantage in terms of bioprospection. Regarding the biotechnological potential of the isolated strains, the maximum yield of EPA production in the most promising candidate (*Vibrio* sp. 618 isolated at the 1000 m depth) was 7.33% TFA at 120 h, grown at 10 °C without optimization. This value is similar to that observed for *V. cyclotrophicus* (10% TFA) and *V. pelagius* (8.7% TFA) [12], and it remains to be seen if further optimization could substantially increase this yield. We also highlight the great value of using the *pfa*A-KS molecular marker in order to optimize the screening of potential EPA/DHA producers, and its potential to differentiate clusters with different production yields, which needs to be evaluated in future studies.

4. Materials and Methods

4.1. Isolation and Bioprospection of Omega-3 Fatty Acid Bacterial Isolates and Culture Conditions

Seawater samples were obtained in May 2016 during the Oceanographic cruise TRIENAL in the Bay of Biscay (Spain) onboard the research vessel Ramón Margalef. Samples were collected using a rosette sampler at three coastal stations, from different depths in the same water column: Shallow (5 m), mesopelagic (500 m), and bathypelagic (1,000 m) waters. Seawater samples were stored at 4 °C before isolation, which was close to the environmental temperature in situ at 500 and 1000 m depths; seawater at 5 m depth was approximately 16 °C. Bacteria isolation was performed by plating out the seawater (0.1 mL) in Marine Artificial Seawater (MASW) agar plates containing 25.5 g of Instant Ocean® sea salt, 5 g peptone, 1 g yeast extract, and 15 g of bacteriological agar per liter. The agar plates were incubated in aerobic conditions for 3 days at 10 °C. Isolated colonies (217) were selected based on the phenotypic heterogenicity, streaked over MASW medium and incubated for 5 days at 10 °C for purification and conservation for further analysis.

4.2. TTC Screening of Omega-3 Fatty Acid-Producing Bacteria

Previously to the TTC screening, gram reaction of the marine isolates was tested by the KOH method [58]. The catalase test was performed using 3% hydrogen peroxide and the oxidase test was carried out using Oxidase Strips (Microgen Camberley, UK). Gram-negative isolates were then plated on MASW agar supplemented with 0.1% (w/v) 2,3,5-Triphenyl-Tetrazolium chloride (TTC) (Alfa Aesar, Tewksbury, USA) for 3 days at 10 °C for an initial screening of omega-3 producing bacteria, as previously described [42]. TTC-positive (129) and TTC-negative (2) pure isolates were selected for subsequent work.

4.3. Phylogenetic Analysis of the Isolates Based on the 16S rRNA Gene Marker

Amplification of 16S rDNA partial genes of the isolates was performed by colony PCR using the following universal primers (5′-GGTGGAGCATGTGGTTTAATTCGA-3′ and 5′-CCCGGGAACGTATTCACCG-3′) for an initial taxonomic identification of the isolates. Based on this preliminary analysis, the full-length 16S rDNA of those isolates affiliated with *Vibrio* sp. were amplified using the universal primers 27F: 5′-AGAGGTTGATCMTGGCTCAG-3′ and 1525R: 5′-AAGGAG GTGWTCCARCC-3′ (Lane et al., 1991). PCR amplifications were performed using a C1000 Touch Thermal Cycler (Bio-Rad, Watford, UK) and the 2× PCR Master Mix (Thermo Scientific, Waltham, USA) in a total volume of 50 μL under the following conditions: Initial denaturation at 95 °C for 3 min, followed by 30 cycles of 94 °C for 30s, 50 °C for 30 s, and 72 °C for 2 min, and a final extension at 72 °C for 10 min. ExoSAP-IT™ PCR Product Cleanup Reagent (Thermo Scientific) was used after DNA agarose gel purification of PCR products. DNA sequencing was carried out by STAB VIDA sequencing service (Caparica, Portugal) using 16S-DNA and 27F/1525R primers for obtaining partial and complete sequences according to the Sanger sequencing service's instructions. The software Bioedit [59] was used for quality assessment of the sequences. BLASTn [60] against the non-redundant nucleotide sequence and the 16S ribosomal RNA sequences (Bacteria and Archaea) databases from NCBI (https://blast.ncbi.nlm.nih.gov/Blast.cgi) were performed for taxonomic assignment. 16S rDNA sequences of the nearest strains and some outgroups were obtained from the NCBI database. For the phylogenetic analysis of *Vibrio* sp. isolates, a phylogenetic tree was built using the complete 16S rDNA sequences with those from other marine gamma-Proteobacteria strains experimentally characterized as omega-3 fatty acid producers, such as *Shewanella* (*S. benthica* ATCC 43992, *S. frigidimarina* ACAM 591, *S. baltica* MAC1, *S. violacea* DSS12, *S. halifaxensis* HAW-EB4, *S. japonica* KMM 3299, *S. piezotolerans* WP3, *S. psychrophile* WP2, *S. pneumatohori* SCRC-2738), *Colwellia* (*C. psychrerythraea* ACAM 550), *Moritella* (*M. marina* MP-1), *Psychromonas* (*P. marina* JCM 10501), *Photobacterium* (*P. profundum* SS9), *Vibrio* (*V. cyclitrophicus*, *V. pelagius*, and *V. splendidus* LGP32). A local BLASTn and BLASTx of *pfa* genes was performed against the NCBI database for searching orthologs of the *pfa*A gene. Strains with *pfa*A sequence hits with *e*-value $\leq 1\times10^{-30}$) were analyzed also for the presence of *pfa* (B–E) genes. The complete predicted PKS genetic cluster (*pfa*A–E) was identified in *S. amazonensis* SB2B, *S. marinintestina*, *S. woodyi* ATTCC 51908, *P. ingrahamii* strain 37, *V. tapetis strain* HH6087, *V. crassostreae* strain LGP7, *V. splendidus* BST398, *V. tasmaniensis* strain Carson D39, *V. atlanticus* strain VB 11.11, *V. alginolyticus* strain K08M4, and *V. lentus* strain 4OM4T, and added to the phylogenetic tree. The 16S rDNA of *V. vulnificus* strain 324 was also added as a PKS-negative strain. All sequences were first assembled and aligned using the MUSCLE algorithm. Phylogenetic distance trees were inferred by maximun likehood analysis using MEGA X software [61]. Confidence was assessed using 1000 replicates' bootstrapping. The bootstrap 50% cut-off is indicated in the branches. The phylogenetic tree was edited using iTOL [62]. ClustalO was used for pairwise alignment and identity calculation [63].

4.4. Molecular Screening of the pfa Cluster

The β-ketosynthase (KS) domain present in the *pfa*A gene was used as a molecular marker of the PKS pathway by using the degenerated primers *pfa*A-KS [30]. PCR conditions were the following: Initial denaturation at 95 °C 3 min, then 30 cycles of 30 s at 95 °C, 30 s at 53 °C, 60 s at 72 °C, and a final extension 10 min at 72 °C. KS nucleotide sequences were aligned using MUSCLE, and a maximum likelihood phylogenetic tree was constructed using MEGA X software [61]. *pfa*A-KS sequences of other *Vibrio* species available in GenBank, such as *Vibrio alginolyticus* KM08M4, *V. crassotreae* 9CS106, and *V. tapetis* CECT4600, were also included, even if it has not been experimentally demonstrated that they are omega-3 producers. *Pfa*A-KS amino acid sequences were aligned using Clustal O [63]. Furthermore, delta 9-fatty acid desaturase (Δ9-*des*) gene was used as a molecular marker of the aerobic fatty acid desaturation process by using previously designed degenerated primers DE1-DE2 designed for identification of the Δ9-*des* gene in the gamma-proteobacteria *Psychrobacter urativorans* DSM14009 [64].

4.5. Nucleotide Sequence Accession Numbers

Complete sequences of 16S rDNA and *pfa*A-KS from *Vibrio* sp. isolates were submitted to GenBank under the following accession numbers: MN974023 to MN974047 and MN991210 to MN991224, respectively.

4.6. Screening of LC-PUFAs Production in Bacterial Isolates

A group of 24 strains isolated from 5,500 and 1000 m of depth were selected for lipid analysis. These strains included 22 TTC-positive red isolates affiliated with *Pseudoalteromonas* sp. (10), *Alteromonas* sp. (5), and *Vibrio* sp. (7) and TTC-negative isolates (2). Marine bacterial cultures were grown in 150 mL of MASW medium for 24 h at 10 °C in aerobic conditions and 130 rpm. Exponential phase cultures (OD600nm 0.2–0.6) were harvested by centrifugation at 4000 rpm during 20 min. Total lipid extraction of freeze-dried bacterial lyophilized pellet (100–1000 mg) and fatty acid methyl esters (FAMEs) were conducted by following the Bligh and Dyer method (Bligh & Dyer, 1959) with modifications. Fatty acid extractions were achieved adding 3 mL MeOH, 1.5 mL CH_3Cl, and 1.2 mL water and vortexing for 1 min for homogenization using an IKA™ Ultra-Turrax™ T 10 dispenser (IKA, Staufen, Germany). The same proportions of CH_3Cl and water were added until the final proportion of the solvents was 3 mL MeOH, 3 mL CH_3Cl, and 2.4 mL distilled water. Vortexing was repeated for 1 min for re-extraction and tubes were centrifuged 5 min at 3000 rpm. After discarding the upper and intermediate phases, the lower phase was collected. Humidity and insoluble parts were removed from the lower phase by a column (Pasteur pipette) containing $NaSO_4$ and glass wool. The extracted fatty acids were methylated by evaporation of the CH_3Cl layer with N_2 to dryness and 0.2% (*w/v*) $NaOCH_3$ and pumice stone were added. Samples were boiled with a reflux condenser for 10 min. Two drops of phenolphthalein were added to the chilled samples and HCl-MeOH (2:1) until they reached complete acidification. Samples were boiled again in the same conditions, and when cooled, 5 mL of n-hexane was added, and tubes were agitated for 1 min. A concentrated solution of NaCl was added to increase the rising volume of the n-hexane-phase, allowing the complete recovery of FAMEs. FAMEs were dried and dissolved in a final volume of 200–1000 μL of *n*-hexane. The Bligh and Dyer method was used for quantification of EPA in *Vibrio* sp. 414 and 618 at two different temperatures (10 and 25 °C) in 24 and 120-h cultures. Nonadecanoic acid (C19:0) (19 mg/L) was used as an internal standard (IS) for fatty acid abundance determination (mg/mL). FAMEs were identified with chromatographic standards. FA composition and EPA quantification were performed on an Agilent 5890A model gas chromatograph (GC) equipped with a capillary Agilent D-B 23 p. No 122-2362, J&W Scientific (60 m × 0.25 mm × 0.25 μm) and flame ionization detection (FID) using helium as a carrier gas. The unsaturation index was calculated as the sum of the mean percentages (by weight) of the UFA species multiplied by the number of double bonds [10].

4.7. Analysis of Lipid Content in Vibrio sp. Strains

Due to the low content of EPA observed in some *Vibrio* sp. strains by GC-FID, lipid analyses were performed also using FAMEs identification and quantification by gas chromatography/mass spectrometry (GC/MS). Strains were grown at 10 °C on MBSW medium 24 h at 130 rpm. *E. coli* BL21-AI (Invitrogen, ThermoFisher Scientific, Waltham, MA, USA) was included as a negative control. Bacterial cultures were grown in 150 mL of MASW broth 24 h at 10 °C in aerobic conditions and 130 rpm whereas *E. coli* was grown in LB medium. Mid-late exponential phase cells (OD600nm 0.4–0.8) were harvested by centrifugation at 4000 rpm during 20 min. Total cellular fatty acid extraction and FAMEs preparation were carried out following the protocol MIDI–Microbial Identification System, previously described [65] (www.midi-inc.com). All the organic extractions were concentrated by evaporation and dissolved in *n*-hexane (50 μL). GC/MS (Agilent Technologies GC 5975C, autosampler 7683B coupled to MS 7890A) equipped with an Agilent column J&W 112-88A7 (100 m × 250 μm × 0.25 μm) was used with helium as a carrier and usually run at 1.3 mL/min. Data was processed using Agilent ChemStation 7.0 software (Santa Clara, CA, USA). The percentage of total peak area was used for estimating the relative EPA content. The NIST02.L mass spectral library was used for FAMEs confirmation.

4.8. Statistical Analysis

Cellular fatty acid composition profiles of bacterial strains were calculated as average ± standard deviation of the same isolation depth. The fatty acid profile for EPA production in *Vibrio* sp. bacterial isolates grown for 24 h was represented as the result of three biological replicates ± standard deviation. Student's *t*-test were used for calculating significative differences (p-value < 0.05) between two samples means ($n \geq 3$).

4.9. Whole-Genome Sequencing Using the MinION Platform

The isolates *Vibrio* sp. 414 (isolated at 5 m depth), and 618 (isolated from 1000 m depth) were grown in MB media overnight at 25 °C. Pellet was harvested, and chromosomal DNA was extracted with a QIAamp DNA Mini Kit (QIAGEN) and purified with DNeasy PowerClean Pro Clean Up (QIAGEN, Hilden, Germany). The purified DNA was quantified using a NanoDropTM spectrometer and visualized with an agarose gel. The Qubit dsDNA HS (High Sensitivity) Assay Kit was used for DNA quantification in a Qubit 2.0 fluorometer (Thermo Fisher Scientific). The Nanopore Rapid Sequencing (SQK-RAD004) (Oxford Nanopore Technologies, Oxford, UK) protocol was used for genomic DNA library preparation and sequencing with MinION (Oxford Nanopore Technologies). A flow cell quality control (QC) was performed before loading the library into the device following the manufacturer's recommendations. The data sets were analyzed by automated MinION Data analysis with EPI2ME Software What's-In-My-Pot (WIMP, Oxford, UK) workflow. Genome assembly was performed using the SMARTdenovo software. Sequencing data obtained by MinION was deposited in ENA under the study accession number PRJEB36532. BLASTx was used for searching orthologs of *Vibrio tasmaniensis* (LGP32) PKS genes (*pfa*A–E) in both *Vibrio* sp. genomes. SnapGene® Viewer software (GSL Biotech; available at snapgene.com) was used for PKS gene cluster annotation.

4.10. Phenotypic Assays of Vibrio sp. Strains

Isolates belonging to the *Vibrio* genus were grown in MASW agar for 48 h at 10 °C and then examined in a dark room at room temperature for visualization of bioluminescence phenotypes. Pre-inocula (1/100) in MASW were used for monitoring the growth of two biological replicates of EPA-producing *Vibrio* sp. 618 and non-producing *Vibrio* sp. 414 in parallel at 10 and 25 °C during 8 h. Bacterial growth was measured spectrophotometrically based on the optical density at 600 nm (OD600nm) along a 45-h incubation at 10 and 25 °C. Growth parameters were calculated using the Roberts and Baranyi complete model incorporated in tthe Fdfit online tool in the ComBase database (combined database on predictive microbiology information) (www.combase.cc) [66].

Supplementary Materials: Figure S1: Phylogenetic tree based on partial 16S rRNA nucleotide sequences. *Alteromonas* sp. *Pseudoalteromonas* sp. and *Vibrio* sp. sequences have been incorporated for genus representation. Figure S2: Microbial phylogenetic diversity based on 16S rDNA sequences (shallow water −5 m, 500 m and 1,000 m). Percentage of TTC-positive (+column) and TTC-negative (−column) colonies are shown according to each bacterial genus and seawater depth. Figure S3: Lipid profile of selected TTC-positive isolates grouped by genera. (**a**) *Pseudoalteromonas* sp. isolates.(**b**) *Alteromonas* sp. isolates. (**c**) *Vibrio* sp. isolates. Data obtained by GC-FID analysis following the Blight&Dyer protocol. Figure S4: Multiple-sequence alignment (MSA) of amino acid sequences of amplified *pfa*A-KS domain in *Vibrio* sp. isolates. Isolate number_Depth isolation (meters). WP_012604512.1 (*Vibrio splendidus* LGP32). Figure S5: Lipid profile of selected TTC-positive isolates belonging to *Vibrio* sp. isolates, which were *pfa*A-KS positive (upper panel) and *pfa*A-KS negative (lower panel). Data obtained by GC/MS analysis following the Sasser protocol. Figure S6: GC/MS of 5,8,11,14,17-eicosapentanoic acid methyl ester peak from library (upper panel) compared with one EPA-positive samples (lower panel). Figure S7: GC-FID fatty acid profile of an EPA-producer isolate (*Vibrio* sp. 618, lower pannel) and a non-producer (*Vibrio* sp. 414, upper pannel). Isolates were grown at 10 °C and C19:0 was used as internal standard. Figure S8: Bacterial growth curve at 10 °C/25 °C of EPA-producer *Vibrio* sp. 618 and non-producer *Vibrio* sp. 414. R1 and R2 represent different biological replicates. Table S1: List of phenotypic and genotypic characteristics of gamma-proteobacteria isolates from Biscay Bay. Table S2: Data derived from MinION genome sequencing.

Author Contributions: Conceptualization, M.E. and L.A.-S.; methodology, M.E., L.A.-S., I.H., I.M., J.F., M.E.B., and E.S.; software, M.E. and L.A.-S; formal analysis M.E.; validation, M.E. and L.A.-S.; M.E. and L.A.-S.; investigation, M.E. and L.A.-S.; resources, L.A.-S.; data curation, M.E., I.H. and L.A.-S.; writing—original draft preparation M.E.; writing—review and editing, M.E. and L.A.-S; visualization, M.E.; supervision, L.A.-S.; project administration, L.A.-S. and I.H.; funding acquisition, L.A.-S. and I.H. All authors have read and agreed to the published version of the manuscript.

Acknowledgments: We acknowledge the contribution of M. Cuesta, B. Beldarrain, C. Perez, E. Erauskin, and M. Santesteban who collaborated and performed the oceanographic sampling. L.A.-S. was funded by a 'Ramón y Cajal' contract (RYC-2012-11404) from the Spanish Ministry of Economy and Competitiveness. This paper is contribution number 953 from AZTI (Marine Research Division).

References

1.   Swanson, D.; Block, R.; Mousa, S.A. Omega-3 Fatty Acids EPA and DHA: Health Benefits Throughout Life. *Adv. Nutr.* **2012**, *3*, 1–7. [CrossRef] [PubMed]

2.   Abedi, E.; Sahari, M.A. Long-chain polyunsaturated fatty acid sources and evaluation of their nutritional and functional properties. *Food Sci. Nutr.* **2014**, *2*, 443–463. [CrossRef] [PubMed]

3.   Burdge, G.C. Metabolism of α-linolenic acid in humans. *Prostaglandins Leukotrienes Essent. Fatty Acids* **2006**, *75*, 161–168. [CrossRef] [PubMed]

4.   Lee, J.H.; O'Keefee, J.H.; Levie, C.J.; Marchioli, R.; Harris, W.S. Omega-3 Fatty Acids for Cardioprotection. *Mayo Clin. Proceed.* **2008**, *83*, 324–332. [CrossRef]

5.   Reimers, A.; Ljung, H. The emerging role of omega-3 fatty acids as a therapeutic option in neuropsychiatric disorders. *Ther. Adv. Psychopharmacol.* **2019**, *9*, 2045125319858901. [CrossRef]

6.   Calder, P.C. Marine omega-3 fatty acids and inflammatory processes: Effects, mechanisms and clinical relevance. *Biochimica Biophysica Acta (BBA) – Mol. Cell Biol. Lipids* **2015**, *1851*, 469–484. [CrossRef]

7.   Astrup, A.V.; Bazinet, R.; Brenna, T.; Calder, P.C.; Crawford, M.A.; Dangour, A.; Donahoo, W.T.; Elmadfa, I.; Galli, C.; Gerber, M.; et al. Fats and fatty acids in human nutrition. Report of an expert consultation. *FAO Food Nutr. Paper* **2008**, 91.

8. Finco, A.M.d.O.; Mamani, L.D.G.; Carvalho, J.C.; de Melo Pereira, G.V.; Thomaz-Soccol, V.; Soccol, C.R. Technological trends and market perspectives for production of microbial oils rich in omega-3. *Crit. Rev. Biotechnol.* **2017**, *37*, 656–671. [CrossRef]

9. Rubio-Rodríguez, N.; Beltrán, S.; Jaime, I.; de Diego, S.M.; Sanz, M.T.; Carbadillo, J.R. Production of omega-3 polyunsaturated fatty acid concentrates: A review. *Inn. Food Sci. Emerg. Technol.* **2010**, *2010*. *11*, 1–12. [CrossRef]

10. Delong, E.F.; Yayanos, A.A. Biochemical function and ecological significance of novel bacterial lipids in deep-sea procaryotes. *Appl. Environ. Microbiol.* **1986**, *51*, 730–737. [CrossRef]

11. Adarme-Vega, T.C.; Thomas-Hall, S.R.; Schenk, P.M. Towards sustainable sources for omega-3 fatty acids production. *Curr. Opin. Biotechnol.* **2014**, *2014*. *26*, 14–18. [CrossRef]

12. Moi, I.M.; Leow, A.T.C.; Ali, M.S.M.; Rahman, R.; Salleh, A.A.; Sabri, S. Polyunsaturated fatty acids in marine bacteria and strategies to enhance their production. *Appl. Microbiol. Biotechnol.* **2018**, *102*, 5811–5826. [CrossRef] [PubMed]

13. Gladyshev, M.I.; Sushchik, N.N.; Makhutova, O.N. Production of EPA and DHA in aquatic ecosystems and their transfer to the land. *Prostaglandins Lipid Mediators* **2013**, *107*, 117–126. [CrossRef] [PubMed]

14. Freese, E.; Rutters, H.; Koster, J.; Rullkotter, J.; Sass, H. Gammaproteobacteria as a possible source of eicosapentaenoic acid in anoxic intertidal sediments. *Microb. Ecol.* **2009**, *57*, 444–454. [CrossRef] [PubMed]

15. Russell, N.J.; Nichols, D.S. Polyunsaturated fatty acids in marine bacteria—a dogma rewritten. *Microbiology* **1999**, *145*, 767–779. [CrossRef] [PubMed]

16. Ringø, E.; Jøstensen, J.P.; Olsen, R.E. Production of eicosapentaenoic acid by freshwater Vibrio. *Lipids* **1992**, *27*, 564–566. [CrossRef]

17. Dailey, F.E.; MacGraw, J.E.; Jensen, B.J.; Bishop, S.S.; Lokken, J.P.; Dorff, K.J.; Ripley, M.P.; Munro, J.B. The Microbiota of Freshwater Fish and Freshwater Niches Contain Omega-3 Fatty Acid-Producing Shewanella Species. *Appl. Environ. Microbiol.* **2016**, *82*, 218–231. [CrossRef]

18. Allen, E.E.; Facciotti, D.; Bartlett, D.H. Monounsaturated but not polyunsaturated fatty acids are required for growth of the deep-sea bacterium Photobacterium profundum SS9 at high pressure and low temperature. *Appl. Environ. Microbiol.* **1999**, *65*, 1710–1720. [CrossRef]

19. Yoshida, K.; Hashimoto, M.; Hori, R.; Adachi, T.; Okuyama, H.; Orikasa, Y.; Nagamine, T.; Shimizu, S.; Ueno, A.; Morita, N. Bacterial Long-Chain Polyunsaturated Fatty Acids: Their Biosynthetic Genes, Functions, and Practical Use. *Mar. Drugs* **2016**, *14*, 94. [CrossRef]

20. Kawamoto, J.; Kurihara, T.; Yamamoto, K.; Nagayasu, M.; Tani, Y.; Mihara, H.; Baba, T.; Sato, S.B.; Esaki, N. Eicosapentaenoic acid plays a beneficial role in membrane organization and cell division of a cold-adapted bacterium, Shewanella livingstonensis Ac10. *J. Bacteriol.* **2009**, *191*, 632–640. [CrossRef]

21. Okuyama, H.; Orikasa, Y.; Nishida, T.; Watanabe, K.; Morita, N. Bacterial genes responsible for the biosynthesis of eicosapentaenoic and docosahexaenoic acids and their heterologous expression. *Appl. Environ. Microbiol.* **2007**, *73*, 665–670. [CrossRef]

22. Okuyama, H.; Orikasa, Y.; Nishida, T. Significance of antioxidative functions of eicosapentaenoic and docosahexaenoic acids in marine microorganisms. *Appl. Environ. Microbiol.* **2008**, *74*, 570–574. [CrossRef]

23. Nishida, T.; Morita, N.; Yano, Y.; Orisaka, Y.; Okuyama, H. The antioxidative function of eicosapentaenoic acid in a marine bacterium, Shewanella marinintestina IK-1. *FEBS Lett.* **2007**, *581*, 4212–4216. [CrossRef]

24. Nishida, T.; Hori, R.; Morita, N.; Okuyama, H. Membrane eicosapentaenoic acid is involved in the hydrophobicity of bacterial cells and affects the entry of hydrophilic and hydrophobic compounds. *FEMS Microbiol. Lett.* **2010**, *306*, 91–96. [CrossRef]

25. Wang, M.; Chen, H.; Gu, Z.; Zhang, H.; Wein, C.; Yong, Q.C. ω3 fatty acid desaturases from microorganisms: structure, function, evolution, and biotechnological use. *Appl. Microbiol. Biotechnol.* **2013**, *97*, 10255–10262. [CrossRef]

26. Metz, J.G.; Roessler, P.; Facciotti, C.; Levering, C.; Dittrich, F.; Lassner, M.; Valentine, R.; Lardizabal, K.; Domerque, F.; Yamada, A. Production of Polyunsaturated Fatty Acids by Polyketide Synthases in Both Prokaryotes and Eukaryotes. *Science* **2001**, *293*, 290–293. [CrossRef]

27. Yazawa, K. Production of eicosapentaenoic acid from marine bacteria. *Lipids* **1996**, *31*, S297–S300. [CrossRef]

28. Allen, E.E.; Bartlett, D.H. Structure and regulation of the omega-3 polyunsaturated fatty acid synthase genes from the deep-sea bacterium Photobacterium profundum strain SS9. *Microbiology* **2002**, *148*, 1903–1913.

[CrossRef]

29. Shulse, C.N.; Allen, E.E. Widespread occurrence of secondary lipid biosynthesis potential in microbial lineages. *PLoS One* **2011**, *6*, e20146. [CrossRef]

30. Shulse, C.N.; Allen, E.E. Diversity and distribution of microbial long-chain fatty acid biosynthetic genes in the marine environment. *Environ. Microbiol.* **2011**, *13*, 684–695. [CrossRef]

31. ICES. *Bay of Biscay and the Iberian Coast. Ecoregion—Ecosystem overview*; ICES Book: Copenhagen, Denmark, 2018.

32. Joint, I.; Muhling, M.; Querellou, J. Culturing marine bacteria - an essential prerequisite for biodiscovery. *Microb. Biotechnol.* **2010**, *3*, 564–575. [CrossRef] [PubMed]

33. Bouhlel, Z.; Arnold, A.A.; Warschawski, D.E.; Lemarchand, K.; Tremblay, R.; Marcotte, I. Labelling strategy and membrane characterization of marine bacteria Vibrio splendidus by in vivo(2)H NMR. *Biochim. Biophys. Acta Biomembr.* **2019**, *1861*, 871–878. [CrossRef] [PubMed]

34. Bianchi, A.C.; Olazábal, L.; Torre, A.; Loperena, L. Antarctic microorganisms as source of the omega-3 polyunsaturated fatty acids. *World J. Microbiol. Biotechnol.* **2014**, *30*, 1869–1878. [CrossRef] [PubMed]

35. Koga, Y. Thermal adaptation of the archaeal and bacterial lipid membranes. *Archaea (Vancouver, B.C.)* **2012**, 789652. [CrossRef] [PubMed]

36. De Carvalho, C.C.C.R.; Fernandes, P. Production of metabolites as bacterial responses to the marine environment. *Mar. Drugs* **2010**, *8*, 705–727. [CrossRef] [PubMed]

37. Abd Elrazak, A.; Ward, A.C.; Glassey, J. Polyunsaturated fatty acid production by marine bacteria. *Bioprocess. Biosys. Eng.* **2013**, *36*, 1641–1652. [CrossRef]

38. Hamamoto, T.; Takata, N.; Kudo, T.; Horikoshi, K. Characteristic presence of polyunsaturated fatty acids in marine psychrophilic vibrios. *FEMS Microbiol. Lett.* **1995**, *129*, 51–56. [CrossRef]

39. Ringø, E.; Jøstensen, J.P.; Olsen, R.E. Production of Eicosapentaenoic Acid (20:5 n-3) by Vibrio pelagius Isolated from Turbot *(Scophthalmus maximus (L.)) Larvae. Appl. Environ. Microbiol.* **1992**, *58*, 3777–3778. [CrossRef]

40. Ivanova, E.P.; Zhukova, N.V.; Svetashev, V.I.; Gorshkova, N.M.; Kurilenko, V.V.; Frolova, G.M.; Mikhailov, V.V. Evaluation of Phospholipid and Fatty Acid Compositions as Chemotaxonomic Markers of Alteromonas-Like Proteobacteria. *Curr. Microbiol.* **2000**, *41*, 341–345. [CrossRef]

41. Skerratt, J.H.; Bowman, J.P.; Nichols, P.D. Shewanella olleyana sp. nov.; a marine species isolated from a temperate estuary which produces high levels of polyunsaturated fatty acids. *Int. J. Syst. Evol. Microbiol.* **2002**, *52*, 2101–2106.

42. Ryan, J.; Farr, H.; Visnovsky, S.; Vyssotski, M.; Visnovsky, M. A rapid method for the isolation of eicosapentaenoic acid-producing marine bacteria. *J. Microbiol. Methods* **2010**, *82*, 49–53. [CrossRef] [PubMed]

43. Abd El Razak, A.; Ward, A.C.; Glassey, J. Screening of marine bacterial producers of polyunsaturated fatty acids and optimisation of production. *Microb. Ecol.* **2014**, *67*, 454–464. [CrossRef] [PubMed]

44. Alagarsamy, S.; Sabeena Farvin, K.H.; Fakhraldeen, S.; Kooramattom, M.R.; Al-Yamani, F. Isolation of Gram-positive Firmibacteria as major eicosapentaenoic acid producers from subtropical marine sediments. *Lett. Appl. Microbiol.* **2019**, *69*, 121–127. [CrossRef]

45. Hamamoto, T.; Takata, N.; Kudo, T.; Horikoshi, K. Effect of temperature and growth phase on fatty acid composition of the psychrophilic Vibrio sp. strain no. 5710. *FEMS Microbiol. Lett.* **1994**, *119*, 77–81.

46. Day, A.P.; Oliver, J.D. Changes in membrane fatty acid composition during entry of Vibrio vulnificus into the viable but nonculturable state. *J. Microbiol.* **2004**, *42*, 69–73.

47. Ibragimova, M.Y.; Salafutdinov, I.I.; Sahin, F.; Zhdanoz, R.I. Biomarkers of Bacillus subtilis total lipids FAME profile under various temperatures and growth phases. *Dokl. Biochem. Biophys.* **2012**, *443*, 109–112. [CrossRef]

48. Jia, J.; Chen, Y.; Jiang, Y.; Tang, J.; Yang, L.; Liang, C.; Jia, Z.; Zhao, L. Visualized analysis of cellular fatty acid profiles of Vibrio parahaemolyticus strains under cold stress. *FEMS Microbiol. Lett.* **2014**, *357*, 92–98. [CrossRef]

49. Nogi, Y.; Masui, N.; Kato, C. *Photobacterium profundum* sp. nov.: A new, moderately barophilic bacterial species isolated from a deep-sea sediment. *Extremophiles* **1998**, *2*, 1–7. [CrossRef]

50. Allen, E.E.; Bartlett, D.H. FabF Is Required for Piezoregulation of *cis*-Vaccenic Acid Levels and Piezophilic Growth of the Deep-Sea Bacterium *Photobacterium profundum* Strain SS9. *J. Bacteriol.* **2000**, *182*, 1264–1271. [CrossRef]

51. Le Roux, F.; Zouine, M.; Chakroun, N.; Binesse, J.; Saulnier, D.; Bouchier, C.; Zidane, N.; Ma, L.; Rusniok, C.; Laius, A. Genome sequence of Vibrio splendidus: an abundant planctonic marine species with a large genotypic diversity. *Environ. Microbiol.* **2009**, *11*, 1959–1970. [CrossRef]

52. Urdaci, M.C.; Marchand, M.; Grimont, P.A.D. Grimont, Characterization of 22 Vibrio species by gas chromatography analysis of their cellular fatty acids. *Res. Microbiol.* **1990**, *141*, 437–452. [CrossRef]

53. Hoffmann, M.; Fischer, M.; Whittaker, P. Evaluating the use of fatty acid profiles to identify deep-sea Vibrio isolates. *Food Chem.* **2010**, *122*, 943–950. [CrossRef]

54. Xu, M.; Wang, J.; Mou, H. Fatty acid profiles of Vibrio parahaemolyticus and its changes with environment. *J. Basic. Microbiol.* **2015**, *55*, 112–120. [CrossRef]

55. Garba, L.; Mohamad Ali, M.S.; Oslan, S.N. Molecular Cloning and Functional Expression of a Delta9- Fatty Acid Desaturase from an Antarctic *Pseudomonas sp. A3*. *PLoS One* **2016**, *11*, e0160681.

56. Li, Y.; Dietrich, M.; Schmid, R.D.; He, B.; Ouyang, P.; Urlacher, V.B. Identification and Functional Expression of a Δ9-Fatty Acid Desaturase from Psychrobacter urativorans in *Escherichia coli*. *Lipids* **2008**, *43*, 207–213. [CrossRef]

57. Noor Zalih, R.; Garba, L.; Shukuri Mo, M.; Nurbava Os, S. Review on Fatty Acid Desaturases and their Roles in Temperature Acclimatisation. *J. Appl. Sci.* **2017**, *17*, 282–295. [CrossRef]

58. Buck, J.D. Nonstaining (KOH) method for determination of gram reactions of marine bacteria. *Appl. Environ. Microbiol.* **1982**, *44*, 992–993. [CrossRef]

59. Hall, T.A. BioEdit: a user-friendly biological sequence alignment editor and analysis program for Windows 95/98/NT. *Nucl. Acids. Symp. Ser.* **1999**, *41*, 95–98.

60. Altschul, S.F.; Gish, W.; Miller, W.; Myers, E.W.; Lipman, D.J. Basic local alignment search tool. *J. Mol. Biol.* **1990**, *215*, 403–410. [CrossRef]

61. Kumar, S.; Stecher, G.; Li, M.; Knyaz, C.; Tamura, K. MEGA X: Molecular Evolutionary Genetics Analysis across Computing Platforms. *Mol. Biol. Evol.* **2018**, *35*, 1547–1549. [CrossRef]

62. Letunic, I.; Bork, P. Interactive tree of Life (iTOL) v4: recent updates and new developments. *Nucleic Acids Res.* **2019**, *47*, W256–W259. [CrossRef] [PubMed]

63. Sievers, F.; Wilm, A.; Dineen, D.; Gibson, T.J.; Karplus, K.; Li, W.; Lopez, R.; McWilliam, H.; Remmert, M.; Söding, J.; et al. Fast, scalable generation of high-quality protein multiple sequence alignments using Clustal Omega. *Mol. Syst. biol.* **2011**, *7*, 539. [CrossRef] [PubMed]

64. Li, Y.; Xu, X.; Dietrich, M.; Urlacher, V.B.; Schmid, R.D.; Ouyang, P.; He, B. Identification and functional expression of a Δ9 fatty acid desaturase from the marine bacterium *Pseudoalteromonas* sp. *MLY15*. *J. Mol. Catal. B: Enzym.* **2009**, *56*, 96–101. [CrossRef]

65. Sasser, M. *Identification of Bacteria By Gas Chromatography of Cellular Fatty Acids*; MIDI Technical Note 101; MIDI: Newark, NJ, USA, 1990; pp. 1–7.

66. Baranyi, J.; Tamplin, M.L. ComBase: A Common Database on Microbial Responses to Food Environments. *J. Food Prot.* **2004**, *67*, 1967–1971. [CrossRef]

3

Dietary Bioactive Fatty Acids as Modulators of Immune Function: Implications on Human Health

Naren Gajenthra Kumar [1]⊚, Daniel Contaifer [2], Parthasarathy Madurantakam [3], Salvatore Carbone [4,5]⊚, Elvin T. Price [2], Benjamin Van Tassell [2], Donald F. Brophy [2]⊚ and Dayanjan S. Wijesinghe [2,6,7,*]⊚

[1] Department of Microbiology and Immunology, School of Medicine, Virginia Commonwealth University, Richmond, VA 23298, USA; ngajenthrakum@mymail.vcu.edu
[2] Department of Pharmacotherapy and Outcomes Sciences, School of Pharmacy, Virginia Commonwealth University, Richmond, VA 23298, USA; dcontaifer@vcu.edu (D.C.); etprice@vcu.edu (E.T.P.); bvantassell@vcu.edu (B.V.T.); dbrophy@vcu.edu (D.F.B.)
[3] Department of General Practice, School of Dentistry, Virginia Commonwealth University, Richmond, VA 23298, USA; madurantakap@vcu.edu
[4] Department of Kinesiology & Health Sciences, College of Humanities & Sciences, Virginia Commonwealth University, Richmond, VA 23220, USA; salvatore.carbone@vcuhealth.org
[5] VCU Pauley Heart Center, Department of Internal Medicine, Virginia Commonwealth University, Richmond, VA 23298, USA
[6] da Vinci Center, Virginia Commonwealth University, Richmond, VA 23220, USA
[7] Institute for Structural Biology, Drug Discovery and Development, Virginia Commonwealth University School of Pharmacy, Richmond, VA 23298, USA
* Correspondence: wijesingheds@vcu.edu.

Abstract: Diet is major modifiable risk factor for cardiovascular disease that can influence the immune status of the individual and contribute to persistent low-grade inflammation. In recent years, there has been an increased appreciation of the role of polyunsaturated fatty acids (PUFA) in improving immune function and reduction of systemic inflammation via the modulation of pattern recognition receptors (PRR) on immune cells. Extensive research on the use of bioactive lipids such as eicosapentaenoic acid (EPA) and docosahexaenoic acid (DHA) and their metabolites have illustrated the importance of these pro-resolving lipid mediators in modulating signaling through PRRs. While their mechanism of action, bioavailability in the blood, and their efficacy for clinical use forms an active area of research, they are found widely administered as marine animal-based supplements like fish oil and krill oil to promote health. The focus of this review will be to discuss the effect of these bioactive fatty acids and their metabolites on immune cells and the resulting inflammatory response, with a brief discussion about modern methods for their analysis using mass spectrometry-based methods.

Keywords: EPA; DHA; FDA regulations; Immune function; toll like receptors; essential fatty acids; non-essential fatty acids; PPAR

1. Introduction

Dietary fatty acids, either by themselves or via their metabolites, have the capacity to influence human health and health outcomes [1]. A detailed dissection of the components of lipids associated with poor cardiovascular health in the past decade has enabled the identification of putative lipid biomarkers predictive of poor cardiovascular health. Lipidomic analysis to study models of dyslipidemia have shown an accumulation of saturated fatty acids and omega-6 fatty acids-associated lipids [2] and are considered to be inflammatory in nature [3]. Increase in disorders like type II diabetes, cardiovascular diseases, and atherosclerosis, which are highly associated with an unhealthy diet, have brought forth the importance of lipid homeostasis in health and disease. Furthermore, with the advent of lipidomics, an increasing grasp on the diversity of lipid species suggests that the relative abundance of lipids influence outcomes [4], rather than the mere presence or absence of a lipid species. The prevailing dogma suggests that an increase in free omega-3 polyunsaturated fatty acids and associated lipids (e.g., omega-3: omega-6 ratio) is known to promote health in humans and is correlated with lower levels of systemic inflammation. Bioactive lipids, specifically polyunsaturated long chain fatty acids, are classified based on their degree of unsaturation, which is insufficient to infer their biological function, and it is important to discuss the ways in which fatty acids play a role in inflammation and immune function.

Studies on the human lipidome, not limited to the classes of phospholipids, cholesterol esters, triacyl glycerol, and fatty acids have been implicated as an area of vital research for diet and lifestyle-associated disorders. Fatty acids differing in their position of desaturation (omega 3 vs. omega 6) play distinct roles in the body and are the primary focus of our discussion (Figure 1). These fatty acids are considered essential fatty acids as humans are unable to synthesize the basic precursors. The most basic dietary precursor for omega-3 fatty acid is the alpha-linolenic (ALA) acid or linolenic acid which is a fatty acid (FA) containing 18 carbons with three double bonds (18:3)with the first double bond from the non-carboxyl end beginning at the third carbon (n-3) and abbreviated in whole as ALA FA 18:3 n-3. This lipid is converted to the anti-inflammatory eicosapentaenoic acid (EPA FA 20:5 n-3), and docosahexaenoic acid (DHA FA 22:6 n-3) (Figure 1). Dietary omega- 6 fatty acids like linoleic acid (FA 18:2 n-6) are converted to gamma-LA (FA 18:3 n-6) and arachidonic acid (AA FA 20:4 n-6), and have distinct roles in inflammation (Figure 1). These fatty acids serve as precursors to many bioactive lipids. When taken via diet, they are converted to monoglycerides and free fatty acids in the intestinal lumen, followed by incorporation into chylomicrons and lipoproteins for circulation within the bloodstream. Omega-3 fatty acids are anti-inflammatory, whereas omega-6 fatty acids are pro-inflammatory, and this association depends on the lipid metabolites produced downstream from these precursors. Biochemically, higher concentrations of dietary bioactive lipids like EPA and DHA compete with AA for synthesis of lipid mediators and can tip the balance towards less inflammatory/pro-resolution phenotypes [5–7]. Resolution may occur when the conversion of arachidonic acid to inflammatory mediators by cyclooxygenase-2 (COX-2) is competed off by EPA and DHA to produce pro-resolution lipids (reviewed in [6]). In addition to their metabolic flux, these fatty acids are known to competitively modulate signaling through pattern recognition receptors and G protein coupled receptors (GPR40) [7,8] on leukocytes [9–11] and thus reduce the risk of inflammation-mediated cardiovascular disease progression. Metabolites of long chain fatty acids, also known as eicosanoids, can interact with G-protein-coupled receptors GPCRs [8] and have been implicated in the development of atherosclerosis. Thus it may be possible to ultimately allow for targeted, personalized applications of lipid formulations for managing systemic inflammation perpetrated by particular cell types of the immune system (T-cells, B cells, and dendritic cells) and the treatment of disorders associated with unhealthy diet [12,13]. While this is an exciting area of research, the narrow window of physiological response demands accurate quantitation of lipid species which is now possible with advances in the field of lipidomics.

Advances in liquid chromatography coupled to high resolution mass spectrometry, and more recently ion mobility methods [14], have enabled comprehensive characterization of the mammalian lipidome. Chromatographic and ion mobility separation prior to mass spectrometry is required to overcome the challenge of isobaric overlap when analyzing lipids by mass spectrometry. In this regard, reverse phase chromatographic methods enable the separation of lipids based on hydrophobicity (Figure 3b) primarily resulting from the fatty acid composition of the lipid species. On the other hand, hydrophilic interaction liquid chromatography (HILIC) [15] provides class-based separation of lipids based on the lipid molecule head groups, the primary determinants of lipid classes. The use of such complimentary methods has enabled the quantification of thousands of lipid species which can be used for targeted monitoring of lipids for personalized therapeutic approaches [12,13]. While liquid chromatography-based separation suffices for most lipid separations, to date, separation in the gas phase (gas chromatography) prior to mass spectrometric analysis remains the most reliable method free fatty acid analysis. It should be noted that newer separation methods like supercritical fluid chromatography coupled to mass spectrometry is also demonstrating great promise towards quantitative analysis of bioactive lipids.

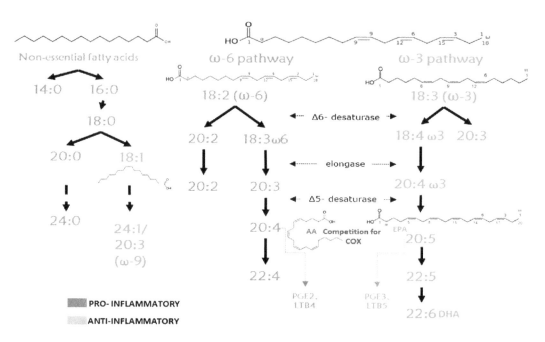

Figure 1. Fatty acid configuration and their role in inflammation-related pathways. n- omega-position of unsaturation, fatty acid nomenclature XX:Y; XX—number of carbon, Y—number of unsaturated bonds. AA—arachidonic acid, EPA—eicosapentaenoic acid, DHA—docosahexaenoic acid, COX—cyclooxygenase, LTB—leukotriene, PGE—prostaglandin.

In the following review, we seek to summarize the role of lipids in immunopathology while discussing the relevant advances in analytical and statistical methods in lipidomic research for studying a broad variety of inflammation-associated diseases.

2. Fatty Acids and Immune Function

2.1. Fatty Acids Influence Inflammatory Repertoire

Fatty acids are classified as short chain, medium chain, and long chain fatty acids. There are three primary means by which fatty acids can influence the inflammatory repertoire of the host; substrates for biosynthesis of inflammatory mediators, activation of cell receptors [16,17], and modulation of membrane fluidity to alter cell function. Bulk of the work in this field has focused around the mechanisms by which polyunsaturated fatty acids (PUFA) act as substrates for the biosynthesis of inflammatory eicosanoid mediators (Figure 1). While some work has focused on delineating the mechanism by which fatty acids interact with cell surface receptors or even modulate cellular function through the production of oxylipins [18], its implications on overall health, and their efficacy as interventions remain elusive. Dietary intake of omega-3 unsaturated fatty acids provides precursors for the production of anti-inflammatory lipids like five series leukotrienes (LTx_5) or three series prostanoids (PGx_3). On the other hand, an increase in omega-6 fatty acids leads to the production of pro-inflammatory mediators like four series leukotrienes(LTA_4, LTB_4, LTC_4, LTD_4, LTE_{44}) and two series prostanoids (PGx_2) [19] (Figure 1). The metabolites produced, further regulate inflammation by feedback inhibition of biosynthetic enzymes [7,20]. This autocrine mechanism of regulation by the two series of prostaglandins has been well studied [3,21]. In the body, fatty acids and their metabolites [22] are usually present together and influence the end result by their relative concentration in the milieu. The initial inflammatory response, as represented by the production of LTx_4, is important for the infiltration of neutrophils to the site of infection and thereby beginning the cascade for the production of pro-inflammatory cytokines [23]. The incorporation of these bioactive lipids into membrane phospholipids affects membrane fluidity and surface receptor expression and regulates the function of immune cells. Enrichment of T cells and neutrophils with omega-3/omega-6 fatty acid supplemented in the media provides an evidence for the incorporation of exogenous fatty acids into membrane phospholipids of immune effector cells [24,25]. Taken together it suggests that immune cells are capable of incorporating exogenous fatty acids into the cell membrane. The incorporation of fatty acids changes the membrane architecture and signaling, thereby altering the function of cell surface pattern recognition receptors [26,27].

Reports investigating this phenomenon have provided some insights into the incorporation of fatty acids into the cell membrane of macrophages [28]. This effect is seen specifically in activated polymorphonuclear neutrophils (PMN's) where there is a loss of membrane incorporated unsaturated fatty acids resulting from heightened intracellular phospholipase activity ($cPLA_2$) due to the activation of leukocytes [29,30] downstream of cell surface receptors like Toll-like receptors (TLRs).

Lipids enriched in dietary fatty acids can influence the inflammatory profile during episodes of sterile or infectious inflammation. Enrichment of omega-3 fatty acids in the media or diet improves the function of lymphocytes by improving mitogen-mediated activation of immune cells [31–33]. An omega-3 rich diet further promotes the development of a T_H2-type immune response [34,35] promoting the production of associated anti-inflammatory cytokines like IL-4 [34,35], and reduction of pro-inflammatory TNF-α [20,27,32–34]. In contrast, a coconut oil-based diet (rich in saturated fatty acids) led to the development of an IFN-γ dominant cytokine profile, characteristic of a T_H1 immune response. While a T_H1 response may be beneficial in a parasitic infection, a T_H2 response is preferred in most cases. Expanding on the effect of omega-3 PUFA on T cell subtypes, it is interesting to note that not only a fish oil-enriched diet, but also purified EPA and DHA suppressed IL-17 production from T_H17 cells resulting in reduced STAT-3 phosphorylation and ROR-$\gamma\tau$ expression. This concomitant decrease in T_H1/T_H17 in response to omega-3 fatty acids (EPA and DHA) may have implications on the ability to reduce enteric inflammation [36]. Similarly, a diet rich in monounsaturated fatty acids (MUFA) like oleic acid (Mediterranean diet) has been shown to improve high density lipoprotein (HDL) function [37] and is protective in patients for high risk for cardiovascular disease [37–39] (Tables 1 and 2).

2.2. Fatty Acids Influence Immune Functions by Interacting with Cell Surface Receptors on Immune Cells

In an effort to dispel the confusion about the mechanism of action of polyunsaturated fatty acids and find cognate receptor interaction, some landmark studies have determined a few receptors that fatty acids interact with on host cells like the peroxisome proliferation activating receptor (PPAR) and Toll-like receptors (TLRs) [40,41]. Interaction with these receptors results in the activation of signaling cascades activating the transcription of anti-inflammatory cytokines while suppressing the transcription of pro-inflammatory cytokines (Figure 2). Studies investigating the interaction of fatty acids with Toll-like receptors (TLRs), particularly TLR2 and TLR4, on leukocytes have shown that saturated fatty acids (C12:0 and C16:0) cause an increased expression of cyclooxygenase-2 (COX-2) and phosphorylation of ERK (p-ERK) in a MyD88 independent manner [16]. Consistent with reported literature, it was also found that DHA and other n-3 fatty acids caused suppression of COX-2 and p-ERK. This study dispels any contention of the bacterial contaminants or the influence of bovine serum albumin (BSA) in eliciting the activation of TLRs [16]. It was also found that monocytic cells (THP-1) and macrophages (RAW 264.7) had a heightened response to saturated fatty acids under serum-starved conditions which was influenced by the reactive oxygen species (ROS) status of the microenvironment, while EPA and DHA suppressed this response [40]. It must also be noted that administration of high fat diet rich in saturated fats has been shown to increase endotoxin (LPS) production by gram negative bacteria [4] and thus TLR4 activation. Saturated fatty acids also activate an additional number of pro-inflammatory pathways, such as the one related to the intracellular macromolecular complex Nod-like receptor pyrin domain-containing protein (NLRP)-3 inflammasome, primarily responsible for the production of the pro-inflammatory cytokines IL-1β and IL-18 [42–45]. Conversely, unsaturated fatty acids inhibit such detrimental effects, exerting anti-inflammatory properties.

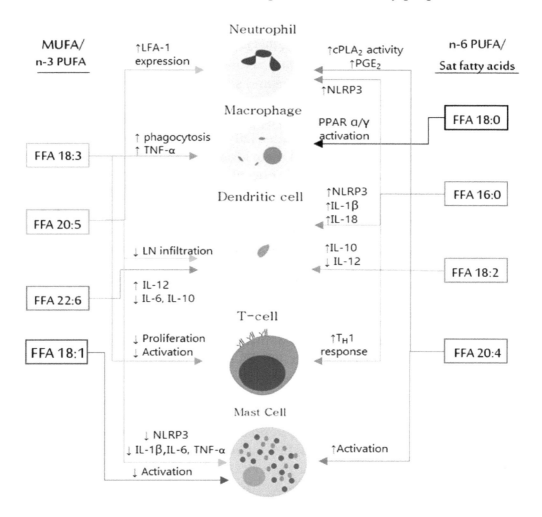

Influence of fatty acids on immune function

Figure 2. Influence of fatty acids on immune cells as summarized in Table 1. Abbreviations; LFA-1—leukocyte factor antigen-1, IL—interleukin, Th1—T helper type I, TNF-α—tumor necrosis factor alpha, LN—lymph node, cPLA$_2$—cytosolic phospholipase A2, PPAR—peroxisome proliferator-activated receptor, NLRP3—nucleotide-binding and oligomerization domain-like receptor, leucine-rich repeat and pyrin domain–containing 3, ↑—levels increase, ↓—levels decrease.

Another kind of membrane-associated receptor other than TLR is the PPAR-gamma (PPAR-γ). It is trafficked to the cell surface in membrane lipid rafts and influences the inflammatory response in an NF-κB dependent manner. Some disease presentations like atherosclerosis are seen due to a PPAR-γ dependent expression of oxLDL uptake receptors that lead to the formation of foam cells in fatty streaks (plaques) seen in arteries. In most cases, this is a result of downstream activation of a cell surface receptor like PPAR-γ [10,46,47], which initiates the cascade required for the immediate production of inflammatory lipid mediators like prostaglandins and cytokines (i.e., IL-2) prior to the infiltration of the tissue by immune cells. Research on this front has provided information that fatty acids and other eicosanoids, like PGJ$_2$, are ligands for PPAR-γ and may increase the transcription of the PPAR-γ [48,49]. However, the suppression of NF-κB-mediated pro-inflammatory responses by EPA and DHA are independent of PPAR-γ [50]. This modulation of the immune response has also been elicited in the requirement of PPAR-γ for the maturation of dendritic cells and the suppression of macrophage-mediated inflammation by n-3 fatty acids [51]. A more responsive counter part of the PPAR family of receptors to n-3 supplementation are PPAR-α and PPAR-δ. The expression of these genes was observed to be upregulated in mice fed with high fat diets and this increase was suppressed when mice were fed with a high fat diet supplemented with DHA and EPA [52,53]. Thus, n-3 fatty acids are antagonistic to pro-inflammatory pathways regulated by TLR (i.e., saturated fatty acids) and PPAR cell surface receptors.

Similar to the interaction of eicosanoids with membrane spanning G protein coupled receptors, long chain fatty acids also interact with G protein coupled receptors on the surface of cells. An upcoming area of research is the interaction of short chain [54–56] and long chain fatty acids with yet uncharacterized G protein coupled receptors and their role in modulating inflammation. For a comprehensive overview of the specific interactions of the free fatty acids with their cognate G protein receptors, the readers are directed to some excellent reviews in this emerging field [57,58].

2.3. Fatty Acids Influence Lymphocyte Proliferation and Cytokine Profiles

It has been reported as early as 1988 that inhibition of the lipoxygenase pathway of inflammation prevents the differentiation of monocytes to macrophages after supplementation with arachidonic acid (n-6) [31]. A concomitant increase in PC specific phospholipase activity is observed in instances where linoleic acid (n-6) is found esterified in the sn2 position of phospholipids [31]. Depending on the stage of development of the macrophage, enrichment of cells with n-3 fatty acids such as linolenate (FA 18:3 n-3) promotes a rapid acute immune response by a reduced production of TNF-alpha when compared to macrophages enriched with n-6 fatty acids such as linolenic acid [20]. A more recent study reiterates the findings that n-3 fatty acids can stimulate the production of IL-4 while saturated fatty acids promote the production of IFN-γ [34]. In addition to influencing cytokine profiles, n-3 fatty acids also promote the expression of the Mac-1 complex (CD-11b/CD-18) on the surface of neutrophils on the cell surface. While the exact mechanism of action remains unclear to date, the expression of Mac-1 on the surface is not dependent on the metabolic production of pro-inflammatory lipids like PGE_2 or LTB_4 [59]. Thus, suggesting that while arachidonic acid metabolites do not play a role in the expression of surface markers and differentiation of monocyte derived cell types, they are important in initiating the cascade responsible for terminal differentiation. Arachidonic acid, an n-6 fatty acid previously shown to be enriched in atherosclerotic plaques, has also been suggested to induce the expression of CD36 and scavenger receptor A [49] on the surface of macrophages. An expression of these receptors promotes prolonged residence of macrophages in bloodstream and the development of foam cells due to increased receptor-mediated oxLDL uptake. A review summarizes the connection between dietary nutrients and immune function [35,60] and we elaborate on some of the mechanisms by which dietary lipids influence immune function at the cellular level.

Fatty acids may also regulate the production of pro-inflammatory cytokines by interacting with the NLRP3 inflammasome and modulating the production of IL-1 family of cytokines (IL-1β and IL-18) [61,62]. Signaling cascade downstream of pattern recognition cell surface receptor (TLR) provides the initial signal for the production IL-1 family of pro-inflammatory cytokines [63]. The maturation of the cytokines is controlled by a multiprotein complex called the inflammasome. Though interesting, it is not surprising that the saturated/n-3 fatty acids play antagonistic roles to n-6 fatty acids in inflammasome function [64,65]. While saturated fatty acids like palmitic acid (C16:0) induce the production of IL-18 and IL-1β in an NLRP3 dependent manner, monounsaturated fatty acid (MUFA) like oleic acid (C18:1) and n-3 PUFA inhibit the production of these cytokines [64] and are involved with the transcriptional repression of NLRP3. Taken together, an increased proportion of saturated and n-6 fatty acids inhibit immune cell development and promotes inflammation. MUFA [38,66] and PUFA (n-3) suppress the production of the pro-inflammatory cytokines by interacting with the

NLRP3 inflammasome and modulate inflammation at the transcriptional and translational level by suppressing gene expression of the components of the inflammasome and preventing the NLRP3 dependent maturation of IL-1 family of cytokines [43,52,53,62,64,65].

A resurgence of interest in bioactive lipids has been complemented with the characterization of the bioactive lipids composition of M1/M2 polarized macrophages [67]. Studies have shown that the lipidomic composition and the resulting phospholipase subtype mediated activity to influence the development of proinflammatory subtype (M1) over the anti-inflammatory subtype (M2) [67]. While it continues to be determined whether the product of 12-Lipoxygenase (12-HETE) is important for shifting the polarization of macrophages to the M2 subtypes, more work focused on the eicosanoid signaling and its relation to cell types differentiation is required. Resulting from the distinct mechanisms of differentiation to macrophage subtypes, M1/M2 cells have distinct eicosanoid compositions where M1 cells are abundant in LTB4 and PGE2 while M2 macrophages are abundant in pro-resolving mediators from the 5-lipoxygenase pathway (5-LOX) and eicosanoids derived from n-3 fatty acids like EPA and DHA such as resolvin D2 and D5 (RvD2, 5). Even more interesting is that M2 cells produced a higher concentration of PGD_2, the anti-inflammatory metabolite of arachidonic acid [68,69]. Taken together, these reports suggest an overlooked role of bioactive lipids and their mediators in modulating the outcomes of infection and inflammation. Further emphasizing the need for the analysis of localized lipid metabolism at the sight of interest instead of classical systemic evaluation of efficacy of the efficacy of bioactive lipids in health and disease.

Table 1. Influence of fatty acids and their metabolites on lymphocyte functions. Source—Endogenous/supplement—the fatty acid was supplemented in an in vivo study or enriched in formulations in vitro. Synthetic—fatty acid was used in in-vitro studies. PLA2—phospholipase A2, PGE2—prostaglandin E2.

Lipid	Source	Immune Cell	Function	Ref.
Fatty acids FA 20:4, FA 20:5, FA 22:6	Endogenous, supplement	Neutrophil	Adherence to endothelia (CD11a and CD 11b)	[32]
FA 18:3 n-3	Supplement	Alveolar macrophages	Increased phagocytosis, Increased TNF-αproduction	[20,33]
FA 18:3 n-3	Oral	T-cell	Suppress T cell proliferation	[70]
FA 20:4	PLA2-II mediated release of arachidonic acid (only release no metabolism)	Neutrophil	Increased mac-1 (CD-11b/CD18) expression	[59]
FA 18:0, 18:2, 18:3, 20:4	Endogenous	Macrophages and hepatocytes	Ligand binding activators of PPAR-α, PPAR-γ	[41]
FA 18:2 n-6	Dietary source	Dendritic cells	Reduced infiltration of LN and activation of T-cell. Reduced IL-12 increased IL -10	[71,72]
FA 20:5	Synthetic	Mast Cells	Decreased activation	[73]
FA 22:6 n-3	Synthetic	Dendritic cells	Increased IL-12 Reduced IL-6 and IL-10	[74]
FA Metabolites PGE2	Endogenous	Lymphocytes	Inhibitor T_H1 response (IL-12)	[21]
Leukotriene B4	Endogenous, supplement	Neutrophil	Adherence to endothelia (CD11a and CD 11b)	[32]
Inflammasome Palmitic acid (C16:0)	Supplement	NLRP3 inflammasome	Increased IL-1β, IL-18	[44,45]
Oleic acid (C 18:1)	Supplement and dietary sources	NLRP3 inflammasome	Decreased IL-1β, TNF-α, IL-6	[43]

2.4. Maintaining Data Quality and Rigor in Studies Involving Bioactive Lipids

Activity of bioactive lipids is controlled by enzymatic conversion to inactive metabolites or sequestering them in phospholipids and thereby producing oxidized phospholipids. The levels of bioactive lipids in human plasma have been quantified down to the picomolar (pM) range [75], and a reliable way of analyzing modulation in these bioactive lipids is through mass spectrometry-based approaches. For those interested in oxidized phospholipids (i.e., lipids with bioactive lipids esterified to the phospholipid backbone) and their functions, a combination of chromatographic methods and ion mobility methods are required to distinguish phospholipids with traditional fatty acids from phospholipids containing sequestered eicosanoids (Figure 3c).

Prior to determining the abundance and function of phospholipids that contribute to the production of bioactive lipids, the overall lipidomic composition needs to be determined. Infusion-based lipidomics-based approaches such as MS/MS^{All} provide a rapid method to determine lipidomic compositions (Figure 3a). This simplified approach provided a comprehensive coverage of lipids in a short run time [76], enabling the identification of the lipids as well as their constituent components such as the fatty acids. Problems of isobaric overlap can be addressed by liquid chromatography-based tandem mass spectrometry but require significantly longer run times with the advantage of resolving isomeric lipid species, i.e., lipids with similar exact mass, based on their retention time and fragmentation pattern (Figure 3b). Identification and analysis of oxidized phospholipids benefit from the use of liquid chromatography-based separation to determine the lipid classes that sequester the bioactive lipid, followed by the comprehensive characterization of the lipid molecule based on accurate mass and fragmentation by tandem MS (Figure 3c).

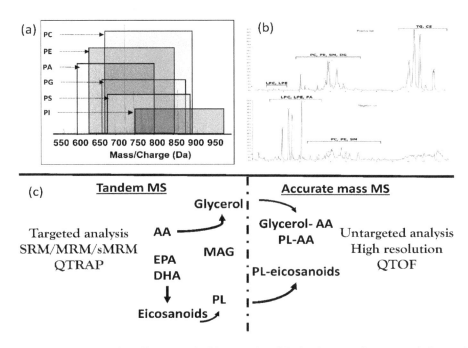

Figure 3. (**a**) Isobaric overlap (kindly provided by Dr. Paul Baker) typically seen while analyzing lipids. (**b**) Separation of lipids based on class in reverse phase chromatography in positive and negative mode. (**c**) Schematic of fatty acid and eicosanoid metabolism and suggested approaches for quantification of bioactive fatty acids and associated oxylipins (PL-eicosanoids).

With respect to studying omega-3 and omega-6 fatty acids involvement in a disease state or changes occurring at cellular level, comprehensive and targeted LC-MS/MS MRM methods enable simultaneous quantification of more than 100 lipid metabolites, including prostaglandins, leukotrienes, and other eicosanoids, resolvins, protectins, and other free fatty acids like arachidonic acid, eicosapentaenoic acid and docosahexaenoic acid [12,77,78] (Figure 3c (Targeted method)).

3. Implications of Bioactive Fatty Acids and Their Metabolites on Human Health

The US Food and Drug Administration (FDA) regulates the requirements for appropriate food labeling in the Title 21CRF101, including labeling of dietary supplements. In 2003, a final rule published in the Federal Register required that trans fatty acids be included in nutrition labeling, based on requests from the Center for Science in the Public Interest, due to the detrimental effects of such nutrients on plasmatic lipoproteins with potential increased risk of cardiovascular and metabolic diseases (i.e., increased low density lipoprotein-cholesterol (LDL-C)) [79,80].

In an effort to regulate the industry and inform the public on food and nutrition, the Food and Agriculture Organization of the United Nations (FAO) and the World Health Organization (WHO) provided a report of an expert consultation on fats and fatty acids in human nutrition. Published in 2008, this guidance is a useful reference on nutritional requirements and recommended dietary lipid intakes. This expert consultation focused on the role of specific fatty acid groups, such as the role of long-chain polyunsaturated fatty acids (LCPUFA) in neonatal and infant mental development, besides their role in maintenance of long-term health and prevention of specific chronic diseases. It was recommended that the n-3 PUFA and n-6 PUFA include more than one fatty acid with distinct attributes and biological function, and labelling them based on category may not be the best path forward. Due to an increasing acknowledgement of the importance of lipids during the initial years of life, this report also describes the requirements and the recommendations for fat and fatty acid for infants of 0–2 years and children of 2–18 years old, respectively [80].

A landmark randomized controlled trial was initially published in 2013 and then republished in 2018, is the 'Primary Prevention of Cardiovascular Disease with a Mediterranean Diet' [39,81] (PREDIMED). In the PREDIMED study, the consumption of energy unrestricted high-fat diets supplemented with extra-virgin olive oil or nuts, foods rich in unsaturated fatty acids, induced an impressive 30% relative risk reduction of major cardiac events compared to subjects who were counseled to follow a relatively low-fat diet. In addition, the subjects randomized to the high-fat diets experience a lower incidence of type 2 diabetes, therefore suggesting that the consumption of healthy fats also improves metabolic outcomes, in addition to cardiovascular outcomes [39]. These results not only support the concept that increasing the consumption of healthy fats (i.e., unsaturated n-3 fatty acids) may prevent cardiovascular diseases, but also that perhaps the prior cut-off of 30%–35% of total calories deriving from fat recommended by the FAO and WHO was still a sub-optimal amount, with the risk of precluding healthy subjects as well as patients with established diseases from the beneficial effects of the high-healthy fats diet. In fact, a recent presidential advisory from the American Heart Association [38] emphasized focusing on replacement of saturated fatty acids with unsaturated fatty acids to reduce the risk of cardiovascular diseases, without recommending a specific goal for total fat intake in terms of total % of calories. Other studies assessing the benefit of using polyunsaturated fatty acids in conditions other than cardiovascular disease can be found as follows in (Table 2). While discussions on the efficacy of the use of bioactive lipids in each of these studies is out of the scope of this review, it can be appreciated that sustained supplementation with high amounts of n-3 polyunsaturated fatty acids promote improved health outcomes.

Table 2. Clinical studies involving polyunsaturated fatty acids.

Study Design	Lipids	Study Endpoints	Results	Ref.
Double blind RCT study	Omega-3 long-chain polyunsaturated fatty acids	Allergic symptoms in children from mothers supplemented with 2.7 g omega-3 LCPUFA daily	Fewer allergies in children whose mothers received high omega-3 LCPUFA supplement.	[39]
REDUCE-IT study—double blind RCT study	Eicosapentaenoic acid	Cardiovascular death	2 g of EPA twice daily reduce risk of ischemic events.	[82]
Double blind RCT study	Eicosapentaenoic acid	Reduction of depressive symptoms	Omega-3 supplementation benefit patients with major depressive episode without comorbid anxiety disorder.	[83]
PREDIMED study random subsample—parallel-group randomized trial	Mediterranean diet	Cardiovascular events	Incidence of cardiovascular events was lower in patients receiving Mediterranean diet supplemented with extra-virgin oil or nuts.	[84]
PREDIMED study random subsample—parallel-group randomized trial	Mediterranean diet	Effect of HDL particles on reverse cholesterol transport	The diet increased cholesterol efflux, decreased cholesteryl ester transfer protein activity and increased HDL ability to esterify cholesterol.	[37]
Double blind RCT study	Fish oil n-3-PUFA	Muscle strength and average isokinetic power	n-3 PUFA therapy slows muscle mass decline and function in older adults	[85]
Double blind RCT study	Fish oil n-3-PUFA	Response of lysophospholipids to obesity	Obesity impact lysophospholipid metabolism abolishing its sensitivity to n-2 PUFA.	[86]
Compassionate protocol	Fish oil-based lipid emulsion	Resolution of cholestasis (plasma conjugated bilirubin <2 mg/dL)	All survival demonstrated resolution of cholestasis, compared with only 10% of non-surviving.	[87]
Double blind RCT study	EPA-DHA intake	Creatinine–cystatin C-based GFR	Long term supplementation with 400 mg/d of EPA-DHA provides slower kidney function decline in CKD patients.	[88]
Open-label randomized study	EPA-DHA intake	Cumulative rate of all-cause death, non-fatal myocardial infarction, and non-fatal stroke	Treatment significantly lowered risk of death and cardiovascular death.	[89]

4. Conclusions

Diet plays a major role in affecting metabolic and cardiovascular health. Particularly, fatty acids such as omega-3 and omega-6 have been subject of intense scrutiny in the last decades. Multiple studies agree on the fact that supplementation with bioactive lipids results in an increased bioavailability; cell type specific and systemic. Acknowledgement and appreciation of the antagonistic roles of n-3 PUFA/MUFA and SFA/n-6 PUFA in inflammation provide avenues for further research into their mechanism of action. While it has been difficult to determine if fatty acids interact with pattern recognition receptors by cognate interactions, there is consensus about their mechanism of action and final targets (reviewed in [90]). Commonality between studies that report on the benefits of n-3 fatty acids on health is that they are long term studies with high doses of supplementation. Thus, suggesting the need for a sustained intake to modify host lipid composition and thereby reduce the severity of diseases involving lipid-mediated signaling and inflammatory cascade. In addition to that, separate reports suggesting that n-3 fatty acids when present in concentration in excess of 40 uM in vitro and in vivo elicit beneficial effects by suppressing pro-inflammatory responses even at the cellular level argue in favor of determining endogenous concentrations of n-3 fatty acids in studies that did not show an affect with supplementation. In addition to their small window of bioactivity and rapid turnover, the field of study is marred by looking at systemic effects in diseases with localized etiology.

Much like immune function, bioactive lipids classically act in localized autocrine and paracrine circuits, the downstream effects of which determine physiological outcomes. Thus, while systemic evaluation of levels of bioactive lipids are informative of circulating levels, they represent the resting physiological levels unless sampling is performed in a state of active inflammation or disease. Furthermore, in supplementation studies, the quantification of a few bioactive lipids outside of the context of its precursors and metabolites provides incomplete information about their efficacy and may be thought to contribute to the variability in studies involving bioactive lipids.

Further research on testing the efficacy of lipid formulations rich in n-3 fatty acids/MUFAs by advanced lipidomics approaches to specifically monitor the flux of bioactive lipids [91] in relation to inflammation in intervention studies are warranted. Such studies hold promise to dispel any ambiguity that may remain with regards to the role of n-3 fatty acids/MUFAs as beneficial interventions in acute and chronic conditions that lead to the presentation of metabolism-associated disorders like cardiovascular diseases.

Author Contributions: Conceptualization, D.S.W.; writing-original draft, N.G.K., D.C., S.C.; visualization, N.G.K. and D.C.; writing-review and editing, P.M., S.C., E.T.P., B.V.T., D.F.B.; supervision, D.S.W.; project administration, N.G.K.; funding acquisition, D.S.W.

References

1. Billingsley, H.E.; Carbone, S.; Lavie, C.J. Dietary Fats and Chronic Noncommunicable Diseases. *Nutrients* **2018**, *10*, 1385. [CrossRef]
2. Yin, W.; Carballo-Jane, E.; McLaren, D.G.; Mendoza, V.H.; Gagen, K.; Geoghagen, N.S.; McNamara, L.A.; Gorski, J.N.; Eiermann, G.J.; Petrov, A.; et al. Plasma lipid profiling across species for the identification of optimal animal models of human dyslipidemia. *J. Lipid Res.* **2012**, *53*, 51–65. [CrossRef]
3. Aronoff, D.M.; Canetti, C.; Peters-Golden, M. Prostaglandin E2 inhibits alveolar macrophage phagocytosis through an E-prostanoid 2 receptor-mediated increase in intracellular cyclic AMP. *J. Immunol.* **2004**, *173*, 559–565. [CrossRef]
4. Kaliannan, K.; Wang, B.; Li, X.; Kim, K.; Kang, J.X. A host-microbiome interaction mediates the opposing effects of omega-6 and omega-3 fatty acids on metabolic endotoxemia. *Nat. Publ. Gr. Nat. Publ. Group* **2015**, 1–17. [CrossRef] [PubMed]
5. Spite, M.; Serhan, C.N. Novel lipid mediators promote resolution of acute inflammation: Impact of aspirin and statins. *Circ. Res.* **2010**, *107*, 1170–1184. [CrossRef] [PubMed]
6. Serhan, C.N. Resolution Phase of Inflammation: Novel Endogenous Anti-Inflammatory and Proresolving Lipid Mediators and Pathways. *Annu. Rev. Immunol.* **2007**, *25*, 101–137. [CrossRef] [PubMed]

7. Di Marzo, V. Arachidonic acid and eicosanoids as targets and effectors in second messenger interactions. *Prostaglandins Leukot. Essent. Fat. Acids* **1995**, *53*, 239–254. [CrossRef]

8. Dennis, E.A.; Norris, P.C. Eicosanoid storm in infection and inflammation. *Nat. Rev. Immunol.* **2015**, *15*, 511–523. [CrossRef] [PubMed]

9. Simopoulos, A.P. The Importance of the Omega-6/Omega-3 Fatty Acid Ratio in Cardiovascular Disease and Other Chronic Diseases. *Exp. Biol. Med.* **2008**, *233*, 674–688. [CrossRef] [PubMed]

10. Lee, J.Y.; Zhao, L.; Hwang, D.H. Modulation of pattern recognition receptor-mediated inflammation and risk of chronic diseases by dietary fatty acids. *Nutr. Rev.* **2010**, *68*, 38–61. [CrossRef]

11. Honda, K.L.; Lamon, S.; Matthan, N.R.; Wu, D.; Lichtenstein, A.H. EPA and DHA Exposure Alters the Inflammatory Response but not the Surface Expression of Toll–Like Receptor 4 in Macrophages. *Lipids* **2015**, *50*, 121–129. [CrossRef] [PubMed]

12. Ekroos, K.; Jänis, M.; Tarasov, K. Lipidomics: A Tool for Studies of Atherosclerosis. *Curr. Atheroscler. Rep.* **2010**, *12*, 273–281. [CrossRef] [PubMed]

13. Stegemann, C.; Pechlaner, R.; Willeit, P.; Langley, S.R.; Mangino, M.; Mayr, U.; Menni, C.; Moayyeri, A.; Santer, P.; Rungger, G.; et al. Lipidomics Profiling and Risk of Cardiovascular Disease in the Prospective Population-Based Bruneck Study. *Circulation* **2014**, *129*, 1821–1831. [CrossRef] [PubMed]

14. Lintonen, T.P.; Baker, P.R.; Suoniemi, M.; Ubhi, B.K.; Koistinen, K.M.; Duchoslav, E.; Campbell, J.L.; Ekroos, K. Differential mobility spectrometry-driven shotgun lipidomics. *Anal. Chem.* **2014**, *86*, 9662–9669. [CrossRef] [PubMed]

15. Cífková, E.; Holčapek, M.; Lísa, M.; Vrána, D.; Melichar, B.; Študent, V. Lipidomic differentiation between human kidney tumors and surrounding normal tissues using {HILIC}-{HPLC}/{ESI}–{MS} and multivariate data analysis. *J. Chromatogr. B* **2015**, *1000*, 14–21. [CrossRef] [PubMed]

16. Huang, S.; Rutkowsky, J.M.; Snodgrass, R.G.; Ono-Moore, K.D.; Schneider, D.A.; Newman, J.W.; Adams, S.H.; Hwang, D.H. Saturated fatty acids activate TLR-mediated proinflammatory signaling pathways. *J. Lipid Res.* **2012**, *53*, 2002–2013. [CrossRef]

17. De Lima-salgado, T.M. Molecular Mechanisms by Which Saturated Fatty Acids Modulate TNF- a Expression in Mouse Macrophage Lineage. *Cell Biochem. Biophys.* **2011**, 89–97. [CrossRef]

18. Calder, P.C. Polyunsaturated fatty acids and inflammation. *Biochem. Soc. Trans.* **2005**, *33*, 423–427. [CrossRef]

19. O'Donnell, V.B.; Murphy, R.C. New families of bioactive oxidized phospholipids generated by immune cells: Identification and signaling actions Review article New families of bioactive oxidized phospholipids generated by immune cells: Identification and signaling actions. *Blood J. Am. Soc. Hematol.* **2012**, *120*, 1985–1992. [CrossRef]

20. Watanabe, S.; Onozaki, K.; Yamamoto, S.; Okuyama, H. Regulation by dietary essential fatty acid balance of tumor necrosis factor production in mouse macrophages. *J. Leukoc. Biol.* **1993**, *53*, 151–156. [CrossRef]

21. Detlnitive, B.; Boeije, L.C.M.; Aarden, L.A.; Blood, C.; Service, T. Prostaglandin-E2 Is a Potent Inhibitor of Human Interleukin 12 Production. *J. Exp. Med.* **1995**, *181*, 775–779.

22. Calder, P.C. Long chain fatty acids and gene expression in inflammation and immunity. *Curr. Opin. Clin. Nutr. Metab. Care* **2013**, *16*, 425–433. [CrossRef] [PubMed]

23. Sadik, C.D.; Luster, A.D. Lipid-cytokine-chemokine cascades orchestrate leukocyte recruitment in inflammation. *J. Leukoc. Biol.* **2012**, *91*, 207–215. [CrossRef] [PubMed]

24. Zeyda, M.; Zlabinger, G.J.; Waldha, W.; Stulnig, T.M. Polyunsaturated fatty acids interfere with formation of the immunological synapse. *J. Leukoc. Biol.* **2005**, *77*, 680–688. [CrossRef]

25. Ojala, P.J.; Hirvonen, T.E.; Hermansson, M.; Somerharju, P.; Parkkinen, J. Acyl chain-dependent effect of lysophosphatidylcholine on human neutrophils. *J. Leukoc. Biol.* **2007**, *82*, 1501–1509. [CrossRef]

26. Gurzell, E.A.; Teague, H.; Harris, M.; Clinthorne, J.; Shaikh, S.R.; Fenton, J.I. DHA-enriched fish oil targets B cell lipid microdomains and enhances ex vivo and in vivo B cell function. *J. Leukoc. Biol.* **2013**, *93*, 463–470. [CrossRef]

27. Monk, J.M.; Hou, T.Y.; Turk, H.F.; McMurray, D.N.; Chapkin, R.S. n3 PUFAs reduce mouse CD4+ T-cell ex vivo polarization into Th17 cells. *J. Nutr.* **2013**, *143*, 1501–1508. [CrossRef]

28. Phillips, A. Changes in the Incorporation of Free Fatty Acids Upon the Stimulation of Human Polymorphonuclear Leukocytes. *J. Leukoc. Biol.* **1986**, *39*, 267–284. [CrossRef]

29. Lindner, S.C.; Köhl, U.; Maier, T.J.; Steinhilber, D.; Sorg, B.L. TLR2 ligands augment cPLA2alpha activity and lead to enhanced leukotriene release in human monocytes. *J. Leukoc. Biol.* **2009**, *86*, 389–399. [CrossRef]

30. Lee, I.T.; Lee, C.W.; Tung, W.H.; Wang, S.W.; Lin, C.C.; Shu, J.C.; Yang, C.M. Cooperation of TLR2 with MyD88, PI3K, and Rac1 in lipoteichoic acid-induced cPLA2/COX-2-dependent airway inflammatory responses. *Am. J. Pathol.* **2010**, *176*, 1671–1684. [CrossRef]

31. Bomalaski, J.S.; Freundlich, B.; Steiner, S.; Clark, M.A. The role of fatty acid metabolites in the differentiation of the human monocyte-like cell line U937. *J. Leukoc. Biol.* **1988**, *44*, 51–57. [CrossRef] [PubMed]

32. Bates, E.J.; Ferrante, A.; Harvey, D.P.; Poulos, A. Polyunsaturated fatty acids increase neutrophil adherence and integrin receptor expression. *J. Leukoc. Biol.* **1993**, *53*, 420–426. [CrossRef] [PubMed]

33. Turek, J.J.; Schoenlein, I.A.; Clark, L.K.; Vanalstine, W.G. Dietary polyunsaturated fatty-acid effects on immune cells of the porcine lung. *J. Leukoc. Biol.* **1994**, *56*, 599–604. [CrossRef] [PubMed]

34. Wallace, F.A.; Miles, E.A.; Evans, C.; Stock, T.E.; Yaqoob, P.; Calder, P.C. Dietary fatty acids influence the production of Th1-but not Th2-type cytokines. *J. Leukoc. Biol.* **2001**, *69*, 449–457.

35. Field, C.J.; Johnson, I.R.; Schley, P.D. Nutrients and their role in host resistance to infection. *J. Leukoc. Biol.* **2002**, *71*, 16–32.

36. Hekmatdoost, A.; Wu, X.; Morampudi, V.; Innis, S.M.; Jacobson, K. Dietary oils modify the host immune response and colonic tissue damage following Citrobacter rodentium infection in mice. *Am. J. Physiol. Gastrointest. Liver Physiol.* **2013**, *304*, G917–G928. [CrossRef]

37. Hernáez, Á.; Castañer, O.; Elosua, R.; Pintó, X.; Estruch, R.; Salas-Salvadó, J.; Corella, D.; Arós, F.; Serra-Majem, L.; Fiol, M.; et al. The Mediterranean Diet improves HDL function in high cardiovascular risk individuals: A randomized controlled trial. *Circulation* **2017**, *135*, 633–643. [CrossRef]

38. Sacks, F.M.; Lichtenstein, A.H.; Wu, J.H.; Appel, L.J.; Creager, M.A.; Kris-Etherton, P.M.; Miller, M.; Rimm, E.B.; Rudel, L.L.; Robinson, J.G.; et al. Dietary Fats and Cardiovascular Disease: A Presidential Advisory From the American Heart Association. *Circulation* **2017**, *136*, e1–e23. [CrossRef]

39. Estruch, R.; Ros, E.; Salas-Salvadó, J.; Covas, M.I.; Corella, D.; Arós, F.; Gómez-Gracia, E.; Ruiz-Gutiérrez, V.; Fiol, M.; Lapetra, J.; et al. Primary Prevention of Cardiovascular Disease with a Mediterranean Diet. *N. Engl. J. Med.* **2013**, *368*, 1279–1290. [CrossRef]

40. Wong, S.W.; Kwon, M.; Choi, A.M.K.; Kim, H.; Nakahira, K.; Hwang, D.H. Fatty Acids Modulate Toll-like Receptor 4 Activation through Regulation of Receptor Dimerization and Recruitment into Lipid Rafts in a Reactive Oxygen Species-dependent Manner. *J. Biol. Chem.* **2009**, *284*, 27384–27392. [CrossRef]

41. Kliewer, S.A.; Sundseth, S.S.; Jones, S.A.; Brown, P.J.; Wisely, G.B.; Koble, C.S.; Devchand, P.; Wahli, W.; Willson, T.M.; Lenhard, J.M.; et al. Fatty acids and eicosanoids regulate gene expression through direct interactions with peroxisome proliferator-activated receptors alpha and gamma. *Proc. Natl. Acad. Sci. USA* **1997**, *94*, 4318–4323. [CrossRef]

42. van de Veerdonk, F.L.; Netea, M.G.; Dinarello, C.A.; Joosten, L.A. Inflammasome activation and IL-1β and IL-18 processing during infection. *Trends Immunol.* **2011**, *32*, 110–116. [CrossRef]

43. Wen, H.; Gris, D.; Lei, Y.; Jha, S.; Zhang, L.; Huang, M.T.H.; Brickey, W.J.; Ting, J.P. Fatty acid–Induced NLRP3-ASC inflammasome activation interferes with insulin signaling. *Nat. Immunol.* **2011**, *12*, 408–415. [CrossRef]

44. Mills, K.H.G.; Roche, H.M. Dietary saturated fatty acids prime the NLRP3 inflammasome via TLR4 in dendritic cells—Implications for diet-induced insulin resistance. *Mol. Nutr. Food Res.* **2012**, *56*, 1212–1222. [CrossRef]

45. Sui, Y.; Luo, W.; Xu, Q.; Hua, J. Dietary saturated fatty acid and polyunsaturated fatty acid oppositely affect hepatic NOD-like receptor protein 3 inflammasome through regulating nuclear factor-kappa B activation. *World J. Gastroenterol.* **2016**, *22*, 2533–2544. [CrossRef]

46. Zhao, L.; Lee, J.Y.; Hwang, D.H. Inhibition of pattern recognition receptor-mediated inflammation by bioactive phytochemicals. *Nutr. Rev.* **2011**, *69*, 310–320. [CrossRef]

47. Villamón, E.; Roig, P.; Gil, M.L.; Gozalbo, D. Toll-like receptor 2 mediates prostaglandin E(2) production in murine peritoneal macrophages and splenocytes in response to Candida albicans. *Res. Microbiol.* **2004**, *156*, 115–118. [CrossRef]

48. Unoda, K.; Doi, Y.; Nakajima, H.; Yamane, K.; Hosokawa, T.; Ishida, S.; Kimura, F.; Hanafusa, T.

Eicosapentaenoic acid (EPA) induces peroxisome proliferator-activated receptors and ameliorates experimental autoimmune encephalomyelitis. *J. Neuroimmunol.* **2013**, *256*, 7–12. [CrossRef]

49. Ricote, M.; Huang, J.T.; Welch, J.S.; Glass, C.K. The peroxisome proliferator-activated receptorϒ (PPARϒ) as a regulator of monocyte/macrophage function. *J. Leukoc. Biol.* **1999**, *66*, 733–739. [CrossRef]

50. Draper, E.; Reynolds, C.M.; Canavan, M.; Mills, K.H.; Loscher, C.E.; Roche, H.M. Omega-3 fatty acids attenuate dendritic cell function via NF-kB independent of PPARϒ. *J. Nutr. Biochem.* **2011**, *22*, 784–790. [CrossRef]

51. Valledor, A.F.; Ricote, M. Nuclear receptor signaling in macrophages. *Biochem. Pharmacol.* **2004**, *67*, 201–212. [CrossRef] [PubMed]

52. Garay-Lugo, N.; Domínguez-Lopez, A.; Miliar Garcia, A.; Aguilar Barrera, E.; Gomez Lopez, M.; Gómez Alcalá, A.; Martínez Godinez, M.D.L.A.; Lara-Padilla, E. n-3 fatty acids modulate the mRNA expression of the Nlrp3 inflammasome and Mtor in the liver of rats fed with high-fat or high-fat/fructose diets. *Immunopharmacol. Immunotoxicol.* **2016**, *38*, 353–363. [CrossRef] [PubMed]

53. Collino, M.; Benetti, E.; Rogazzo, M.; Mastrocola, R.; Yaqoob, M.M.; Aragno, M.; Thiemermann, C.; Fantozzi, R. Reversal of the deleterious effects of chronic dietary HFCS-55 intake by PPAR—δ agonism correlates with impaired NLRP3 inflammasome activation. *Biochem. Pharmacol.* **2013**, *85*, 257–264. [CrossRef] [PubMed]

54. Yang, G.; Chen, S.; Deng, B.; Tan, C.; Deng, J.; Zhu, G.; Yin, Y.; Ren, W. Implication of G Protein-Coupled Receptor 43 in Intestinal Inflammation: A Mini-Review. *Front. Immunol.* **2018**, *9*, 1434. [CrossRef] [PubMed]

55. Vinolo, M.A.R.; Rodrigues, H.G.; Nachbar, R.T.; Curi, R. Regulation of Inflammation by Short Chain Fatty Acids. *Nutrients* **2011**, *3*, 858–876. [CrossRef]

56. Bhutia, Y.D.; Ganapathy, V. Short, but Smart: SCFAs Train T Cells in the Gut to Fight Autoimmunity in the Brain. *Immunity* **2015**, *43*, 629–631. [CrossRef]

57. Husted, A.S.; Trauelsen, M.; Rudenko, O.; Hjorth, S.A.; Schwartz, T.W. GPCR-Mediated Signaling of Metabolites. *Cell Metab.* **2017**, *25*, 777–796. [CrossRef]

58. Yonezawa, T.; Kurata, R.; Yoshida, K.; Murayama, M.A.; Cui, X. Free Fatty Acids-Sensing G Protein-Coupled Receptors in Drug Targeting and Therapeutics. *Curr. Med. Chem.* **2013**, *20*, 3855–3871. [CrossRef]

59. Takasaki, J.; Kawauchi, Y.; Yasunaga, T.; Masuho, Y. Human type II phospholipase A2-induced Mac-1 expression on human neutrophils. *J. Leukoc. Biol.* **1996**, *60*, 174–180. [CrossRef]

60. Dahan, S.; Segal, Y.; Shoenfeld, Y. Dietary factors in rheumatic autoimmune diseases: A recipe for therapy? *Nat. Rev. Rheumatol.* **2017**, *13*, 348–358. [CrossRef]

61. De Nardo, D.; Latz, E. NLRP3 inflammasomes link inflammation and metabolic disease. *Trends Immunol.* **2011**, *32*, 373–379. [CrossRef] [PubMed]

62. Ralston, J.C.; Lyons, C.L.; Kennedy, E.B.; Kirwan, A.M.; Roche, H.M. Fatty Acids and NLRP3 Inflammasome—Mediated Inflammation in Metabolic Tissues. *Annu. Rev. Nutr.* **2017**, *37*, 77–102. [CrossRef] [PubMed]

63. Guo, H.; Callaway, J.B.; Ting, J.P. Inflammasomes: Mechanism of action, role in disease, and therapeutics. *Nat. Med.* **2015**, *21*, 677–687. [CrossRef] [PubMed]

64. Shen, L.; Yang, Y.; Ou, T.; Key, C.C.; Tong, S.H.; Sequeira, R.C.; Nelson, J.M.; Nie, Y.; Wang, Z.; Boudyguina, E.; et al. Dietary PUFAs attenuate NLRP3 inflammasome activation via enhancing macrophage autophagy. *J. Lipid Res.* **2017**, *58*, 1808–1821. [CrossRef]

65. L'homme, L.; Esser, N.; Riva, L.; Scheen, A.; Paquot, N.; Piette, J.; Legrand-Poels, S. Unsaturated fatty acids prevent activation of NLRP3 inflammasome in human monocytes/macrophages. *J. Lipid Res.* **2013**, *54*. [CrossRef]

66. de Pablo, M.A.; Alvarez de Cienfuegos, G. Modulatory effects of dietary lipids on immune system functions. *Immunol. Cell Biol.* **2000**, *78*, 31–39. [CrossRef]

67. Ashley, J.W.; Hancock, W.D.; Nelson, A.J.; Bone, R.N.; Hubert, M.T.; Wohltmann, M.; Turk, J.; Ramanadham, S. Polarization of Macrophages toward M2 Phenotype Is Favored by Reduction in iPLA2beta (Group VIA Phospholipase A2). *J. Biol. Chem.* **2016**, *291*, 23268–23281. [CrossRef]

68. Werz, O.; Gerstmeier, J.; Libreros, S.; De la Rosa, X.; Werner, M.; Norris, P.C.; Chiang, N.; Serhan, C.N. Human macrophages differentially produce specific resolvin or leukotriene signals that depend on bacterial pathogenicity. *Nat. Commun.* **2018**, *9*, 59. [CrossRef]

69. Sorgi, C.A.; Zarini, S.; Martin, S.A.; Sanchez, R.L.; Scandiuzzi, R.F.; Gijón, M.A.; Guijas, C.; Flamand, N.;

Murphy, R.C.; Faccioli, L.H. Dormant 5-lipoxygenase in inflammatory macrophages is triggered by exogenous arachidonic acid. *Sci. Rep.* **2017**, *7*, 10981. [CrossRef]

70. Seiler, M.; Rossettl, G.; Zurier, R.B. Oral administration of unsaturated fatty acids: Effects human on peripheral blood T lymphocyte proliferation. *J. Leukoc. Biol.* **1997**, *62*, 438–443.

71. Draper, E.; DeCourcey, J.; Higgins, S.C.; Canavan, M.; McEvoy, F.; Lynch, M.; Keogh, B.; Reynolds, C.; Roche, H.M.; Mills, K.H.; et al. Conjugated linoleic acid suppresses dendritic cell activation and subsequent Th17 responses. *J. Nutr. Biochem.* **2014**, *25*, 741–749. [CrossRef] [PubMed]

72. Penedo, L.A.; Nunes, J.C.; Gama, M.A.S.; Leite, P.E.C.; Quirico-Santos, T.F.; Torres, A.G. Intake of butter naturally enriched with cis9,trans11 conjugated linoleic acid reduces systemic inflammatory mediators in healthy young adults. *J. Nutr. Biochem.* **2013**, *24*, 2144–2151. [CrossRef] [PubMed]

73. Wang, X.; Ma, D.W.L.; Kang, J.X.; Kulka, M. N-3 Polyunsaturated fatty acids inhibit Fc ε receptor I-mediated mast cell activation. *J. Nutr. Biochem.* **2015**, *26*, 1580–1588. [CrossRef] [PubMed]

74. Zapata-Gonzalez, F.; Rueda, F.; Petriz, J.; Domingo, P.; Villarroya, F.; Diaz-Delfin, J.; de Madariaga, M.A.; Domingo, J.C. Human dendritic cell activities are modulated by the omega-3 fatty acid, docosahexaenoic acid, mainly through PPARγ: RXR heterodimers: Comparison with other polyunsaturated fatty acids. *J. Leukoc. Biol.* **2017**, *84*, 1172–1182. [CrossRef] [PubMed]

75. Burla, B.; Arita, M.; Arita, M.; Bendt, A.K.; Cazenave-Gassiot, A.; Dennis, E.A.; Ekroos, K.; Han, X.; Ikeda, K.; Liebisch, G.; et al. MS-based lipidomics of human blood plasma: A community-initiated position paper to develop accepted guidelines 1. *J. Lipid Res.* **2018**, *59*, 2001–2007. [CrossRef]

76. Simons, B.; Kauhanen, D.; Sylvänne, T.; Tarasov, K.; Duchoslav, E.; Ekroos, K. Shotgun Lipidomics by Sequential Precursor Ion Fragmentation on a Hybrid Quadrupole Time-of-Flight Mass Spectrometer. *Metabolites* **2012**, *2*, 195–213. [CrossRef]

77. Ejsing, C.S.; Duchoslav, E.; Sampaio, J.; Simons, K.; Bonner, R.; Thiele, C.; Ekroos, K.; Shevchenko, A. Automated identification and quantification of glycerophospholipid molecular species by multiple precursor ion scanning. *Anal. Chem.* **2006**, *78*, 6202–6214. [CrossRef]

78. Astarita, G.; Kendall, A.C.; Dennis, E.A.; Nicolaou, A. Targeted lipidomics strategies for oxygenated metabolites of polyunsaturated fatty acids. *Biochim. Biophys. Acta* **2015**, *1851*, 456–468. [CrossRef]

79. Federal {Register}. Available online: https://www.gpo.gov/fdsys/pkg/FR-2003-07-11/html/FR-2003-07-11-FrontMatter.htm (accessed on 11 July 2003).

80. Food and Agriculture Organization of the United Nations. Fats and fatty acids in human nutrition: report of an expert consultation: Geneva. Rome, 10–14 November 2008.

81. Gómez-gracia, E.; Ruiz-gutiérrez, V.; Fiol, M. Primary prevention of cardiovascular disease with a mediterranean diet. *N. Engl. J. Med.* **2013**, 1279–1290. [CrossRef]

82. Bhatt, D.L.; Steg, P.G.; Miller, M.; Brinton, E.A.; Jacobson, T.A.; Ketchum, S.B.; Doyle, R.T., Jr.; Juliano, R.A.; Jiao, L.; Granowitz, C.; et al. Cardiovascular Risk Reduction with Icosapent Ethyl for Hypertriglyceridemia. *N. Engl. J. Med.* **2019**, *380*, 11–22. [CrossRef]

83. Lesperance, F.; Frasure-Smith, N.; St-Andre, E.; Turecki, G.; Lesperance, P.; Wisniewski, S.R. The efficacy of omega-3 supplementation for major depression: A randomized controlled trial. *J. Clin. Psychiatry.* **2011**, *72*, 1054–1062. [CrossRef] [PubMed]

84. Estruch, R.; Ros, E.; Salas-Salvadó, J.; Covas, M.I.; Corella, D.; Arós, F.; Gómez-Gracia, E.; Ruiz-Gutiérrez, V.; Fiol, M.; Lapetra, J.; et al. Primary Prevention of Cardiovascular Disease with a Mediterranean Diet Supplemented with Extra-Virgin Olive Oil or Nuts. *N. Engl. J. Med.* **2018**, *378*, e34. [CrossRef]

85. Smith, G.I.; Julliand, S.; Reeds, D.N.; Sinacore, D.R.; Klein, S.; Mittendorfer, B. Fish oil-derived n-3 PUFA therapy increases muscle mass and function in healthy older adults. *Am. J. Clin. Nutr.* **2015**, *102*, 115–122. [CrossRef] [PubMed]

86. del Bas, J.M.; Caimari, A.; Rodriguez-Naranjo, M.I.; Childs, C.E.; Paras Chavez, C.; West, A.L.; Miles, E.A.; Arola, L.; Calder, P.C. Impairment of lysophospholipid metabolism in obesity: Altered plasma profile and desensitization to the modulatory properties of n-3 polyunsaturated fatty acids in a randomized controlled trial. *Am. J. Clin. Nutr.* **2016**, *104*, 266–279. [CrossRef] [PubMed]

87. Premkumar, M.H.; Carter, B.A.; Hawthorne, K.M.; King, K.; Abrams, S.A. High rates of resolution of cholestasis in parenteral nutrition-associated liver disease with fish oil-based lipid emulsion monotherapy. *J. Pediatr.* **2013** *162*, 793–798. [CrossRef]

88. Hoogeveen, E.K.; Geleijnse, J.M.; Kromhout, D.; Stijnen, T.; Gemen, E.F.; Kusters, R.; Giltay, E.J. Effect of omega-3 fatty acids on kidney function after myocardial infarction: The Alpha Omega Trial. *Clin. J. Am. Soc. Nephrol.* **2014**, *9*, 1676–1683. [CrossRef]

89. GISSI-Prevenzione Investigators. Dietary supplementation with n-3 polyunsaturated fatty acids and vitamin E after myocardial infarction: Results of the GISSI-Prevenzione trial. Gruppo Italiano per lo Studio della Sopravvivenza nell'Infarto miocardico. *Lancet (Lond. Engl.)* **1999**, *354*, 447–455. [CrossRef]

90. Yin, J.; Peng„ Y.; Wu, J.; Wang, Y.; Yao, L. Toll-like receptor 2/4 links to free fatty acid-induced inflammation and β-cell dysfunction. *J. Leukoc. Biol.* **2013**, *95*, 47–52. [CrossRef]

91. Ecker, J.; Liebisch, G. Application of stable isotopes to investigate the metabolism of fatty acids, glycerophospholipid and sphingolipid species. *Prog. Lipid Res.* **2014**, *54*, 14–31. [CrossRef]

4

Preliminary Estimations of Insect Mediated Transfers of Mercury and Physiologically Important Fatty Acids from Water to Land

Sydney Moyo

Department of Oceanography and Coastal Sciences, Louisiana State University, Baton Rouge, LA 70803, USA; sydmoyo@gmail.com

Abstract: Aquatic insects provide an energy subsidy to riparian food webs. However, most empirical studies have considered the role of subsidies only in terms of magnitude (using biomass measurements) and quality (using physiologically important fatty acids), negating an aspect of subsidies that may affect their impact on recipient food webs: the potential of insects to transport contaminants (e.g., mercury) to terrestrial ecosystems. To this end, I used empirical data to estimate the magnitude of nutrients (using physiologically important fatty acids as a proxy) and contaminants (total mercury (Hg) and methylmercury (MeHg)) exported by insects from rivers and lacustrine systems in each continent. The results reveal that North American rivers may export more physiologically important fatty acids per unit area (93.0 ± 32.6 Kg Km^{-2} year^{-1}) than other continents. Owing to the amount of variation in Hg and MeHg, there were no significant differences in MeHg and Hg among continents in lakes (Hg: 1.5×10^{-4} to 1.0×10^{-3} Kg Km^{-2} year^{-1}; MeHg: 7.7×10^{-5} to 1.0×10^{-4} Kg Km^{-2} year^{-1}) and rivers (Hg: 3.2×10^{-4} to 1.1×10^{-3} Kg Km^{-2} year^{-1}; MeHg: 3.3×10^{-4} to 8.9×10^{-4} Kg Km^{-2} year^{-1}), with rivers exporting significantly larger quantities of mercury across all continents than lakes. Globally, insect export of physiologically important fatty acids by insect was estimated to be $\sim43.9 \times 10^6$ Kg year^{-1} while MeHg was ~649.6 Kg year^{-1}. The calculated estimates add to the growing body of literature, which suggests that emerging aquatic insects are important in supplying essential nutrients to terrestrial consumers; however, with the increase of pollutants in freshwater systems, emergent aquatic insect may also be sentinels of organic contaminants to terrestrial consumers.

Keywords: aquatic ecosystems; subsidies; eicosapentaenoic acid; docosahexaenoic acid; food webs

1. Introduction

The movement of materials between juxtaposed habitats has received much attention by food web and landscape ecologists in the last four decades (reviewed by Richardson and Sato [1]). Freshwater ecologists have long documented that exogenous organic matter (e.g., terrestrial leaves) fuels rivers via inputs of nutrients and organic matter [2], but in recent decades, the importance of aquatic insect subsidies to riparian predators (e.g., bats; [3]) has been emphasized [4–6]. These aquatic subsidies are known to affect the behaviour, productivity, and diversity of riparian predators [7,8].

One such subsidy is in the form of physiologically important fatty acids (eicosapentaenoic acid (EPA; 20:5ω3) and docosahexaenoic acid (DHA; 22:6ω3)), both of which are of fundamental physiological importance to all organisms [5,9] because most consumers do not possess the necessary enzymes to synthesize them in the required quantities, so they must obtain them from their diet. These physiologically important fatty acids are required for the maintenance of cell membrane structure and function [10,11], regulating hormonal processes and preventing cardiovascular diseases [12].

Aquatic insects are one group of organisms known to be key exporters of physiologically important fatty acids to terrestrial systems [13], and because many adult insects do not return to the water [14], they represent a net loss of organic nutrients from the aquatic system, and potential food for consumers in adjacent terrestrial ecosystems. A plethora of studies on fatty acids in aquatic systems generally support the premise that aquatic insects are richer in physiologically important fatty acids [15–17] than their terrestrial counterparts [13]. Aquatic insects lay their eggs in freshwaters, where the larvae then develop and accumulate physiologically important fatty acids [18]. Subsequently, owing to their complex life cycles, aquatic insects can effectively transfer physiologically important fatty acids to the terrestrial system when they emerge and fall prey to terrestrial predators [19]. As such, knowledge of fatty acids in food sources and consumers is important both for obtaining basic dietary information on consumers within one habitat and for assessing the nutritional implications of reciprocal fluxes in juxtaposed habitats.

Further to providing critical nutrients to terrestrial consumers, aquatic insects can also supply unwanted contaminants to recipient food webs [20]. One such contaminant is mercury, a metal that has become a global concern because of its toxicity. Specifically, methylmercury (MeHg) is of concern as it concentrates at the base of aquatic food webs (e.g., algae) and is subsequently biomagnified, resulting in high concentrations of MeHg in the tissues of predators (e.g., spiders; [21]). The potential of MeHg to be biomagnified presents a health hazard to aquatic organisms and terrestrial wildlife with trophic linkages to aquatic food webs (e.g., those that consume emergent aquatic insects; [22–24]). While many studies have examined the movement of contaminants between habitats (e.g., Du et al. [25]), few studies have concurrently measured the fluxes of contaminants and fatty acids from streams to riparian zones; even though stream contamination is widespread [26].

Great strides have been made by individual researchers on the potential export of fatty acids from water to land (e.g., [13,27]), however, studies looking into the potential export by insects are scanty. Furthermore, our current knowledge of transfer of fatty acids and contaminants extends only to site-specific studies (with many being biased toward the Northern Hemisphere), effectively limiting our ability to understand the global effects of stream-derived contaminants and nutrients across aquatic–terrestrial boundaries.

Through the seminal works of Gladyshev and others [18], the first global estimate of physiologically important fatty acids by emerging insects was estimated to be between 0.1 Kg km^{-2} year^{-1} to as high as 672.2 Kg km^{-2} year^{-1}. One would expect that with new studies documenting fatty acids in insects, these estimates may have changed significantly. To date, no global estimates are available for the global estimate of mercury from water to land. To this end, the aim of this study was to build on past works by Gladyshev et al. [18] and estimate the continental and global export of contaminants (methylmercury) and nutrients (physiologically important fatty acids) from freshwater systems to land and to determine the extent of coupling between contaminants and nutrients.

2. Material and Methods

2.1. Literature Search and Data Extraction

To quantify export of physiologically important fatty acids and mercury (Aim: estimate continental and global export of mercury and physiologically important fatty acids via insects) from freshwater systems to land, I quantified subsidies (using physiologically important fatty acids; DHA + EPA) and the potential export of contaminants (methylmercury and total mercury) from freshwater to terrestrial ecosystems by carrying out an extensive search of the scientific literature. To identify relevant studies, a comprehensive literature search was conducted using papers from scientific databases (Google Scholar©, Scholars Portal© and Thomson Reuters Web of Science©) using the search algorithm: fatty acids OR mercury*AND benthic invert*aquatic insects* OR insect emergence. I also included papers from the first global estimates of insect emergence and fatty acids listed in works by Gladyshev et al. [18]. These initial searches yielded >400 articles published up to October 2019. From this initial set, the final dataset (Tables 1–5) was chosen based on the following criteria: (1) emergence reported in mg m^{-2} year (or comparable units) for the year, (2) fatty acids and mercury were reported in mg g^{-1} and ng g^{-1}, respectively (or comparable units e.g., ug g^{-1}), for benthic insects, (3) only emergence traps were used to collect emergent insects, (4) studies that did not use allometric equations (length-weight regressions) to estimate the dry weight of emergent insects (e.g., [13]) that may overestimate emergence rates [28], and (5) only studies published in English, were included in literature surveys. Criterion 2 excluded studies that reported fatty acid and mercury data as relative proportions or percentages (%).

In several cases, fatty acid, mercury and emergence data were available for different seasons or from different locations. Within a single location, a grand mean was calculated from the fatty acid data from that location, regardless of season; thus, the values represented the average values for a location. Data from different studies were combined to provide a grand mean for each type of data (fatty acid, methylmercury, total mercury, emergence).

To standardise values with those reported in the broader scientific literature, I ensured that all units were converted to match those reported in the literature by other authors [27].

2.2. Calculation of Surface Area

Total surface area (Km^2) was estimated by calculating areas of lakes and rivers for six of the world's continents (Africa, Asia, Australia, Europe, North America and South America; Table S1 in Supporting information). I used estimates from the Global Lakes and Wetlands Database (GLWD; [29]), Digital Chart of the World (DCW; [30]), HydroSheds (basins and stream networks; [31]) and HydroK1 (US Geological Survey. [32], empirical data supplied by authors [33]) to calculate the total surface area of lakes and rivers. All Shapefiles (.shp) were visualized and surface areas measured using GRASS GIS [34] and QGIS (version 3.10, [35]). For global estimates of surface areas of lakes and rivers, theoretical calculations from several models in the literature were used (see Supplementary information; Table S2).

Aquatic insects develop and live in only a small portion of aquatic habitats. For instance, over 72% of insects only live in the littoral area of lakes near the shore [36]. Similarly, littoral zones can make up anywhere from 3.4% to 30.3% of the surface area of lakes [36]. As such, I adjusted the measurements of all areas to account for the littoral zone to be between 3.4% to 30.3% (average of 18.6% for all Lakes).

2.3. Emergence of Insects

Data for emerging aquatic insects (dry weight; $g \, m^{-2} \, year^{-1}$) were extracted from diverse literature data (Figure 1; Tables 1 and 2). Because only a very small percentage of emergent aquatic return to the stream, I used the average calculations of return of insect to freshwaters. For instance, Jackson and Fisher [14] enumerated the return of adult aquatic insects to be only 3.1% of the emerged biomass returned to the stream. Elsewhere, Gray [37] found that less than 1% of aquatic insects in a prairie stream returned to the aquatic system, whereas other researchers have documented larger (9.2%) returns by biomass in lacustrine systems [38,39]. As such, I corrected the net export to account for the return of between 1% to 9.2% for lakes and rivers (average of 4.43% return rate).

2.4. Estimates of Physiologically Important Fatty Acids in Aquatic Insects

Available data on physiologically important fatty acids (Figure 1; mg g^{-1} of dry mass) were obtained based on studies that quantitatively determined the fatty acids content of insects using standard fatty acid extraction methods (e.g., [40,41]). Some data reported were for aquatic insect larvae and these were included in the analysed dataset. Fatty acid content of insect differs with life stages from larvae to adults [41], however, the life-stage differences in physiologically fatty acids are minor. For example, some mosquito (Culicidae) larvae and adults have been observed to contain approximately similar quantities of physiologically important fatty acids [41]. Where data were reported as wet weight, I used the moisture content given by the authors to calculate the dry mass. Taxa included were from Europe and Asia (Table 3). Most data collected indicated that Diptera are the most dominant order in most emergence data sets.

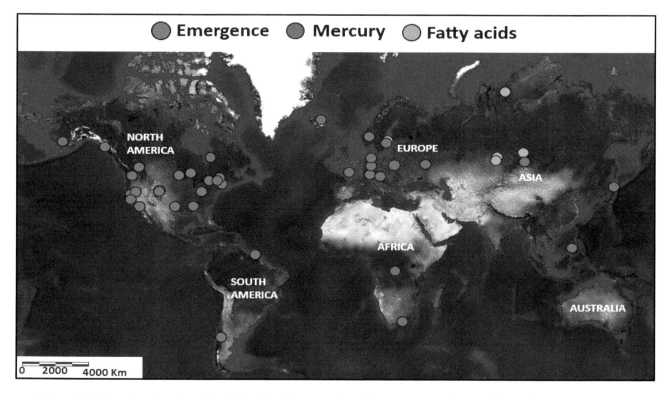

Figure 1. Map showing locality of studies documenting the emergence fatty acids and mercury content in six continents.

Table 1. Insect emergence from lakes (g DM m^{-2} year^{-1}) for available continents. 'Community' denotes instances where whole taxa values are reported. Average and coefficient of variation in bold represents the grand average that was used to calculate emergence for Africa, South America, Asia, Australia.

Continent	Taxa	Emergence	Reference
Europe			
	Chironomidae, Ephemeroptera, Trichoptera	4.0	[42]
	Community	1.8	[43]
	Community	1.4	[43]
	Community	1.1	[43]
	Community	2.4	[44]
	Chironomidae	1.9	[45]
	Chironomidae	0.2	[46]
	Community	0.2 [a]	[47]
Average ± SD		1.6 ± 1.2	
Coefficient of variation (%)		70.9	
North America			
	Chironomidae	1.5 [b]	[38]
	Community	1.1 [c]	[48]
	Chironomidae	0.2 [d]	[49]
	Chironomidae	1.9	[14]
Average ± SD		1.2 ± 0.6	
Coefficient of variation (%)		53.8	
Average ± SD		**1.5 ± 1.0**	
Coefficient of variation (%)		**70**	

[a] average values calculated from Table 3 of the reference. [b] averaged author's data. [c] Recalculated from authors data. [d] average value calculated from Table 2 of the reference.

Table 2. Insect emergence from rivers (g DM m^{-2} year^{-1}) for available continents. 'Community' denotes instances where whole taxa values are reported. Values in 'bold' denote the grand means and standard deviation for all available data. Average and coefficient of variation in bold represents the grand average that was used to calculate emergence for Australia and South America.

Continent	Taxa	Emergence	Reference
Africa			
	Trichoptera	0.5 [e]	[14]
	Community	4.0 [e]	[14]
Average ± SD		2.2 ± 1.7	
Coefficient of variation (%)		78.6	
Asia			
	Community	2.1 [f]	[50]
	Community	1.2 [g]	[51]
Average		1.7 ± 0.5	
Coefficient of variation (%)		27.3	
Europe			
	Diptera, Trichoptera, Ephemeroptera	1.7	[52]
	Ephemeroptera, Plecoptera, Trichoptera	3.6 [h]	[14]
	Ephemeroptera, Plecoptera, Trichoptera	5.0 [h]	[14]
	Community	5.4 [h]	[14]
	Community	2.6 [h]	[14]
	Community	2.6 [h]	[14]
	Community	3.7 [h]	[14]
	Community	3.7 [h]	[14]
	Community	2.0 [h]	[14]
	Community	2.6 [h]	[14]
	Community	3.2 [h]	[14]
	Chironomidae	1.9 [h]	[14]
Average		3.2 ± 1.1	
Coefficient of variation (%)		35.7	
North America			
	Diptera, Chironomidae	1.2 [i]	[53]
	Trichoptera, Ephemeroptera, Plecoptera, Diptera	6.6 [j]	[54]
	Ephemeroptera, Plecoptera, Trichoptera	0.3	[39]
	Chironomidae, Ephemeroptera, Trichopetra	23.1 [h]	[14]
	Community	5.3	[14]
	Community	7.1	[14]
Average ± SD		7.8 ± 9.2	
Coefficient of variation (%)		117.4	
Average ± SD		**4.5 ± 4.5**	
Coefficient of variation (%)		**100.4**	

[e] data for Democratic republic of Congo (formerly Zaire) stream from Table 5 of the reference; [f] averaged from using average weight of insect specimen dry mass 150 µg; [g] recalculated from Figure 1C of the reference; [h] data for Europe from Table 5 of the reference; [i] averaged author's data; [j] recalculated from authors data.

Table 3. Physiologically important fatty acids (EPA+DHA, mg g^{-1} of dry mass) in emergent aquatic insects in lakes and rivers. Taxa in italics represent fatty acids measured in insect larvae. Average and coefficient of variation in bold represents the grand average that was used to calculate emergence for all six continents.

Continent	Taxa	EPA +DHA	Reference
Lentic			
	Odonata	8.27 [k]	[55]
	Chironomidae	11.9	[46]
	Community	17.8 [l]	[56]
	Chironomidae	4.0	[40]
	Chironomidae	7.0	[40]
	Ephemeroptera	11.3	[27]
	Chironomidae	10.1	[57]
	Culicidae	6.77	[41]
Average ± SD		**9.6 ± 3.9**	
Coefficient of variation (%)		**41**	
Lotic			
	Trichoptera [m]	11.6	[58]
	Ephemeroptera [m]	12.8	[58]
	Chironomidae [m]	7.7	[58]
	Chironomidae	18.1	[18]
	Trichoptera	9.4	[27]
Average ± SD		**11.9 ± 3.6**	
Coefficient of variation (%)		**30**	

[k] converted wet weight to dry weight based on authors data of moisture of ~71.7%; [l] average estimated from Figure 3 of the reference; [m] dry weight estimated from the reference using moisture contents of 83.8% Trichoptera, Chironomidae 78.0%, Ephemeroptera (80%).

Table 4. Total mercury (Hg, mg g^{-1} of dry mass) and methylmercury (MeHg, mg g^{-1}) in emergent aquatic insects in lakes. 'Community' denotes instances where whole taxa values are reported. Average and coefficient of variation in bold represents the grand average that was used to calculate emergence for Africa, Asia, Australia, Europe.

Continent	Taxa	Total Mercury	Methylmercury	Reference
Lentic				
North America				
	Trichoptera, Diptera	[n] 4.2×10^{-4}	[n] 1.6×10^{-4}	[48]
	Coleoptera	1.8×10^{-4}	1.1×10^{-4}	[59]
	Ephemeroptera	1.3×10^{-4}	1.4×10^{-5}	[59]
	Hemiptera	2.6×10^{-4}	1.2×10^{-4}	[59]
	Odonata	1.4×10^{-4}	1.0×10^{-4}	[59]
	Trichoptera	1.3×10^{-4}	4.9×10^{-5}	[59]
	Trichoptera	4.9×10^{-4}	2.5×10^{-5}	[60]
	Odonata	1.1×10^{-4}	5.7×10^{-5}	[60]
	Ephemeroptera	1.1×10^{-4}	2.1×10^{-5}	[60]
	Coleoptera	1.5×10^{-4}	2.0×10^{-5}	[60]
	Trichoptera	3.8×10^{-5}	1.6×10^{-5}	[60]
	Odonata	7.1×10^{-5}	4.8×10^{-5}	[60]
	Ephemeroptera	7.5×10^{-5}	1.9×10^{-5}	[60]
	Odonata	9.7×10^{-5}	1.1×10^{-4}	[61]
	Ephemeroptera	1.1×10^{-4}	7.9×10^{-5}	[61]
	Trichoptera	5.0×10^{-5}	3.7×10^{-5}	[61]
	Diptera	6.9×10^{-5}	3.6×10^{-5}	[61]
	Odonata	-	1.3×10^{-4}	[62]
	Diptera	-	7.9×10^{-5}	[62]
	Trichoptera	-	8.9×10^{-5}	[62]
Average ± SD		$1.3 \times 10^{-4} \pm 8.9 \times 10^{-5}$	$6.6 \times 10^{-5} \pm 4.3 \times 10^{-5}$	

Table 4. *Cont.*

Continent	Taxa	Total Mercury	Methylmercury	Reference
Coefficient of variation (%)		70	65	
South America				
	Diptera	[o] 1.3×10^{-3}	-	[63]
	Ephemeroptera	5.7×10^{-4}	-	[63]
	Odonata	1.7×10^{-4}	-	[63]
	Plecoptera	2.0×10^{-3}	-	[63]
	Trichoptera	3.1×10^{-4}	-	[63]
	Community	2.0×10^{-4}	3.4×10^{-5}	[64]
	Community	2.8×10^{-4}	1.9×10^{-4}	[65]
Average ± SD		$6.9 \times 10^{-4} \pm 6.4 \times 10^{-4}$	$7.0 \times 10^{-5} \pm 4.9 \times 10^{-5}$	
Coefficient of variation (%)		93	68	
Average ± SD		**$2.9 \times 10^{-4} \pm 4.4 \times 10^{-4}$**	**$7.0 \times 10^{-5} \pm 4.9 \times 10^{-5}$**	
Coefficient of variation (%)		**150**	**70**	

[n] mean from data presented in Table 3 in authors data; [o] units converted from ug g to mg g^{-1}.

Table 5. Total mercury (Hg, mg g^{-1} of dry mass) and methylmercury (MeHg, mg g^{-1}) in emergent aquatic insects in rivers. 'Community' denotes instances where whole taxa values are reported. Average and coefficient of variation (in bold) represents the grand average that was used to calculate emergence for Africa, Asia, Australia, Europe, and South America.

Continent	Taxa	Total Mercury	Methylmercury	Reference
Lotic				
North America				
	Diptera	[p] 4.5×10^{-4}	[p] 2.0×10^{-4}	[66]
	Ephemeroptera	[q] 3.4×10^{-5}	[q] 1.8×10^{-5}	[67]
	Trichoptera	5.1×10^{-5}	*	[68]
	Community	2.7×10^{-4}	*	[69]
	Ephemeroptera	8.1×10^{-5}	*	[70]
	Plecoptera	6.1×10^{-5}	7.3×10^{-5}	
	Diptera	2.0×10^{-5}	*	[22]
Average ± SD		$1.4 \times 10^{-4} \pm 1.5 \times 10^{-4}$	$9.6 \times 10^{-5} \pm 7.5 \times 10^{-5}$	
Coefficient of variation (%)		108	78	
South America				
	Community	5.7×10^{-4}	5.0×10^{-4}	[65]
Average ± SD		**$1.9 \times 10^{-4} \pm 2.0 \times 10^{-4}$**	**$2.0 \times 10^{-4} \pm 1.9 \times 10^{-4}$**	
Coefficient of variation (%)		**104**	**95**	

[p] based on average from authors data; [q] based on means of authors data. * Asterisks denote instance where data were not recorded cited reference.

2.5. Estimates of Mercury and Methylmercury Content in Aquatic Insects

Data on Hg and MeHg (mg g^{-1} of dry mass; Tables 4 and 5) were obtained based on studies that quantitatively determined the content of the two forms of mercury in aquatic insects using advanced mercury analyzers like amalgamation-thermal atomic absorption spectrometers [48,66]. While original data were presented by most authors in ng g^{-1}, I converted the values to mg g^{-1} (by multiplying all ng g^{-1} values by 1×10^{-6}) for all analyses to match the values reported for emergence data.

2.6. Data Analyses

Initially, content for fatty acids (mg g^{-1}) was multiplied by the emergence to obtain the export of fatty acids in (Kg Km^{-2} year^{-1}). Mercury content data were converted from ug g^{-1} to mg g^{-1} and subsequently multiplied by emergence to obtain methylmercury (MeHg) and total mercury (Hg) as Kg Km^{-2} year^{-1}.

To estimate the total net export (Kg year^{-1}) of mercury and fatty acids from water to land, export of mercury and fatty acids (Kg km^{-2} year^{-1}) were multiplied by the estimate of areas of lakes and rivers (Km^2) globally and by continent. Because some continents had no available emergence and mercury data for lakes (e.g., Africa, Australia, Asia and South America) and rivers (e.g., South America, I used the grand mean calculated for all available data for each ecosystem type (Lake or River).

All means and coefficients to variations (CV) were calculated for each data type. All mean values for data were compared using MedCalc® (statistical software version 14.8.1, software bvba, Ostend, Belgium; http://www.medcalc.org; 2018) and following procedures described in Altman [71].

3. Results

All literature survey data for fatty acids, Hg, MeHg are presented in Tables 1–5. Overall, the data, as evidenced by high coefficients of variation depict that there is a lot of variation in fatty acid and mercury data recorded in the literature. For example, Hg (Table 5) has a coefficient of variation of over 100 percent. Similarly, the grand means for fatty acids and mercury also show large variations across datasets.

3.1. Continental Exports of Physiologically Important Fatty Acids

Considering export of physiologically important fatty acids per unit area, lentic systems export similar quantities of fatty acids across all six continents in this study (range: 11.3 to 14.2 Kg Km^{-2} year^{-1}; Figure 2).

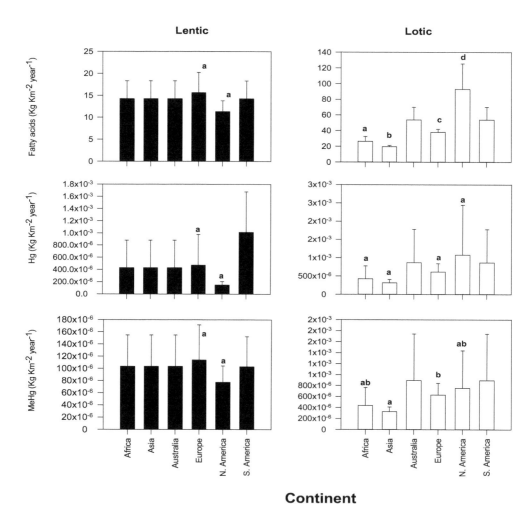

Figure 2. Estimate (±SD) of physiologically important fatty acids, methylmercury (MeHg) and total mercury (Hg) calculated for each continent. The letters depict results from Medcalc® comparison of means calculator within each continent, where values with the same letters depict no significant difference between the export values. Note that only continents where emergence data are available are statistically compared.

In rivers (Figure 2), North America exports a larger amount of fatty acids (93.0 ± 32.6 Kg Km^{-2} year^{-1}; Figure 2) compared to all other continents (range: 19.7 to 53.8 Kg Km^{-2} year^{-1}) per unit area. The lowest exports of fatty acids per unit area exported from river to land by aquatic insects were in Asia (19.7 Kg Km^{-2} year^{-1}).

Considering the total area of rivers and lakes by continent reveals that the quantity of fatty acids (Kg year^{-1}) exported from lakes to land are highest in Asia (2.2×10^6 Kg year^{-1}; Figure 3) and North America (2.2×10^6 Kg year^{-1}), with Australia exporting the lowest amount of fatty acids (3.4×10^4 Kg year^{-1}; Figure 3).

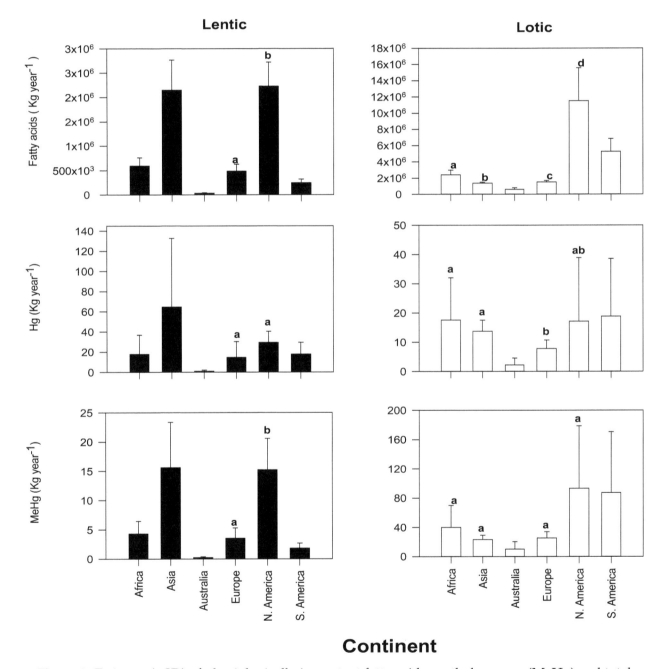

Figure 3. Estimate ($\pm$SD) of physiologically important fatty acids, methylmercury (MeHg) and total mercury (Hg) calculated for each continent. The letters depict results from Medcalc® comparison of means calculator within each continent, where values with the same letters depict no significant difference between the export values. Note that only continents were emergence data are available are statistically compared.

In rivers, North America contributes more to the export of fatty acids (11.5 $\times 10^6$ Kg year^{-1}) than all the other continents (range: 62.4 $\times 10^4$ to 52.7 $\times 10^5$ Kg year^{-1}). South America is the second largest exporter of fatty acids from river to land (52.7 $\times$ 105 Kg year^{-1}), with Australia exporting the lowest (62.4 $\times 10^4$ Kg year^{-1}). Overall, rivers across all continents contribute more to export of fatty acids than lakes.

3.2. Continental Exports of Mercury and Methylmercury

Regarding the export of Hg and MeHg from lakes to land per unit area, there are no significant differences among the exports of Hg (range: 1.5 $\times 10^{-4}$ to 1.0 $\times 10^{-3}$ Kg Km^{-2} year^{-1}; Figure 2) and MeHg (range: 77.2 $\times 10^{-6}$ to 103 $\times 10^{-6}$ Kg Km^{-2} year^{-1}) in lentic systems.

In rivers, there were no significant differences in flow of Hg from water to land among continents per unit area (mean range: 3.2 $\times 10^{-4}$ to 1.1 $\times 10^{-3}$ Kg Km^{-2} year^{-1}; $p > 0.05$). Similarly, there were no significant differences among exports of MeHg by continent. The only exception was between Europe and Asia, where Europe (6.4 $\times 10^{-4}$ Kg Km^{-2} year^{-1}) exported more MeHg per unit area from land to water than Asia (3.3 $\times 10^{-4}$ Kg Km^{-2} year^{-1}).

By considering the total area of rivers and lakes at each continent, I was able to calculate the amount of Hg and MeHg exported from water to land per year (Kg year^{-1}). The results from these calculations reveal that there are no significant differences in export of Hg from lakes (Figure 3). Australia was the only exception as it had significantly lower exports of Hg (2 Kg year^{-1}) from lake compared to all the other continents. Methylmercury exported from lake to land is greatest in Asia (15.6 Kg year^{-1}) and North America (15.2 Kg year^{-1}) compared to the other continents (mean range: 0.3 to 4.33 Kg year^{-1}).

In rivers, there were no significant difference in exports of Hg and MeHg from river to land, with exceptions occurring between some continents (e.g., export of Hg is significantly higher in Europe than in Australia).

3.3. Global Exports of Physiologically Important Fatty Acids and Mercury

Global export of fatty acids per year are higher in rivers (35.4 $\times 10^6$ Kg year^{-1}) than in lakes (85.1 $\times 10^5$ Kg year^{-1}; Figure 4; $p < 0.001$). Similarly, MeHg exports are higher in rivers (572.1 Kg year^{-1}) than in lakes (255.9 Kg year^{-1}; Figure 4). Congruent to MeHg exports, Hg differs significantly between rivers and lakes globally (587.7 Kg year^{-1} for rivers versus 61.9 Kg year^{-1} for lakes; Figure 4; $p < 0.05$).

Overall, global estimates reveal that there is some coupling between mercury and fatty acid exports; when fatty export and emergence are high, the values are synchronous to mercury exports by insects (Figures 2–4).

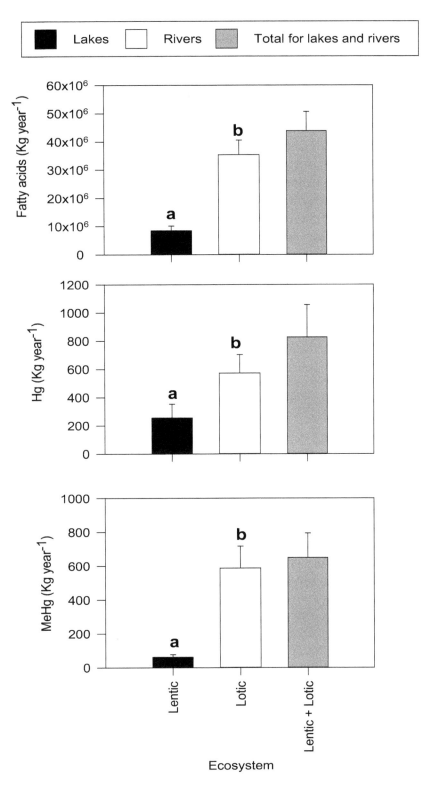

Figure 4. Global estimate (±SD) of physiologically important fatty acids, methylmercury (MeHg) and total mercury (Hg) calculated from diverse ecosystems. The letters depict results from Medcalc® comparison of means calculator between lentic and lotic systems, where values with the same letters depict no significant difference between the export values.

4. Discussion

Subsidies are known to affect terrestrial consumers in recipient systems, but these cross-boundary fluxes also transport persistent mercury [26]. Here, the first global perspective of the potential synchrony between export of physiologically fatty acids is presented using a plethora of data from different systems. The estimates build on general ideas originally formulated for rivers and lakes as donors of aquatic subsidies via emergent insects [18,72], which have demonstrated the importance of exports of nutrients from water to adjacent land [18,51,72]. One key finding from this this work is that there is synchrony between physiologically important fatty acids and mercury; because of emergence rates. Congruent to previous research (e.g., [54,73]), the results also demonstrate how the export of physiologically important fatty acids and mercury values vary spatially (by continents), with the North American continent exporting more fatty acids from water to land than all other continents.

The estimate of fatty acids exported from water to land (11.3–93.0 Kg km^{-2} year^{-1}; Figure 2) are within the range of the first estimate documented to date (0.1 to 672 Kg km^{-2} year^{-1}) [18]. The differences in the values obtained may be driven by the availability of more emergence data from other ecosystems. Presently, there are no estimates for export of mercury by aquatic insects to compare with these findings (Figure 1), mainly as a result of prior studies being focused on one aspect on the export of subsidies (nutrients). More studies on the potential export are thus warranted and should yield more fascinating results on the effects of subsidy type on consumers. Considering that hundreds of thousands of miles of streams and lakes are impaired by persistent mercury [74], the results suggest that aquatic insects are likely key movers of mercury from freshwater to terrestrial systems at a global scale. While these estimates are cursory, they may have huge implications for the ecology of terrestrial consumers and humans.

5. Implications

5.1. Wildlife

Terrestrial consumers are known to benefit from aquatic subsidies [7]. For example , quality of fatty acids can affect the fitness of tree swallows [75]. Assuming the trophic transfer efficiency of physiologically important fatty acids through the food web to be 10% (i.e., 90% of energy lost at each trophic level; Figure 5) [76,77] in a presumed three-trophic-level food web, aquatic insects can contribute between 0.4×10^6 to 4×10^6 Kg year^{-1} to terrestrial consumers. It is worth noting that while there may be a 10% dissipation with increased trophic level, other researchers have shown that physiologically important fatty acids are retained and are not dissipated by changing trophic positions [78]. To this end, assuming no dissipation of fatty acids happens up the terrestrial food chain implies that fatty acid production of the third level consumers may be equated to the initial contribution of physiologically important fatty acids with insect emergence (Figure 5). The no dissipation scenario is also tenable considering that physiologically important fatty acids are moved through trophic chains at about double the efficiency of biomolecules such as organic carbon and are effectively bioaccumulated (with no dilution) in higher trophic level consumers [79]. However, it must be emphasized that demand by terrestrial consumers for physiological fatty acids is sparse and further studies are warranted to assess terrestrial consumer dietary needs [80].

Terrestrial consumers that depend on aquatic subsidies may suffer irreversible behavioral, physiological, and reproductive effects [81,82] from exposure to MeHg. For example, some birds (e.g., belted kingfisher (*Ceryle alcyon*) and bald eagle (*Haliaeetus leucocephalus*)) and small mammals (e.g., American mink; *Neovison vison*) have been observed to suffer from visual, cognitive, and neurobehavioral effects [82], and even death within a year when exposed to MeHg concentrations of 1 µg g^{-1} [74]. Because MeHg increases in concentration as it progresses up the food chain, one can predict that organisms consuming prey at higher trophic levels are exposed to higher concentrations of total Hg and MeHg (Figure 4; [83,84]). Assuming that MeHg does not change significantly up the food chain suggests that consumers accumulate 649.6 Kg year^{-1}. However, the absolute assimilation efficiencies of MeHg vary with trophic level, uptake pathway, and water chemistry conditions; therefore, the estimates need to be interpreted with caution.

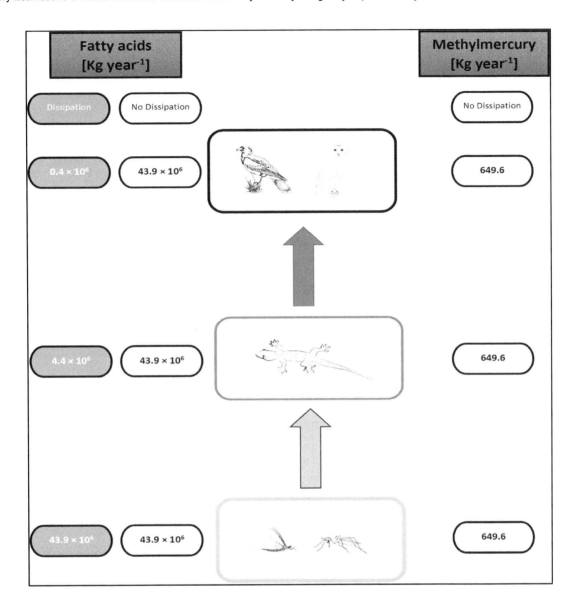

Figure 5. Depiction of the movement of physiologically important fatty acids and methylmercury (MeHg) as mediated my aquatic insects. Transfer efficiency based on traditional trophic pyramid concept of 10% dissipation at each trophic level.

5.2. Climate Change

Climate warming decreases the production of physiologically important fatty acids by decreasing polyunsaturated fatty acid membrane content while simultaneously increasing saturated fatty acids via homeoviscous adaptation [85]. Specifically, climate warming of 2.5 °C is predicted to reduce physiologically important fatty acid in algae by 8.2% to 27.8% (estimated to reduce physiologically important fatty acids from 240 to 225 tonnes [9]. This reduction under climate change will result in many aquatic insects receiving fewer fatty acids and this may subsequently have major effects on terrestrial consumers that often rely on aquatic subsidies to meet their dietary needs. However, some studies show that temperature does not have an effect on the quantity of physiologically fatty acids in consumers. For instance, Gladyshev et al. [86] found that contrasting temperatures have no effect on physiologically important fatty acids (EPA and DHA) with significant effects only observable in C18 saturated and polyenoic acids. As such, it is plausible that the temperature-dependent decrease in

EPA and DHA quantities happens mostly due to changes in the taxonomic composition of aquatic communities as a response to temperature changes [86].

6. Additional Considerations and Conclusions

In any study, there are caveats in protocols that can include trap design and other collection tools [87], so some caution is necessary for interpreting any results. I investigated fluxes from river to land using data collected by several authors in different ecosystems, as such, some variation can be expected in these estimates. For example, Different collection methods and traps may overestimate or underestimate fluxes for a variety of reasons [88,89]. Specifically, emergence traps may underestimate the fluxes of odonates from rivers, as some odonates crawl onto vegetation and rocks rather than fly out [87,90]. Additionally, Odonates, individually, have very high biomasses relative to other aquatic insects [90], and their contributions to outward subsidies may be underestimated in all our calculations. I recommend that additional studies incorporate the capture of crawling insects, as this aspect would improve the estimates of aquatic invertebrate flow from water to land.

Additionally, it is worth noting that the values expressed here for annual export of physiologically important fatty acids and MeHg via insects are preliminary estimates, based on averaging data from different ecosystems, and merely represent an initial attempt to calculate the order of magnitude of exports that are mediated my insects. I am cognisant that there are many limitations and sources of error in this type of global extrapolation, including the fact that fatty acids and mercury concentrations may vary depending on region, growth phase, climate, light regime and local nutrient conditions. For example, various authors have shown that mercury varies substantially over space and time [91,92]. Nevertheless, these kinds of data using a global perspective are needed to give a broader scale (*sensu* Gladyshev et al. [18]), which, in the future, may be refined further to create models to predict how environmental perturbations like climate change may affect the spatial and temporal dynamics of subsidies and methylmercury exported from water to land.

Summarily, these results underscore the need to view freshwater systems as just not nutrient exporters but lateral exporters of harmful contaminants [64] that can potentially be biomagnified within the food web. This view departs from the traditional viewpoint of streams being exporters of nutrients alone. Riparian insectivores (e.g., birds and small mammals) facilitate the transfer of aquatic mercury to higher trophic levels, thus serving as conduits in the dispersal of aquatic contaminants to the broader terrestrial food web [82]. Given the widespread contamination of streams, the ubiquity of stream insects, and the importance of insect subsidies to riparian predators, more research is needed to quantify the magnitude and risk of exposure to riparian food webs.

Author Contributions: S.M. conceived, designed, and executed this study and wrote the manuscript. No other person is entitled to authorship.

Acknowledgments: I am grateful for all the authors whose hard work and data were fundamental to this study. I am also grateful for the assistance of Valerie Motsumi who provided helpful comments on an earlier version of the manuscript. I thank the Louisiana State University Libraries Open Access Author Fund for providing the funds to publish open access.

References

1. Richardson, J.S.; Sato, T. Resource Subsidy Flows across Freshwater–Terrestrial Boundaries and Influence on Processes Linking Adjacent Ecosystems. *Ecohydrology* **2015**, *8*, 406–415. [CrossRef]
2. Brett, M.T.; Bunn, S.E.; Chandra, S.; Galloway, A.W.E.; Guo, F.; Kainz, M.J.; Kankaala, P.; Lau, D.C.P.; Moulton, T.P.; Power, M.E.; et al. How Important Are Terrestrial Organic Carbon Inputs for Secondary Production in Freshwater Ecosystems? *Freshw. Biol.* **2017**, *62*, 833–853. [CrossRef]

3.   Lam, M.M.Y.; Martin-Creuzburg, D.; Rothhaupt, K.O.; Safi, K.; Yohannes, E.; Salvarina, I. Tracking Diet Preferences of Bats Using Stable Isotope and Fatty Acid Signatures of Faeces. *PLoS ONE* **2013**, *8*, e83452. [CrossRef] [PubMed]

4.   Baxter, C.V.; Fausch, K.D.; Saunders, W.C. Tangled Webs: Reciprocal Flows of Invertebrate Prey Link Streams and Riparian Zones. *Freshw. Biol.* **2005**, *50*, 201–220. [CrossRef]

5.   Larsen, S.; Muehlbauer, J.D.; Marti, E. Resource Subsidies between Stream and Terrestrial Ecosystems under Global Change. *Glob. Chang. Biol.* **2016**, *22*, 2489–2504. [CrossRef]

6.   Harris, H.E.; Baxter, C.V.; Davis, J.M. Wildfire and Debris Flows Affect Prey Subsidies with Implications for Riparian and Riverine Predators. *Aquat. Sci.* **2018**, *80*, 37. [CrossRef]

7.   Sabo, J.L.; Power, M.E. River–Watershed Exchange: Effects of Riverine Subsidies on Riparian Lizards and their Terrestrial Prey. *Ecology* **2002**, *83*, 1860–1869.

8.   Sabo, J.L.; Power, M.E. Numerical Response of Lizards to Aquatic Insects and Short-Term Consequences for Terrestrial Prey. *Ecology* **2002**, *83*, 3023–3036. [CrossRef]

9.   Hixson, S.M.; Arts, M.T. Climate Warming Is Predicted to Reduce Omega-3, Long-chain, Polyunsaturated Fatty Acid Production in Phytoplankton. *Glob. Chang. Biol.* **2016**, *22*, 2744–2755. [CrossRef]

10. Twining, C.W.; Brenna, J.T.; Lawrence, P.; Shipley, J.R.; Tollefson, T.N.; Winkler, D.W. Omega-3 Long-Chain Polyunsaturated Fatty Acids Support Aerial Insectivore Performance More than Food Quantity. *Proc. Natl. Acad. Sci. USA* **2016**, *113*, 10920–10925. [CrossRef]

11. Twining, C.W.; Shipley, J.R.; Winkler, D.W. Aquatic Insects Rich in Omega-3 Fatty Acids Drive Breeding Success in a Widespread Bird. *Ecol. Lett.* **2018**, *21*, 1812–1820. [CrossRef] [PubMed]

12. Arts, M.T.; Ackman, R.G.; Holub, B.J. "Essential Fatty Acids" in Aquatic Ecosystems: A Crucial Link between Diet and Human Health and Evolution. *Can. J. Fish. Aquat. Sci.* **2001**, *58*, 122–137. [CrossRef]

13. Moyo, S.; Chari, L.D.; Villet, M.H.; Richoux, N.B. Decoupled Reciprocal Subsidies of Biomass and Fatty Acids in Fluxes of Invertebrates between a Temperate River and the Adjacent Land. *Aquat. Sci.* **2017**, *79*, 689–703. [CrossRef]

14. Jackson, J.K.; Fisher, S.G. Secondary Production, Emergence, and Export of Aquatic Insects of a Sonoran Desert Stream. *Ecology* **1986**, *67*, 629–638. [CrossRef]

15. Bell, J.G.; Ghioni, C.; Sargent, J.R. Fatty Acid Compositions of 10 Freshwater Invertebrates Which Are Natural Food Organisms of Atlantic Salmon Parr (*Salmo salar*): A Comparison with Commercial Diets. *Aquaculture* **1994**, *128*, 301–313. [CrossRef]

16. Ghioni, C.; Bell, J.G.; Sargent, J.R. Polyunsaturated Fatty Acids in Neutral Lipids and Phospholipids of Some Freshwater Insects. *Comp. Biochem. Physiol. B Biochem. Mol. Biol.* **1996**, *114*, 161–170. [CrossRef]

17. Torres-Ruiz, M.; Wehr, J.D.; Perrone, A.A. Trophic Relations in a Stream Food Web: Importance of Fatty Acids for Macroinvertebrate Consumers. *J. North. Am. Benthol. Soc.* **2007**, *26*, 509–522. [CrossRef]

18. Gladyshev, M.I.; Arts, M.T.; Sushchik, N.N. Preliminary Estimates of the Export of Omega-3 Highly Unsaturated Fatty Acids (EPA+DHA) from Aquatic to Terrestrial Ecosystems. In *Lipids in Aquatic Ecosystems*; Kainz, M., Brett, M.T., Arts, M.T., Eds.; Springer New York: New York, NY, USA, 2009; pp. 179–210. [CrossRef]

19. Burdon, F.J.; Harding, J.S. The Linkage between Riparian Predators and Aquatic Insects across a Stream-Resource Spectrum. *Freshw. Biol.* **2008**, *53*, 330–346. [CrossRef]

20. Walters, D.M.; Fritz, K.M.; Otter, R.R. The Dark Side of Subsidies: Adult Stream Insects Export Organic Contaminants to Riparian Predators. *Ecol. Appl.* **2008**, *18*, 1835–1841. [CrossRef]

21. Speir, S.L.; Chumchal, M.M.; Drenner, R.W.; Cocke, W.G.; Lewis, M.E.; Whitt, H.J. Methyl Mercury and Stable Isotopes of Nitrogen Reveal That a Terrestrial Spider Has a Diet of Emergent Aquatic Insects. *Environ. Toxicol. Chem.* **2014**, *33*, 2506–2509. [CrossRef]

22. Lavoie, R.A.; Jardine, T.D.; Chumchal, M.M.; Kidd, K.A.; Campbell, L.M. Biomagnification of Mercury in Aquatic Food Webs: A Worldwide Meta-Analysis. *Environ. Sci. Technol.* **2013**, *47*, 13385–13394. [CrossRef]

23. Chumchal, M.M.; Drenner, R.W.; Greenhill, F.M.; Kennedy, J.H.; Courville, A.E.; Gober, C.A.A.; Lossau, L.O. Recovery of Aquatic Insect-Mediated Methylmercury Flux from Ponds Following Drying Disturbance. *Environ. Toxicol. Chem.* **2017**, *36*, 1986–1990. [CrossRef] [PubMed]

24. Chumchal, M.M.; Drenner, R.W.; Hall, M.N.; Polk, D.K.; Williams, E.B.; Ortega-Rodriguez, C.L.; Kennedy, J.H. Seasonality of Dipteran-Mediated Methylmercury Flux from Ponds. *Environ. Toxicol. Chem.* **2018**, *37*, 1846–1851. [CrossRef]

25. Du, H.; Ma, M.; Igarashi, Y.; Wang, D. Biotic and Abiotic Degradation of Methylmercury in Aquatic Ecosystems: A Review. *Bull. Environ. Contam. Toxicol.* **2019**. [CrossRef]

26. Walters, D.M.; Rosi-Marshall, E.; Kennedy, T.A.; Cross, W.F.; Baxter, C.V. Mercury and Selenium Accumulation in the Colorado River Food Web, Grand Canyon, USA. *Environ. Toxicol. Chem.* **2015**, *34*, 2385–2394. [CrossRef]

27. Gladyshev, M.I.; Gladysheva, E.E.; Sushchik, N.N. Preliminary Estimation of the Export of Omega-3 Polyunsaturated Fatty Acids from Aquatic to Terrestrial Ecosystems in Biomes via Emergent Insects. *Ecol. Complex.* **2019**, *38*, 140–145. [CrossRef]

28. Wesner, J.S. Seasonal Variation in the Trophic Structure of a Spatial Prey Subsidy Linking Aquatic and Terrestrial Food Webs: Adult Aquatic Insects. *Oikos* **2010**, *119*, 170–178. [CrossRef]

29. Lehner, B.; Döll, P. Development and Validation of a Global Database of Lakes, Reservoirs and Wetlands. *J. Hydrol.* **2004**, *296*, 1–22. [CrossRef]

30. U.S. Defense Mapping Agency. *Development of the Digital Chart of the World*; Government Printing Office: Washington, DC, USA, 1992.

31. Lehner, B.; Verdin, K.; Jarvis, A. New Global Hydrography Derived from Spaceborne Elevation Data. *Eos Trans. Am. Geophys. Union* **2008**, *89*, 93. [CrossRef]

32. US Geological Survey. *Long Term Archive*; HYDRO1K [Internet]; U.S. Geological Survey: Reston, VA, USA, 2015.

33. Allen, G.H.; Pavelsky, T.M. Global Extent of Rivers and Streams. *Science* **2018**, *361*, 585–588. [CrossRef] [PubMed]

34. Neteler, M.; Bowman, M.H.; Landa, M.; Metz, M. GRASS GIS: A Multi-Purpose Open Source GIS. *Environ. Model. Softw.* **2012**, *31*, 124–130. [CrossRef]

35. QGIS Development Team. *QGIS Geographic Information System*; Open Source Geospatial Foundation Project: Chicago, IL, USA, 2019; Available online: http://Qgis.Osgeo.Org (accessed on 12 January 2020).

36. Vadeboncoeur, Y.; McIntyre, P.B.; Vander Zanden, M.J. Borders of Biodiversity: Life at the Edge of the World's Large Lakes. *BioScience* **2011**, *61*, 526–537. [CrossRef]

37. Gray, L.J. Emergence Production and Export of Aquatic Insects from a Tallgrass Prairie Stream. *Southwest. Nat.* **1989**, *34*, 313–318. [CrossRef]

38. Stagliano, D.M.; Benke, A.C.; Anderson, D.H. Emergence of Aquatic Insects from 2 Habitats in a Small Wetland of the Southeastern USA: Temporal Patterns of Numbers and Biomass. *J. North. Am. Benthol. Soc.* **1998**, *17*, 37–53. [CrossRef]

39. Francis, T.B.; Schindler, D.E.; Moore, J.W. Aquatic Insects Play a Minor Role in Dispersing Salmon-Derived Nutrients into Riparian Forests in Southwestern Alaska. *Can. J. Fish. Aquat. Sci.* **2006**, *63*, 2543–2552. [CrossRef]

40. Goedkoop, W.; Sonesten, L.; Ahlgren, G.; Boberg, M. Fatty Acids in Profundal Benthic Invertebrates and Their Major Food Resources in Lake Erken, Sweden: Seasonal Variation and Trophic Indications. *Can. J. Fish. Aquat. Sci.* **2000**, *57*, 2267–2279. [CrossRef]

41. Sushchik, N.N.; Yurchenko, Y.A.; Gladyshev, M.I.; Belevich, O.E.; Kalachova, G.S.; Kolmakova, A.A. Comparison of Fatty Acid Contents and Composition in Major Lipid Classes of Larvae and Adults of Mosquitoes (Diptera: Culicidae) from a Steppe Region. *Insect Sci.* **2012**, *20*, 585–600. [CrossRef]

42. Raitif, J.; Plantegenest, M.; Agator, O.; Piscart, C.; Roussel, J.-M. Seasonal and Spatial Variations of Stream Insect Emergence in an Intensive Agricultural Landscape. *Sci. Total Environ.* **2018**, *644*, 594–601. [CrossRef]

43. Salvarina, I.; Gravier, D.; Rothhaupt, K.-O. Seasonal Insect Emergence from Three Different Temperate Lakes. *Limnologica* **2017**, *62*, 47–56. [CrossRef]

44. Silina, A.E. Emergence of Amphibiotic Insects from a Floodplain Lake in the Usman Forest in the Central Russian Forest Steppe. *Contemp. Probl. Ecol.* **2016**, *9*, 421–436. [CrossRef]

45. Dreyer, J.; Townsend, P.A.; Iii, J.C.H.; Hoekman, D.; Vander Zanden, M.J.; Gratton, C. Quantifying Aquatic Insect Deposition from Lake to Land. *Ecology* **2015**, *96*, 499–509. [CrossRef] [PubMed]

46. Borisova, E.V.; Makhutova, O.N.; Gladyshev, M.I.; Sushchik, N.N. Fluxes of Biomass and Essential Polyunsaturated Fatty Acids from Water to Land via Chironomid Emergence from a Mountain Lake. *Contemp. Probl. Ecol.* **2016**, *9*, 446–457. [CrossRef]

47. Djomina, I.V.; Yermokhin, M.V.; Polukonova, N.V. Substance and Energy Flows Formed by the Emergence of Amphibiotic Insects across the Water–Air Boundary on the Floodplain Lakes of the Volga River. *Contemp. Probl. Ecol.* **2016**, *9*, 407–420. [CrossRef]

48. Tremblay, A.; Cloutier, L.; Lucotte, M. Total Mercury and Methylmercury Fluxes via Emerging Insects in Recently Flooded Hydroelectric Reservoirs and a Natural Lake. *Sci. Total Environ.* **1998**, *219*, 209–221. [CrossRef]

49. Sherk, T.; Rau, G. Emergence of Chironomidae from Findley Lake and Two Ponds in the Cascade Mountains, U.S.A. *Netherland J. Aquat. Ecol.* **1992**, *26*, 321–330. [CrossRef]

50. Freitag, H. Composition and Longitudinal Patterns of Aquatic Insect Emergence in Small Rivers of Palawan Island, the Philippines. *Int. Rev. Hydrobiol.* **2004**, *89*, 375–391. [CrossRef]

51. Nakano, S.; Murakami, M. Reciprocal Subsidies: Dynamic Interdependence between Terrestrial and Aquatic Food Webs. *Proc. Natl. Acad. Sci. USA* **2001**, *98*, 166–170. [CrossRef]

52. Poepperl, R. Benthic Secondary Production and Biomass of Insects Emerging from a Northern German Temperate Stream. *Freshw. Biol.* **2001**, *44*, 199–211. [CrossRef]

53. Whiles, M.R.; Goldowitz, B.S. Hydrologic Influences on Insect Emergence Production from Central Platte River Wetlands. *Ecol. Appl.* **2001**, *11*, 1829–1842. [CrossRef]

54. Rundio, D.E.; Lindley, S.T. Reciprocal Fluxes of Stream and Riparian Invertebrates in a Coastal California Basin with Mediterranean Climate. *Ecol. Res.* **2012**, *27*, 539–550. [CrossRef]

55. Popova, O.N.; Haritonov, A.Y.; Sushchik, N.N.; Makhutova, O.N.; Kalachova, G.S.; Kolmakova, A.A.; Gladyshev, M.I. Export of Aquatic Productivity, Including Highly Unsaturated Fatty Acids, to Terrestrial Ecosystems via Odonata. *Sci. Total Environ.* **2017**, *581–582*, 40–48. [CrossRef]

56. Martin-Creuzburg, D.; Kowarik, C.; Straile, D. Cross-Ecosystem Fluxes: Export of Polyunsaturated Fatty Acids from Aquatic to Terrestrial Ecosystems via Emerging Insects. *Sci. Total Environ.* **2017**, *577*, 174–182. [CrossRef]

57. Makhutova, O.N.; Borisova, E.V.; Shulepina, S.P.; Kolmakova, A.A.; Sushchik, N.N. Fatty Acid Composition and Content in Chironomid Species at Various Life Stages Dominating in a Saline Siberian Lake. *Contemp. Probl. Ecol.* **2017**, *10*, 230–239. [CrossRef]

58. Sushchik, N.N.; Gladyshev, M.I.; Moskvichova, A.V.; Makhutova, O.N.; Kalachova, G.S. Comparison of Fatty Acid Composition in Major Lipid Classes of the Dominant Benthic Invertebrates of the Yenisei River. *Comp. Biochem. Physiol. B Biochem. Mol. Biol.* **2003**, *134*, 111–122. [CrossRef]

59. Tremblay, A.; Lucotte, M.; Rheault, I. Methylmercury in a Benthic Food Web of Two Hydroelectric Reservoirs and a Natural Lake of Northern Québec (Canada). *Water. Air. Soil Pollut.* **1996**, *91*, 255–269. [CrossRef]

60. Gorski, P.R.; Cleckner, L.B.; Hurley, J.P.; Sierszen, M.E.; Armstrong, D.E. Factors Affecting Enhanced Mercury Bioaccumulation in Inland Lakes of Isle Royale National Park, USA. *Sci. Total Environ.* **2003**, *304*, 327–348. [CrossRef]

61. Cremona, F.; Planas, D.; Lucotte, M. Assessing the Importance of Macroinvertebrate Trophic Dead Ends in the Lower Transfer of Methylmercury in Littoral Food Webs. *Can. J. Fish. Aquat. Sci.* **2008**, *65*, 2043–2052. [CrossRef]

62. Tweedy, B.N. Effects of Fish on Emergent Insects and Their Transport of Methyl Mercury from Ponds [Electronic Resource]. UMI Thesis, Texas Christian University, Fort Worth, TX, USA, 2012.

63. Rizzo, A.; Arcagni, M.; Arribére, M.A.; Bubach, D.; Guevara, S.R. Mercury in the Biotic Compartments of Northwest Patagonia Lakes, Argentina. *Chemosphere* **2011**, *84*, 70–79. [CrossRef]

64. Arcagni, M.; Juncos, R.; Rizzo, A.; Pavlin, M.; Fajon, V.; Arribére, M.A.; Horvat, M.; Ribeiro Guevara, S. Species and Habitat-Specific Bioaccumulation of Total Mercury and Methylmercury in the Food Web of a Deep Oligotrophic Lake. *Sci. Total Environ.* **2018**, *612*, 1311–1319. [CrossRef]

65. Dominique, Y.; Maury-Brachet, R.; Muresan, B.; Vigouroux, R.; Richard, S.; Cossa, D.; Mariotti, A.; Boudou, A. Biofilm and Mercury Availability as Key Factors for Mercury Accumulation in Fish (*Curimata cyprinoides*) from a Disturbed Amazonian Freshwater System. *Environ. Toxicol. Chem.* **2009**, *26*, 45–52. [CrossRef]

66. Harding, K.M.; Gowland, J.A.; Dillon, P.J. Mercury Concentration in Black Flies *Simulium* Spp. (Diptera, Simuliidae) from Soft-Water Streams in Ontario, Canada. *Environ. Pollut.* **2006**, *143*, 529–535. [CrossRef]

67. Nagorski, S.A.; Engstrom, D.R.; Hudson, J.P.; Krabbenhoft, D.P.; Hood, E.; DeWild, J.F.; Aiken, G.R. Spatial Distribution of Mercury in Southeastern Alaskan Streams Influenced by Glaciers, Wetlands, and Salmon. *Environ. Pollut.* **2014**, *184*, 62–72. [CrossRef] [PubMed]

68. Cain, D.J.; Carter, J.L.; Fend, S.V.; Luoma, S.N.; Alpers, C.N.; Taylor, H.E. Metal Exposure in a Benthic Macroinvertebrate, *Hydropsyche californica*, Related to Mine Drainage in the Sacramento River. *Can. J. Fish. Aquat. Sci.* **2000**, *57*, 380–390. [CrossRef]

69. Sullivan, S.M.P.; Boaz, L.E.; Hossler, K. Fluvial Geomorphology and Aquatic-to-Terrestrial Hg Export Are Weakly Coupled in Small Urban Streams of Columbus, Ohio. *Water Resour. Res.* **2016**, *52*, 2822–2839. [CrossRef]

70. Dukerschein, J.T.; Wiener, J.G.; Rada, R.G.; Steingraeber, M.T. Cadmium and Mercury in Emergent Mayflies (*Hexagenia Bilineata*) from the Upper Mississippi River. *Arch. Environ. Contam. Toxicol.* **1992**, *23*, 109–116. [CrossRef] [PubMed]

71. Altman, D.G. *Practical Statistics for Medical Research*, 1st ed.; Chapman and Hall/CRC: Boca Raton, FL, USA, 1990.

72. Polis, G.A.; Anderson, W.B.; Holt, R.D. Toward an Integration of Landscape and Food Web Ecology: The Dynamics of Spatially Subsidized Food Webs. *Annu. Rev. Ecol. Syst.* **1997**, *28*, 289–316. [CrossRef]

73. Kautza, A.; Sullivan, S.M.P. Shifts in Reciprocal River-Riparian Arthropod Fluxes along an Urban-Rural Landscape Gradient. *Freshw. Biol.* **2015**, *60*, 2156–2168. [CrossRef]

74. Driscoll, C.T.; Han, Y.-J.; Chen, C.Y.; Evers, D.C.; Lambert, K.F.; Holsen, T.M.; Kamman, N.C.; Munson, R.K. Mercury Contamination in Forest and Freshwater Ecosystems in the Northeastern United States. *BioScience* **2007**, *57*, 17–28. [CrossRef]

75. Martinez del Rio, C.; McWilliams, S.R. How Essential Fats Affect Bird Performance and Link Aquatic Ecosystems and Terrestrial Consumers. *Proc. Natl. Acad. Sci. USA* **2016**, *113*, 11988–11990. [CrossRef]

76. Lindeman, R.L. The Trophic-Dynamic Aspect of Ecology. *Ecology* **1942**, *23*, 399–417. [CrossRef]

77. Gladyshev, M.I.; Sushchik, N.N.; Makhutova, O.N. Production of EPA and DHA in Aquatic Ecosystems and Their Transfer to the Land. *Prostaglandins Other Lipid Mediat.* **2013**, *107*, 117–126. [CrossRef] [PubMed]

78. Kainz, M.J.; Hager, H.H.; Rasconi, S.; Kahilainen, K.K.; Amundsen, P.-A.; Hayden, B. Polyunsaturated Fatty Acids in Freshwater Fishes Increase with Total Lipids Irrespective of Feeding Sources and Trophic Position. *UiT Munin* **2017**. [CrossRef]

79. Gladyshev, M.I.; Sushchik, N.N.; Anishchenko, O.V.; Makhutova, O.N.; Kolmakov, V.I.; Kalachova, G.S.; Kolmakova, A.A.; Dubovskaya, O.P. Efficiency of Transfer of Essential Polyunsaturated Fatty Acids versus Organic Carbon from Producers to Consumers in a Eutrophic Reservoir. *Oecologia* **2011**, *165*, 521–531. [CrossRef] [PubMed]

80. Hixson, S.M.; Sharma, B.; Kainz, M.J.; Wacker, A.; Arts, M.T. Production, Distribution, and Abundance of Long-Chain Omega-3 Polyunsaturated Fatty Acids: A Fundamental Dichotomy between Freshwater and Terrestrial Ecosystems. *Environ. Rev.* **2015**, *23*, 414–424. [CrossRef]

81. Bernhoft, R.A. Mercury Toxicity and Treatment: A Review of the Literature. *J. Environ. Public Health* **2012**, *2012*. [CrossRef] [PubMed]

82. Evers, D. The Effects of Methylmercury on Wildlife: A Comprehensive Review and Approach for Interpretation. *Encycl. Anthr.* **2018**, *5*, 181–194. [CrossRef]

83. Tsui, M.T.K.; Finlay, J.C.; Nater, E.A. Mercury Bioaccumulation in a Stream Network. *Environ. Sci. Technol.* **2009**, *43*, 7016–7022. [CrossRef]

84. Tsui, M.T.K.; Blum, J.D.; Kwon, S.Y.; Finlay, J.C.; Balogh, S.J.; Nollet, Y.H. Sources and Transfers of Methylmercury in Adjacent River and Forest Food Webs. *Environ. Sci. Technol.* **2012**, *46*, 10957–10964. [CrossRef]

85. Fuschino, J.R.; Guschina, I.A.; Dobson, G.; Yan, N.D.; Harwood, J.L.; Arts, M.T. Rising Water Temperatures Alter Lipid Dynamics and Reduce N-3 Essential Fatty Acid Concentrations in *Scenedesmus obliquus* (Chlorophyta)1. *J. Phycol.* **2011**, *47*, 763–774. [CrossRef]

86. Gladyshev, M.I.; Sushchik, N.N.; Dubovskaya, O.P.; Buseva, Z.F.; Makhutova, O.N.; Fefilova, E.B.; Feniova, I.Y.; Semenchenko, V.P.; Kolmakova, A.A.; Kalachova, G.S. Fatty Acid Composition of Cladocera and Copepoda from Lakes of Contrasting Temperature. *Freshw. Biol.* **2015**, *60*, 373–386. [CrossRef]

87. Malison, R.L.; Benjamin, J.R.; Baxter, C.V. Measuring Adult Insect Emergence from Streams: The Influence of Trap Placement and a Comparison with Benthic Sampling. *J. North. Am. Benthol. Soc.* **2010**, *29*, 647–656. [CrossRef]

88. Edwards, E.D.; Huryn, A.D. Effect of Riparian Land Use on Contributions of Terrestrial Invertebrates to Streams. *Hydrobiologia* **1996**, *337*, 151–159. [CrossRef]

89. Wipfli, M.S. Terrestrial Invertebrates as Salmonid Prey and Nitrogen Sources in Streams: Contrasting Old-Growth and Young-Growth Riparian Forests in Southeastern Alaska, U.S.A. *Can. J. Fish. Aquat. Sci.* **1997**, *54*, 1259–1269. [CrossRef]

90. Merritt, R.W.; Cummins, K.W. *Introduction to the Aquatic Insects of North America*, 3rd ed.; Kendall/Hunt: Dubuque, IA, USA, 1996.

91. Shia, R.-L.; Seigneur, C.; Pai, P.; Ko, M.; Sze, N.D. Global Simulation of Atmospheric Mercury Concentrations and Deposition Fluxes. *J. Geophys. Res. Atmos.* **1999**, *104*, 23747–23760. [CrossRef]

92. Li, P.; Feng, X.B.; Qiu, G.L.; Shang, L.H.; Li, Z.G. Mercury Pollution in Asia: A Review of the Contaminated Sites. *J. Hazard. Mater.* **2009**, *168*, 591–601. [CrossRef]

5

Assessment of Fatty Acid Desaturase (Fads2) Structure-Function Properties in Fish in the Context of Environmental Adaptations and as a Target for Genetic Engineering

Zuzana Bláhová [1,*], Thomas Nelson Harvey [2], Martin Pšenička [1] and Jan Mráz [1]

[1] South Bohemian Research Center of Aquaculture and Biodiversity of Hydrocenoses, Faculty of Fisheries and Protection of Waters, University of South Bohemia in České Budějovice, Zátiší 728/II, 389 25 Vodňany, Czech Republic

[2] Centre for Integrative Genetics (CIGENE), Department of Animal and Aquacultural Sciences, Faculty of Biosciences, Norwegian University of Life Sciences, 1430 Ås, Norway

* Correspondence: zblahova@frov.jcu.cz

Abstract: Fatty acid desaturase 2 (Fads2) is the key enzyme of long-chain polyunsaturated fatty acid (LC-PUFA) biosynthesis. Endogenous production of these biomolecules in vertebrates, if present, is insufficient to meet demand. Hence, LC-PUFA are considered as conditionally essential. At present, however, LC-PUFA are globally limited nutrients due to anthropogenic factors. Research attention has therefore been paid to finding ways to maximize endogenous LC-PUFA production, especially in production species, whereby deeper knowledge on molecular mechanisms of enzymatic steps involved is being generated. This review first briefly informs about the milestones in the history of LC-PUFA essentiality exploration before it focuses on the main aim—to highlight the fascinating Fads2 potential to play roles fundamental to adaptation to novel environmental conditions. Investigations are summarized to elucidate on the evolutionary history of fish Fads2, providing an explanation for the remarkable plasticity of this enzyme in fish. Furthermore, structural implications of Fads2 substrate specificity are discussed and some relevant studies performed on organisms other than fish are mentioned in cases when such studies have to date not been conducted on fish models. The importance of Fads2 in the context of growing aquaculture demand and dwindling LC-PUFA supply is depicted and a few remedies in the form of genetic engineering to improve endogenous production of these biomolecules are outlined.

Keywords: fatty acyl desaturase; $\Delta 6$ - desaturase; long-chain polyunsaturated fatty acid; LC-PUFA; $\omega 3$; $\omega 6$; EPA; DHA; AA; essential fatty acid; health; fish; transgene

1. Introduction

Fatty acid desaturase 2 (Fads2) is an endoplasmic reticulum membrane bound protein which acts as the first enzyme in the biosynthesis of long chain ($\geq C_{20}$) polyunsaturated fatty acids (LC-PUFA). This pathway includes physiologically important eicosapentaenoic acid (EPA, $\omega 3$-20:5[5,8,11,14,17]), docosahexaenoic acid (DHA, $\omega 3$-22:6[4,7,10,13,16,19]), and arachidonic acid (AA, $\omega 6$-20:4[5,8,11,14]) which are produced from the shorter and lower level polyunsaturated fatty acids (PUFA) α-linolenic acid (ALA, $\omega 3$-18:3[6,9,12]) and linoleic acid (LA, $\omega 6$-18:2[9,12]). Human and many fish genomes encode for Fads2 as well as for some other enzymes acting in the LC-PUFA biosynthetic pathway, namely Fads1, elongase 5 (Elovl5), elongase 4 (Elovl4) or elongase 2 (Elovl2). LC-PUFA are often referred as conditionally essential nutrients, meaning that however the organism could be capable to produce them, this endogenous production is insufficient to meet demand, hence, LC-PUFA biomolecules must be obtained through the diet. Endogenous production is hypothesised to serve as a compensation apparatus, which helps the organisms to maintain homeostasis under fluctuating environmental conditions and LC-PUFA availability. This could be the reason why marine fish, unlike freshwater fish, do not have the capability to produce LC-PUFA at a significant level [1] as a consequence of living in nutritionally rich oceans. In contrast to the conditional-essentiality of LC-PUFA, precursors of LC-PUFA, LA and ALA, cannot be created de novo in nearly any living animal, since their genomes do not encode for enzymes capable to create them (such would be methyl-end desaturases with $\Delta 12$ and $\Delta 15$ activity converting oleic acid (18:1n-9) into LA (18:2n-6) and ALA (18:3n-3)) [2]. Hence, animals are usually dependent on plants for providing double bonds in the $\Delta 12$ and $\Delta 15$ positions of the two major precursors of the $\omega 6$ and $\omega 3$ fatty acids LA and ALA [3]. These two fatty acids, therefore, are called essential fatty acids. In the literature, however, most often, the conditional-essentiality of AA, EPA and DHA for vertebrates and humans is not considered and with LA and ALA, these biomolecules are altogether called essential fatty acids (EFA).

The exogenous supply of EFA for many animals, including some omnivorous terrestrial animals and humans, is from aquatic ecosystems. In aquatic ecosystems, substantial amounts of EPA and DHA are provided by primary producers. Historically, the primary production of these biomolecules has been associated exclusively to single-cell microorganisms such as photosynthetic microalgae, heterotrophic protists and bacteria. Recently, however, multiple invertebrates, many of them representing abundant groups in aquatic ecosystems, have been confirmed to be able to produce PUFA de novo and farther biosynthesize them into $\omega 3$ LC-PUFA similarly to single-cell microorganisms [4,5]. Once synthesized by microalgae or invertebrates, these biomolecules are transferred through trophic webs to organisms of higher trophic levels. Fish are considered as the best source of $\omega 3$ LC-PUFA for humans. However, anthropogenic factors such as pollution, eutrophication, climate change or biological invasions threaten the LC-PUFA production by primary producers at present. World capture fishery production cannot be increased and aquaculture is expected to be continuously growing to deliver food to humans. Here, this could be seen a paradox. While aquaculture has increasingly become the major source of EPA and DHA for humans, it has also, at the same time, become the greatest consumer of the world's available supply of EPA and DHA. The problem of bridging the gap between supply and demand of LC-PUFA was very recently excellently reviewed by Tocher et al. [1]. Since LC-PUFA have been identified as globally limited nutrients, the ability of an organism to compensate for dietary deficiencies of LC-PUFA by enhanced activity of its endogenous biosynthesis is of great importance for human and animal health as well as for the maintenance of fish as an EFA source for the human diet and aquaculture food.

Although it has been confirmed that alterations in activities of elongases Elovl4 and Elovl5 catalysing subsequent steps in LC-PUFA biosynthesis can alter EPA, DHA and AA production by promoting various disease states [6], Fads2 is still commonly considered as the rate-limiting and the most important enzyme of LC-PUFA biosynthesis. Moreover, in light of some recent significant publications, Fads2 appears as an enzyme with far-reaching implications for environmental sustainability.

2. Significance and Essentiality of LC-PUFA Biomolecules

LC-PUFA are important components of fat and in higher eukaryotes confer fluidity, flexibility and selective permeability to cellular membranes. They greatly influence many physiological processes. Participation of LC-PUFA in several major human pathologies (inflammatory-autoimmune diseases, cardiovascular diseases, cancer and neurodegenerative disorders) has been reviewed recently by Zárate et al. [7]. The pivotal role of lipids as an essential dietary component is now widely accepted; however, many decades of research have gone into this conclusion, which has recently been very well reviewed by Spector and Kim [8]. Fish played an essential role in coming to this conclusion. The essentiality of any fatty acid biomolecule was, for the very first time, reported in what was at the time highly controversial scientific work of George Oswald Burr in 1929 [9,10]. He and co-workers demonstrated that LA and ALA rescued the growth retardance phenotype in rats fed fat-free diet and found the first clue that LA is a precursor of AA. The concept of essential fatty acids appeared [11]. But it was before chain desaturation or elongation of fatty acids had been demonstrated and the authors wondered how two double bonds + two double bonds could equal four [12], as was elucidated later in series of studies conducted by Mead et al. [13]. The linkage between LA, its ω6 fatty acid desaturation and elongation products, and the formation of prostaglandins as biomediators was reported by Bergström et al. [14,15], who stated" . . . *the symptoms of essential fatty acid deficiency at least partly are due to an inadequate biosynthesis of the various members of the prostaglandin hormone system.*" The pathway through which ALA is converted into EPA and DHA was determined in 1960 by Klenk et al. [16]. Noteworthily, no important functions were attributed to ω3 fatty acids for a long time. However, this changed in 1968 when Dr. Jørn Dyerberg made the remarkable discovery that fats in the diet of Greenland Eskimos comprised mainly of fish were associated with a lowered risk of cardiovascular diseases, concretely, that plasma of these people contained large amount of ω3 fatty acids and their phospholipids contained high levels of EPA, but very little AA [17]. Their conclusion was that EPA protects against cardiovascular diseases [18]. Without a doubt, the paradigm had been changed and fatty acids were no more considered to fulfil the only function in energy storage. Widespread interest was awoken in investigations of unsaturated fatty acids as biomolecules indispensable for health. Numerous global and national health agencies and associations and government bodies have produced many recommendations for EFA intake for a healthy human diet through fish consumption. With the advent of molecular and genetic technologies, there appeared much evidence that a balanced abundance of EFA is a prerequisite for health and disease prevention in humans [19–21]. Meeting the dietary demands of a burgeoning human population with a correct dietary balance of EFA and at levels required for normal health and development has become a major challenge. It has been clear that understanding the molecular basis of LC-PUFA biosynthesis would underpin efforts to meet this challenge. Various strategies of human populations regarding EFA metabolism have been shown by Gladyshev and Sushchik this year [22]. Studies performed in fish are advantageous mainly because there is wide variation between fish species in their ability to biosynthesize LC-PUFA, probably as a consequence of inhabiting widely different environments. Comparisons of their genomes and expression levels of genes encoding key elements in the LC-PUFA biosynthetic pathway between species have been promising to increase knowledge of the molecular components of the pathway and of the molecular genetic basis of phenotypic variation in LC-PUFA biosynthesis.

3. Fads2 in LC-PUFA Biosynthesis

The biosynthesis of C_{20-22} LC-PUFA involves alternating steps of desaturation (introduction of an additional double bond) and elongation (addition of two carbons) of the dietary essential C_{18} fatty acids LA and ALA [1]. Firstly, in the biosynthesis of LC-PUFA, Δ6 Fads2 desaturase converts dietary obtained LA (18:2n-6) and ALA (18:3n-3) into gamma-linoleic acid (GLA) and stearidonic acid, respectively. Subsequently, in the biosynthesis of LC-PUFAs of ω3 series, it converts tetracosapentaenoic acid into tetracosahexaenoic acid which is then converted to DHA. Enzymatic steps in the biosynthesis of LC-PUFAs in vertebrates are shown in Figure 1 [23–30]. AA and EPA are biosynthesized in the same pathway in which LA and ALA substrates compete of the same enzymes, respectively. The pathway revealed from studies in vertebrates are the so-called "Δ6 pathway" (Δ6 desaturation–elongation–Δ5 desaturation) and the "Δ8 pathway" (elongation–Δ8 desaturation–Δ5 desaturation). DHA is achieved downstream in the biosynthesis of LC-PUFA from EPA via two alternative routes. Either, two consecutive elongations of EPA produce tetracosapentaenoic acid (TPA, 24:5n-3), which then undergoes a Δ6 desaturation to tetracosahexaenoic acid (THA, 24:6n-3), the latter being β-oxidised to DHA in peroxisome organelles following the translocation from endoplasmic reticulum, the so-called "Sprecher pathway" identified in mammals [24,31], or, the direct Δ4 desaturation of docosapentaenoic acid (DPA, 22:5n-3) into DHA via the 'Δ4 route'. The first Fads2 gene with Δ4 activity was identified in the marine herbivorous fish *Siganus canaliculatus* [32]. It was not only the first enzyme with this activity among fish, but it was the first discovered case in all vertebrates. The discovery indicated that there exists another possible mechanism for DHA biosynthesis, a direct route involving elongation of EPA to 22:5n-3 followed by Δ4 desaturation. If both DHA routes were coexist, this would represent a clear advantage for satisfying DHA requirements through endogenous production. After further identification of 11 teleost species having a putative Δ4 Fads2 by Oboh et al. [33], it was made clear that the direct Δ4 pathway is more widespread among teleost fish than initially believed.

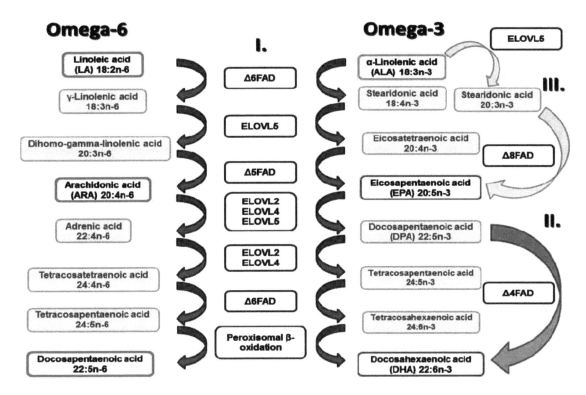

Figure 1. Biosynthetic pathways of long-chain ($\geq C_{20}$) polyunsaturated fatty acids (LC-PUFAs) of ω3 and ω6 families from dietary essential α-linolenic (ALA, 18:3n-3) and linoleic (LA, 18:2n-6) acids, respectively, by elongation and desaturation reactions. Adapted from (Carmona-Antoñanzas et al., 2011; Monroig et al., 2011; Sprecher, 2000; Voss et al., 1991); modified after (Trattner, 2009; Vestergeren, 2014; Yan, 2016).

4. Fads Gene Repertoire in Fish

There is a fundamental difference between fish and mammals regarding the gene repertoire encoding for enzymes performing desaturation activities needed for LC-PUFA biomolecules production. In contrast to mammals, where distinct separate genes *Fads1* and *Fads2* encode enzymes Fads1 and Fads2 with appropriate specificities Δ5 and Δ6 [34], respectively, in fish, *Fads1* gene has been lost during the evolution. As a result, all desaturation steps of the LC-PUFA biosynthetic pathway in fish are catalysed by Fads2 enzymes, exhibiting different Δ activities which can be overlapping to some extent.

Until recently, this scenario was generally accepted with no exceptions. Accordingly, one single Δ6 Fads2 appears most often [35–38]. Less often, a separate Δ6 Fads2 and Δ5 Fads2 paralogues appeared, such as in Atlantic salmon (*Salmo salar*) [39,40], or more than one single Δ6 Fads2 paralog are present such as in common carp (*Cyprinus carpio*) [41] and recently confirmed in numerous Osteoglossomorpha species [42]. In some teleosts studied, Δ6 Fads2 had measurable levels of Δ5 activity [37] or Δ8 activity [42,43]. A single bifunctional Δ6/Δ5 Fads2 acts in zebrafish (*Danio rerio*) [44] which was the first functionally characterized fish desaturase and for some time, it has been considered as an exception, not only in fish but in vertebrates in general. Later, two desaturases from marine rabbitfish (*Siganus canaliculatus*) were functionally characterized, one of which was shown to be Δ6/Δ5 bifunctional and the other Δ5/Δ4 bifunctional [32]. There exist extreme exceptions as well, represented by teleosts lacking *Fads*-like genes in their genomes, namely pufferfish *Takifugu rubripes* and *Tetraodon nigroviridis* [42]. The Atlantic salmon Δ6 and Δ5 *Fads2* cDNAs are very similar, sharing greater than 95% nucleic acid identity, indicating the presence of a recently duplicated locus, probably as the result of the recent salmonid whole genome duplication event [40].

The property of fish Fads2 exhibiting a more varied spectrum of Δ activities towards substrates has been hypothesized by Castro et al. [45] as a result of a functionalization process that occurred in response to dietary availability in natural pray. Functionally characterized Fads2 in numerous teleosts and all their activities determined by heterologous expression in yeast are listed in recent review of Kabeya et al. [46]. However, the persisting lack of information in some teleost lineages, such as Elopomorpha, and other nonteleost lineages, such as Lepisosteiformes, Polypteriformes or Cyclosomata, has hampered the full comprehension of Fads enzymes function in fish for a long time. Current novel insights into the fish LC-PUFA biosynthesis have provided a study on Fads desaturases published by Lopes-Marques et al. [42]. Accordingly, two types of desaturase repertoire are confirmed to appear in teleost fish, separating Elopomorpha from the other living teleost lineages. The orthologous gene to *Fads1* has been found in Japanese eel (*Anguilla* japonica), an Elopomorpha teleost specie, and confirmed by heterologous expression approach in yeast that desaturates the corresponding fatty acid substrates in the Δ5 position as well as sharing the common structural features to mammalian Fads1 enzymes. Farther Fads1 have been identified in some representatives of ancient fish lineages such as the Senegal bichir (*Polypterus senegalus*) and spotted gar (*Lepisosteus oculatus*) by these authors [42].

Based on sequence and phylogenetic data, *Fads2* and *Fads1* genes have been deduced to originate from the vertebrate ancestor and *Fads1* seems to be lost in Teleostei lineages except in Elopomorpha. It could be hypothesized that some teleosts have generated a mechanism to overcome the bottleneck caused by the loss of Δ5 Fads1, since otherwise, they would not be able to convert PUFA to LC-PUFA. Such a mechanism would be *Fads2* gene duplication followed by the process of functionalization as most probably was the case in salmonids, whereby acquisition of Δ5 *Fads2* occurred in one of the several *Fads2* gene copies. Another example would be the zebrafish (*Danio rerio*) in which Δ6 Fads2 acquired the ability to desaturate even in the Δ5 position [42,44]. The loss of canonical *Fads1* gene followed by *Fads2* subfunctionalization that teleosts have undergone during evolution could be linked to and explain the higher plasticity with which fish produce LC-PUFA biomolecules in comparison to other vertebrates.

5. Fads2 Structure and Structural Implications of Substrate Specificity

Fads2 are modular proteins which characteristically have a cytochrome *b5*-like domain on the N-terminus and the main desaturation domain with three histidine-rich regions on the C-terminus [47–49]. The fusion of the cytochrome *b5*-like domain to the main desaturase protein domain enables the NADH cytochrome *b5* reductase to directly transfer electrons to the catalytic site of Fads2 via the cytochrome *b5*-like domain without the requirement for an independent cytochrome *b5* [50,51]. However, solid evidence has been provided that both the cytochrome *b5*-like domain of Fads2 and microsomal cytochrome *b5* are necessary in the process of Δ6 desaturation and that the microsomal cytochrome *b5* does not compensate for the role of cytochrome *b5*-like domain of Fads2, which is accompanied by highly conserved heme-binding HPGG motif of the cytochrome *b5*-like domain. Moreover, protein–protein interactions between Fads2 and microsomal cytochrome *b5* are required for proper Fads2 function [52]. Phylogenetic studies have shown that cytochrome *b5* domain from Δ6 Fads2 proteins form a single cluster which points to a single ancient fusion event that took place in the common ancestor of all eukaryotes [53].

As a hydrophobic membrane-bound protein, Fads2 is extremely recalcitrant to characterization by conventional biochemical methods. A three-dimensional structure of Fads2 by X-ray crystallography is missing to date. The only animal desaturase whose structure is known is the stearoyl-CoA desaturase with Δ9 desaturation activity for which crystal structures have been published in humans [54] and rats [55]. There are some characteristic features common to all desaturases. The amino acid sequences within the substrate binding channel in all membrane desaturases contain the three His-boxes which histidine residues hold two irons in the active site. These histidine residues are of high evolutional conservation [56,57] and take place in very close proximity to the fatty acid substrate, referred to as "contact residues" [51]. Hydropathy analyses have shown that desaturases contain up to three long hydrophobic domains which are long enough to span the membrane bilayer twice whereas the His-boxes have a consistent positioning with respect to these potential membrane spanning domains. Sayanova et al. [56] undertook a massive motif analysis in more than fifty eukaryotic genomes, obtaining 275 desaturases, and reported the sequence logo representations of conserved histidine regions shown in Figure 2.

Figure 2. Sequence logo of histidine boxes in membrane associated front-end desaturases such as Δ6 Fads2. The high of letters corresponds to the occurrence probability. Adapted from (Sayanova, 2001). Modified after (Hashimoto, 2007).

The first report on the structural basis of the substrate specificity of a mammalian front-end fatty acid desaturase was published by Watanabe et al. in 2016 [58]. Using the crystal structure modelling of the human soluble stearoyl-CoA (Δ9) desaturase [54,55], these authors performed homology modelling and revealed that Arg216, Trp244, Gln245, and Leu323 are located near the substrate-binding site. They applied site-directed mutagenesis to create mutations in rat Δ6 Fads2 at those sites they had predicted to influence the enzymatic function. Then, they exchanged these amino acids accordingly to be the same as in the unique bifunctional Δ6/Δ5 Fads2 from zebrafish. They determined amino acid residues responsible for both switching and adding the substrate specificity of rat Δ6 Fads2. Additionally, they predicted tertiary structure of rat Δ6 Fads2. There is a very important outcome from their publication—that one single amino acid in the Δ6 Fads2 desaturase enzyme, when changed, has the potential to switch the specificity towards substrates.

This corresponded to results from investigations on sex pheromones in moths [59], where similarly, a change as small as a single amino acid substitution in a fatty acid desaturase enzyme was sufficient to change the enzymatic function of the whole enzyme, moreover, resulting in huge consequences in reproduction. According to their data delivered, MsexD2 desaturase gene in *Manduca sexta* duplicated during the evolution whereby one copy acquired one amino acid change. Then, in the process of neofunctionalization, this novel gene acquired the ability to introduce another double bond and produce an uncommon sex pheromone with significant implication in species reproduction.

Corresponding data were obtained by the study of Δ6 Fads2 from marine algae *Thalassiosira pseudonana* [60]. Mutation sites in Δ6 Fads2 from *T. pseudonana* were determined which appeared to induce a propensity for the enzyme to favor binding of a particular fatty acid, suggesting that these may be associated with substrate specificity. The focused primarily on desaturation kinetics and assessed molecular mechanisms underlying the catalytic activity of Fads2 in *T. pseudonana*, because this model organism offers the advantage of exhibiting a very high desaturase catalytic activity suitable for such studies. They divided the amino acid sequence of Fads2 into sections and at the same time, they have used Fads2 from *Glossomastix chrysoplasta*, which in the opposite has very low enzymatic activity, and divided it in the same sections. To determine the catalytic activity of each region, the corresponding regions of both Fads2 enzymes were systematically exchanged to construct recombinant swap genes, which were expressed in yeast. Kinetics of enzymatic catalytic activity of recombinant desaturase were measured as well as amino acid residues important for catalytic activity by the use of site-directed mutagenesis were determined. As a result, topology prediction was created depictured in Figure 3. Amino acid substitutions significantly impacted the desaturation catalytic efficiency providing a solid basis for in-depth understanding of catalytic efficiency of Δ6 Fads2 enzyme.

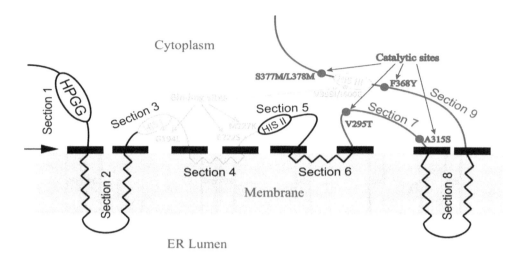

Figure 3. The predicted topology model of Δ6 Fads2. The black solid rectangles indicate boundaries of domains. Four alpha-helices span the membrane. The blue lines and blue dots indicate the areas and sites implicated in substrate specificity, red lines and red dots indicate the areas and sites important for catalytic activity, respectively. ER lumen: endoplasmic reticulum lumen. HIS I, HIS II and HIS III: histidine rich motifs. Modified from (Shi et al., 2018).

The abovementioned studies clearly demonstrated that the strictness of structure-function relationship of Fads2 enzymes might be enormous and acquired changes as small as one single amino acid of enzyme primary structure might have significant consequences for the organism studied which could be a general feature extrapolatable even to more diverse taxa such as fish. Comparative studies of highly effective and minimally effective LC-PUFA biosynthetic machineries either between more or less related species or occurring in one single species (as typically studied in salmonids [29,40,61] have justified, that this is a promising strategy with great potential to gain insight into the challenging LC-PUFA biomolecules research in fish.

The question why some species can survive in EPA and DHA poor environment and other even closely related species not, has been an attractive research topic in the very recent past. An interesting structure-function study performed by Xie et al. [62] has addressed that question by studying the *Fads2* promoter sequence. The binding site for stimulatory protein Sp1 has been found as lacking in the promoter of *Fads2* gene in marine teleost *Epinephelus coioides*. The authors speculated therefore, that the Sp1-binding site absence might be the main cause of the very low Fads2 expression in marine carnivorous teleost species. To test this hypothesis, they inserted the Sp1-binding site from the *Fads2* promoter sequence of the herbivorous *Siganus canaliculatus*, the first marine teleost demonstrated to have LC-PUFA biosynthetic ability, into the corresponding region of *E. coioides Fads2* promoter sequence. As expected, Δ6 desaturation activity increased significantly. The results provided direct evidence for the importance of the Sp1-binding site in determining *Fads2* promoter activity and indicated that its lack may be a reason for very low expression of Fads2 and poor LC-PUFA biosynthetic ability in *E. coioides*. The Sp1-binding site has been found as lacking in marine carnivorous fish *Gadus morhua* [36] as well as in *Dicentrarchus labrax* [63], while in *Oncorhynchus mykiss* its promoter activity was weaker [64].

6. Fads2 Copy Number Variation

The best was yet to come regarding studying marine vs. freshwater fish dealing with LC-PUFA poor food sources. Just a few months ago, Science released an exciting paper from Ishikawa et al. [65]. In this comprehensive study, the authors compared three-spined stickleback (*Gasterosteus aculeatus* species complex) which successfully colonized newly emerged freshwater bodies after glacial retreat with closely related marine Japan Sea stickleback (*G. nipponicus*) which had failed to colonize freshwater. They linked the colonization success to *Fads2* gene copy number, being higher in Pacific Ocean stickleback from the *G. acuelatus* complex. When transgenic Japan Sea stickleback overexpressing

Fads2 was made and fed only DHA-free *Artemia*, the Fads2 transgenics showed a higher survival rate and higher DHA content at 40 days after fertilization than the control GFP-transgenics. Moreover, *Fads2* gene linkage to X chromosome was confirmed, resulting in higher copy number in females. That fact is consistent with higher female survival observed. These results suggested that lower *Fads2* copy number may be a constraint to colonization of DHA-deficient freshwater niches by Japan Sea stickleback. These authors went deeper into the copy number assessing in the context of phylogeny and deduced a general mechanism: Higher *Fads2* copy number contributes to survival with DHA-free diets. Hence, *Fads2* was the metabolic gene important for overcoming the nutritional constraints associated with freshwater colonization in fishes. The authors mentioned the intriguing feature of the *Fads2* gene to make strong signatures of selection even in human such as in Greenland Eskimos which might be farther extrapolated to even more diverse taxa.

7. Fads2 Transgenes

The limited availability of LC-PUFA derived from fish represents the critical bottleneck in food production systems, one that numerous research institutions and aquafeed companies in this field are trying to overcome. Attempts to replace fish-derived LC-PUFA by plant derived alternatives often resulted in low quality products lacking the original content of these health promoting biomolecules [66]. This problem could be minimized by either feeding fish genetically modified plants for enhanced EPA and DHA production or by gene editing fish to be capable to produce endogenous LC-PUFA more effectively. Genetic engineering has been long been utilized as a strategy to increase natural productivity. Genetically engineered organisms could have the potential to reduce pressure on current LC-PUFA natural resources. The efforts and progress to develop transgenic plants as terrestrial sources of ω3 fish oils as well as advances in the field have been reviewed recently by Napier [67]. Transgenic fish have many potential applications in aquaculture, but the research also raised concerns regarding the possible risks to the environment associated with release and escape. A tabulated balance sheet of likely benefits and risks have been published by Maclean and Laight [68]. In this review, we focused on attempts to produce genetically modified fish with an enhanced content of ω3 LC-PUFA.

The first step to modifying the LC-PUFA biomolecules production pathway using genetic engineering was done in zebrafish [69], into which a gene for Δ5 Fads2 from masu salmon was introduced. The result demonstrated that masu salmon (*Oncorhynchus masou*) Δ5 Fads2 is functional in zebrafish and modifies its LC-PUFA metabolic pathway; hence, the technique could be applied to farmed fish to generate a nutritionally richer product for human consumption. The closely relative to zebrafish, the common carp (*Cyprinus carpio*) accounts for about 40% of the total global aquaculture production and could therefore deliver a significant amount of LC-PUFA if they were produced in their body. However, the content of EPA and DHA (mg g^{-1}) in muscle tissue of common carp is relatively low when compared to many other fish species, as revealed by recent meta-analysis data [70].

Some pioneering transgenesis experiments were carried out which reported trends towards increased ω3 LC-PUFA content in muscle of transgenic progeny—Δ5 *Fads2* from masou salmon driven by a β-actin promoter was introduced into common carp [71] and channel catfish (*Ictalurus punctatus*) [72] with the aim to improve ω3 LC-PUFA production. The results have shown promise for future work in this area, when utilizing homozygous transgenic individuals in contrast to the heterozygous individuals utilized in these studies. The effects of the transgene varied between common carp and channel catfish, being higher in common carp [72].

However, only a few month ago, *Haiyouli* construction was published [73], which, in Chinese, means "advantageous carp-like marine fish". This common carp was genetically modified with the aim to elevate production of ω3 LC-PUFA. The transgene used was a fish-codon optimized fatty acid desaturase (*fat1*) coding sequence originally from *Caenorhabditis elegans* driven by the 5′upstream regulatory region of common carp β-actin. Unexpectedly for the authors, under transgene expression fat accumulation of the internal organs decreased and in the liver tissues, *fat1*-transgenic common carp showed less accumulation of lipid droplets when compared with wildtype. However, the quantitative RT-PCR results showed a 10.5-fold increase in Fads2 expression, a 6.5-fold increase in elongase 5 expression and a 3-fold increase in elongase 2 expression in the transgenic tissue, indicating stimulation of LC-PUFA biosynthesis by the expression of exogenous *fat1* desaturase. Interestingly, the transcription of acyl-CoA oxidase 3 increased by 8.2 in transgenic tissue, which perfectly explains the lipid content decrease in internal organs of genetically modified common carp. Intriguingly, the authors stated that the ω6 to ω3 ratio of their transgenic common carp (0.4) was even lower than that of the Atlantic salmon (0.58) reported by Henderson and Tocher [74]. For this reason, *Haiyouli* has been presented as a potentially ideal fish produced in modern society to balance the high ω6 to ω3 ratio of human diets. However, such a conclusion is questionable since they stimulate LC-PUFA biosynthesis by the expression of exogenous *fat1* at the expense of overall lipid biosynthesis. When total mass unit of LC-PUFA per mass unit of filet is calculated and compared to salmon, there may, in fact, be no relative advantage to consuming *Haiyouli*. Hence, it is rather a step forward on the way to constructing the ideal fish.

Successful production of DHA using *Fads2* transgenes has been reported in mammals. In Chinese hamster ovary cells, LC-PUFA-elevated production was achieved by heterologous expression of fish Δ4 Fads2 from *Siganus canaliculatus* with concomitant overexpression of Δ6 Fads2 and Δ5 Fads1 from mice. The authors stated that this new technology has been confirmed as very effective in high-level production of DHA from dietary ALA and provided a potential for the creation of new land animal breeds who could produce DHA abundantly in their related products [75].

If such solutions come into practice, this will have a positive effect in sufficient delivery of health promoting LC-PUFA to humans while at the same time, preserving wild fish populations.

8. Conclusions

Fads2 is a fascinating enzyme with far-reaching implications for both human health and environmental sustainability. It is clear that Fads2 has played an important role in the adaptations to novel environments throughout evolutionary history as differences in both gene expression and copy number have been reported across freshwater and seawater dwelling species. We have demonstrated the importance of this enzyme in the context of growing aquaculture demand and dwindling LC-PUFA supply and outlined a few remedies in the form of genetic engineering to improve endogenous PUFA production. By improving our understanding of Fads2, we can address major environmental concerns and break out of the cycle of exploitation that currently strains our wild fish reserves to feed the growing aquaculture sector.

Author Contributions: Conceptualization, M.P. and J.M.; methodology, Z.B. and T.N.H.; writing—original draft preparation, Z.B. and T.N.H.; writing—review and editing, M.P. and J.M.; supervision, M.P. and J.M.; project administration, M.P. and J.M.; funding acquisition, M.P. and J.M. All authors have read and agreed to the published version of the manuscript.

Acknowledgments: We are grateful to anonymous reviewers for their valuable recommendations to improve the manuscript.

References

1.   Tocher, D.R.; Betancor, M.B.; Sprague, M.; Olsen, R.E.; A Napier, J. Omega-3 Long-Chain Polyunsaturated Fatty Acids, EPA and DHA: Bridging the Gap between Supply and Demand. *Nutrients* **2019**, *11*, 89. [CrossRef]
2.   Bell, M.V.; Tocher, D.R. Biosynthesis of polyunsaturated fatty acids in aquatic ecosystems: General pathways and new directions. In *Lipids in Aquatic Ecosystems*; Springer: New York, NY, USA, 2009; pp. 211–236.

3.   Tinoco, J. Dietary requirements and functions of α-linolenic acid in animals. *Prog. Lipid Res.* **1982**, *21*, 1–45. [CrossRef]

4.   Monroig, Ó.; Kabeya, N. Desaturases and elongases involved in polyunsaturated fatty acid biosynthesis in aquatic invertebrates: A comprehensive review. *Fish. Sci.* **2018**, *84*, 911–928. [CrossRef]

5.   Kabeya, N.; Fonseca, M.M.; Ferrier, D.E.K.; Navarro, J.C.; Bay, L.K.; Francis, D.S.; Tocher, D.R.; Castro, L.F.C.; Monroig, O. Genes for de novo biosynthesis of omega-3 polyunsaturated fatty acids are widespread in animals. *Sci. Adv.* **2018**, *4*, eaar6849. [CrossRef]

6.   Jakobsson, A.; Westerberg, R.; Jacobsson, A. Fatty acid elongases in mammals: Their regulation and roles in metabolism. *Prog. Lipid Res.* **2006**, *45*, 237–249. [CrossRef] [PubMed]

7.   Zárate, R.; El Jaber-Vazdekis, N.; Tejera, N.; Pérez, J.A.; Rodríguez, C. Significance of long chain polyunsaturated fatty acids in human health. *Clin. Transl. Med.* **2017**, *6*, 25. [CrossRef] [PubMed]

8.   Spector, A.A.; Kim, H.Y. Discovery of essential fatty acids. *J. Lipid Res.* **2015**, *56*, 11–21. [CrossRef]

9.   Burr, G.O.; Burr, M.M. A new deficiency disease produced by the rigid exclusion of fat from the diet. *J. Biol. Chem.* **1929**, *82*, 345–367. [CrossRef]

10.   Burr, G.O.; Burr, M.M. On the nature and role of the fatty acids essential in nutrition. *J. Biol. Chem.* **1930**, *86*, 587–621.

11.   Burr, G.O.; Burr, M.M.; Miller, E.S. On the fatty acids essentiality in nutrition III. *J. Biol. Chem.* **1932**, *97*, 1–9.

12.   Holman, R.T.; George, O. Burr and the Discovery of Essential Fatty Acids. *J. Nutr.* **1988**, *118*, 535–540. [CrossRef] [PubMed]

13.   Mead, J. The essential fatty acids: Past, present and future. *Prog. Lipid Res.* **1981**, *20*, 1–6. [CrossRef]

14.   Bergström, S.; Danielsson, H.; Klenberg, D.; Samuelsson, B. The enzymatic conversion of essential fatty acids into prostaglandins. *J. Biol. Chem.* **1964**, *239*, PC4006–PC4008.

15.   Bergström, S.; Danielsson, H.; Samuelsson, B. The enzymatic formation of prostaglandin E2 from arachidonic acid prostaglandins and related factors 32. *Biochim. Biophys. Acta (BBA)—Gen. Subj.* **1964**, *90*, 207–210. [CrossRef]

16.   Klenk, E.; Mohrhauer, H. Metabolism of polyenoic acids in the rat. *Hoppe Seylers Z. Physiol. Chem.* **1960**, *320*, 218–232. [CrossRef]

17.   Dyerberg, J.; Bang, H. Dietary fat and thrombosis. *Lancet* **1978**, *311*, 152. [CrossRef]

18.   Dyerberg, J.; Bang, H.O.; Stoffersen, E.; Moncada, J.; Vane, J.R. Eicosapentaenoic acid and the prevention of thrombosis and atherosclerosis? *Lancet* **1978**, *2*, 117–119. [CrossRef]

19.   Simopoulos, A.P. The omega-6/omega-3 fatty acid ratio: Health implications. Oléagineux, Corps gras. *Lipides* **2010**, *17*, 267–275.

20.   Simopoulos, A.P. Importance of the omega-6/omega-3 balance in health and disease: Evolutionary aspects of diet. *World Rev. Nutr. Diet.* **2011**, *102*, 10–21. [CrossRef]

21.   Scaioli, E.; Liverani, E.; Belluzzi, A. The Imbalance between n-6/n-3 Polyunsaturated Fatty Acids and Inflammatory Bowel Disease: A Comprehensive Review and Future Therapeutic Perspectives. *Int. J. Mol. Sci.* **2017**, *18*, 2619. [CrossRef]

22.   Gladyshev, M.I.; Sushchik, N.N. Long-chain Omega-3 Polyunsaturated Fatty Acids in Natural Ecosystems and the Human Diet: Assumptions and Challenges. *Biomolecules* **2019**, *9*, 485. [CrossRef] [PubMed]

23.   Voss, A.; Reinhart, M.; Sankarappa, S.; Sprecher, H. The metabolism of 7,10,13,16,19-docosapentaenoic acid to 4,7,10,13,16,19-docosahexaenoic acid in rat liver is independent of a 4-desaturase. *J. Boil. Chem.* **1991**, *266*, 19995–20000.

24.   Sprecher, H. Metabolism of highly unsaturated n-3 and n-6 fatty acids. *Biochim. Biophys. Acta (BBA)—Mol. Cell Boil. Lipids* **2000**, *1486*, 219–231. [CrossRef]

25.   Monroig, O.; Li, Y.; Tocher, D.R. Delta-8 desaturation activity varies among fatty acyl desaturases of teleost fish: High activity in delta-6 desaturases of marine species. *Comp. Biochem. Physiol. Part. B Biochem. Mol. Boil.* **2011**, *159*, 206–213. [CrossRef] [PubMed]

26.   Carmona-Antoñanzas, G.; Monroig, O.; Dick, J.R.; Davie, A.; Tocher, D.R. Biosynthesis of very long-chain fatty acids (C > 24) in Atlantic salmon: Cloning, functional characterisation, and tissue distribution of an Elovl4 elongase. *Comp. Biochem. Physiol. Part. B Biochem. Mol. Boil.* **2011**, *159*, 122–129. [CrossRef]

27.   Trattner, S. Quality of lipids in fish fed vegetable oils. Ph.D. Thesis, Acta Universitatis Agriculturae Sueciae, Uppsala, Sweden, May 2009.

28. Vestergeren, A.L.S. Transcriptional regulation in salmonids with emphasis on lipid metabolism. Ph.D. Thesis, Swedish University of Agricultural Sciences, Uppsala, Sweden, December 2014.

29. Yin, Y. Expression of genes involved in regulation of polyunsaturated fatty acid metabolism in liver of Atlantic salmon (*Salmo salar*) undergoing parr-smolt transformation. Master´s Thesis, Norwegian University of Live Sciences, Ås, Norway, September 2016.

30. Ferdinandusse, S.; Denis, S.; Mooijer, P.A.; Zhang, Z.; Reddy, J.K.; Spector, A.A.; Wanders, R.J. Identification of the peroxisomal beta-oxidation enzymes involved in the biosynthesis of docosahexaenoic acid. *J. Lipid Res.* **2001**, *42*, 1987–1995.

31. Sprecher, H. A reevaluation of the pathway for the biosynthesis of 4,7,10,13,16,19-docosahexaenoic acid. *Omega-3 News* **1992**, *7*, 1–3.

32. Li, Y.; Monroig, O.; Zhang, L.; Wang, S.; Zheng, X.; Dick, J.R.; You, C.; Tocher, U.R. Vertebrate fatty acyl desaturase with Δ4 activity. *Proc. Natl. Acad. Sci. USA* **2010**, *107*, 16840–16845. [CrossRef]

33. Oboh, A.; Kabeya, N.; Carmona-Antoñanzas, G.; Castro, L.F.C.; Dick, J.R.; Tocher, U.R.; Monroig, O. Two alternative pathways for docosahexaenoic acid (DHA, 22:6n-3) biosynthesis are widespread among teleost fish. *Sci. Rep.* **2017**, *7*, 3889. [CrossRef]

34. Guillou, H.; Zadravec, D.; Martin, P.G.; Jacobsson, A. The key roles of elongases and desaturases in mammalian fatty acid metabolism: Insights from transgenic mice. *Prog. Lipid Res.* **2010**, *49*, 186–199. [CrossRef]

35. Tocher, D.R. Fatty acid requirements in ontogeny of marine and freshwater fish. *Aquac. Res.* **2010**, *41*, 717–732. [CrossRef]

36. Tocher, D.R.; Zheng, X.; Schlechtriem, C.; Hastings, N.; Dick, J.R.; Teale, A.J. Highly unsaturated fatty acid synthesis in marine fish: Cloning, functional characterization, and nutritional regulation of fatty acyl Δ6 desaturase of Atlantic cod (Gadus morhua L.). *Lipids* **2006**, *41*, 1003–1016. [CrossRef] [PubMed]

37. Zheng, X.; Seiliez, I.; Hastings, N.; Tocher, D.; Panserat, S.; Dickson, C.; Bergot, P.; Teale, A. Characterization and comparison of fatty acyl Δ6 desaturase cDNAs from freshwater and marine teleost fish species. *Comp. Biochem. Physiol. Part. B Biochem. Mol. Boil.* **2004**, *139*, 269–279. [CrossRef] [PubMed]

38. Zheng, X.; King, Z.; Xu, Y.; Monroig, Ó.; Morais, S.; Tocher, D.R. Physiological roles of fatty acyl desaturases and elongases in marine fish: Characterization of cDNAs of fatty acyl Δ6-desaturase and Elovl5 elongase of cobia (*Rachycentron canadum*). *Aquaculture* **2009**, *290*, 122–131. [CrossRef]

39. Leaver, M.J.; Bautista, J.M.; Björnsson, B.T.; Jönsson, E.; Krey, G.; Tocher, D.R.; Torstensen, B.E. Towards Fish Lipid Nutrigenomics: Current State and Prospects for Fin-Fish Aquaculture. *Rev. Fish. Sci.* **2008**, *16*, 73–94. [CrossRef]

40. Gillard, G.; Harvey, T.N.; Gjuvsland, A.; Jin, Y.; Thomassen, M.; Lien, S.; Leaver, M.; Torgersen, J.S.; Hvidsten, T.R.; Vik, J.O.; et al. Life-stage-associated remodelling of lipid metabolism regulation in Atlantic salmon. *Mol. Ecol.* **2018**, *27*, 1200–1213. [CrossRef]

41. Ren, H.-T.; Zhang, G.-Q.; Li, J.-L.; Tang, Y.-K.; Li, H.-X.; Yu, J.-H.; Xu, P. Two Δ6-desaturase-like genes in common carp (Cyprinus carpio var. Jian): Structure characterization, mRNA expression, temperature and nutritional regulation. *Gene* **2013**, *525*, 11–17. [CrossRef]

42. Lopes-Marques, M.; Kabeya, N.; Qian, Y.; Ruivo, R.; Santos, M.M.; Venkatesh, B.; Tocher, D.R.; Castro, L.F.C.; Monroig, Ó. Retention of fatty acyl desaturase 1 (fads1) in Elopomorpha and Cyclostomata provides novel insights into the evolution of long-chain polyunsaturated fatty acid biosynthesis in vertebrates. *BMC Evol. Boil.* **2018**, *18*, 157.

43. Wang, S.; Monroig, Ó.; Tang, G.; Zhang, L.; You, C.; Tocher, D.R.; Li, Y. Investigating long-chain polyunsaturated fatty acid biosynthesis in teleost fish: Functional characterization of fatty acyl desaturase (Fads2) and Elovl5 elongase in the catadromous species, Japanese eel Anguilla japonica. *Aquaculture* **2014**, *434*, 57–65.

44. Hastings, N.; Agaba, M.; Tocher, U.R.; Leaver, M.J.; Dick, J.R.; Sargent, J.R.; Teale, A.J. A vertebrate fatty acid desaturase with Δ5 and Δ6 activities. *Proc. Natl. Acad. Sci. USA* **2001**, *98*, 14304–14309. [CrossRef]

45. Castro, L.F.C.; Monroig, Ó.; Leaver, M.J.; Wilson, J.; Cunha, I.; Tocher, U.R. Functional Desaturase Fads1 (Δ5) and Fads2 (Δ6) Orthologues Evolved before the Origin of Jawed Vertebrates. *PLoS ONE* **2012**, *7*, e31950. [CrossRef] [PubMed]

46. Kabeya, N.; Yoshizaki, G.; Tocher, D.R.; Monroig, Ó. Diversification of Fads2 in Finfish Species: Implications for Aquaculture. In *Investigación y Desarollo en Nutrición Acuícola*; Universidad Autónoma de Nuevo León: San Nicolás de los Garza, Nuevo León, México, 2017; pp. 338–362.

47. Zheng, X.; Tocher, D.R.; Dickson, C.A.; Bell, J.G.; Teale, A.J. Highly unsaturated fatty acid synthesis in vertebrates: New insights with the cloning and characterization of a Δ6 desaturase of Atlantic salmon. *Lipids* **2005**, *40*, 13–24. [CrossRef] [PubMed]

48. Seiliez, I.; Panserat, S.; Kaushik, S.; Bergot, P. Cloning, tissue distribution and nutritional regulation of a Delta6-desaturaselike enzyme in rainbow trout. *Comp. Biochem. Physiol. B Biochem. Mol. Biol.* **2001**, *130*, 83–93. [CrossRef]

49. Cho, H.P.; Nakamura, M.T.; Clarke, S.D. Cloning, expression, and nutritional regulation of the mammalian Delta-6 desaturase. *J. Biol. Chem.* **1999**, *274*, 471–477. [CrossRef] [PubMed]

50. Guillou, H.; D'Andrea, S.; Rioux, V.; Barnouin, R.; Dalaine, S.; Pedrono, F.; Jan, S.; Legrand, P. Distinct roles of endoplasmic reticulum cytochrome *b5* and fused cytochrome *b5*-like domain for rat Delta6-desaturase activity. *J. Lipid Res.* **2004**, *45*, 32–40. [CrossRef]

51. Mitchell, A.G.; Martin, C.E. A novel cytochrome *b5*-like domain is linked to the carboxyl terminus of the Saccharomyces cerevisiae delta-9 fatty acid desaturase. *J. Biol. Chem.* **1995**, *270*, 29766–29772.

52. Dahmen, J.L.; Olsen, R.; Fahy, D.; Wallis, G.J.; Browse, J. Cytochrome b5 coexpression increases Tetrahymena thermophila Δ6 fatty acid desaturase activity in Saccharomyces Cerevisiae. *Eukaryot. Cell.* **2013**, *12*, 923–931. [CrossRef]

53. Gostinčar, C.; Turk, M.; Gunde-Cimerman, N. The evolution of fatty acid desaturases and cytochrome *b5* in eukaryotes. *J. Membr. Biol.* **2010**, *233*, 63–72. [CrossRef]

54. Wang, H.; Klein, M.G.; Zhou, H.; Lane, W.; Snell, G.; Levin, I.; Li, K.; Sang, B.C. Crystal structure of human stearoyl-coenzyme A desaturase in complex with substrate. *Nat. Struct. Mol. Biol.* **2015**, *22*, 581–585. [CrossRef]

55. Bai, Y.; McCoy, J.G.; Levin, E.J.; Sobrado, P.; Rajashankar, K.R.; Fox, B.G.; Zhou, M. X-ray structure of a mammalian stearoyl-CoA desaturase. *Nature* **2015**, *524*, 252–256. [CrossRef]

56. Sayanova, O.; Beaudoin, F.; Libisch, B.; Castel, A.; Shewry, P.R.; Napier, J.A. Mutagenesis and heterologous expression in yeast of a plant Delta6-fatty acid desaturase. *J. Exp Bot.* **2001**, *52*, 1581–1585. [CrossRef] [PubMed]

57. Hashimoto, K.; Yoshizawa, A.C.; Okuda, S.; Kuma, K.; Goto, S.; Kanehisa, M. The repertoire of desaturases and elongases reveals fatty acid variations in 56 eukaryotic genomes. *J. Lipid Res.* **2007**, *49*, 183–191. [CrossRef] [PubMed]

58. Watanabe, K.; Ohno, M.; Taguchi, M.; Kawamoto, S.; Ono, K.; Aki, T. Identification of amino acid residues that determine the substrate specificity of mammalian membrane-bound front-end fatty acid desaturases. *J. Lipid Res.* **2016**, *57*, 89–99. [CrossRef] [PubMed]

59. Buček, A.; Matoušková, P.; Vogel, H.; Šebesta, P.; Jahn, U.; Weissflog, J.; Svatoš, A.; Pichová, I. Evolution of moth sex pheromone composition by a single amino acid substitution in a fatty acid desaturase. *Proc. Natl. Acad. Sci. USA* **2015**, *112*, 12586–12591. [CrossRef]

60. Shi, H.; Wu, R.; Zheng, Y.; Yue, X. Molecular mechanism underlying catalytic activity of delta 6 desaturase from *Glossomastix chrysoplasta* and *Thalassiosira pseudonana*. *J. Lipid Res.* **2018**, *56*, 2309–2321. [CrossRef]

61. Jin, Y.; Olsen, R.E.; Harvey, T.N.; Ostensen, M.A.; Li, K.; Santi, N.; Vadstein, O.; Vik, J.O.; Sandve, S.R.; Olsen, Y. Comparative transcriptomics reveals domestication-associated features of Atlantic salmon lipid metabolism. *bioRxiv* **2019**. Available online: https://www.biorxiv.org/content/10.1101/847848v1.full.pdf (accessed on 20 January 2020). [CrossRef]

62. Xie, D.; Fu, Z.; Wang, S.; You, C.; Monroig, O.; Tocher, D.R.; Li, Y. Characteristics of the *fads2* gene promoter in marine teleost *Epinephelus coioides* and role of Sp1-binding site in determining promoter activity. *Sci. Rep.* **2018**, *8*, 5305. [CrossRef]

63. Geay, F.; Santigosa, I.C.E.; Corporeau, C.; Boudry, P.; Dreano, Y.; Corcos, L.; Bodin, N.; Vandeputte, M.; Zambonino-Infante, J.L.; Mazurais, D.; et al. Regulation of FADS2 expression and activity in European sea bass (*Dicentrarchus labrax L.*) fed a vegetable diet. *Comp. Biochem. Physiol. B Biochem. Mol. Biol.* **2010**, *156*, 237–243. [CrossRef]

64. Geay, F.; Zambonino-Infante, J.; Reinhardt, R.; Kuhl, H.; Santigosa, E.; Cahu, C.; Mazurais, D. Characteristics of fads2 gene expression and putative promoter in European sea bass (*Dicentrarchus labrax*): Comparison with salmonid species and analysis of CpG methylation. *Mar. Genomics.* **2012**, *5*, 7–13. [CrossRef]

65. Ishikawa, A.; Kabeya, N.; Ikeya, K.; Kakioka, R.; Cech, J.N.; Osada, N.; Leal, M.C.; Inoue, J.; Kume, M.; Toyoda, A.; et al. A key metabolic gene for recurrent freshwater colonization and radiation in fishes. *Science* **2019**, *364*, 886–889. [CrossRef]

66. Tocher, D.R. Issues surrounding fish as a source of omega-3 long-chain polyunsaturated fatty acids. *Lipid Technol.* **2009**, *21*, 13–16. [CrossRef]

67. Napier, J.A.; Usher, S.; Haslam, R.P.; Ruiz-Lopez, N.; Sayanova, O. Transgenic plants as a sustainable, terrestrial source of fish oils. *Eur. J. Lipid Sci Technol.* **2015**, *117*, 1317–1324. [CrossRef] [PubMed]

68. Maclean, N.; Laight, R.J. Transgenic fish: An evaluation of benefits and risks. *Fish Fish.* **2001**, *1*, 146–172. [CrossRef]

69. Yoshizaki, G.; Kiron, V.; Satoh, S.; Takeuchi, T. Expression of masou salmon delta5-desaturase-like gene elevated EPA and DHA biosynthesis in zebrafish. *Mar. Biotechnol.* **2007**, *9*, 92–100. [CrossRef] [PubMed]

70. Gladyshev, M.I.; Sushchik, N.N.; Tolomeev, A.P.; Dgebuadze, Y.Y. Meta-analysis of factors associated with omega-3 fatty acid contents of wild fish. *Rev. Fish. Biol Fish.* **2018**, *28*, 277–299. [CrossRef]

71. Cheng, Q.; Su, B.; Qin, Z.; Weng, C.C.; Yin, F.; Zhou, Y.; Fobes, M.; Perera, D.A.; Shang, M.; Soller, F.; et al. Interaction of diet and the masou salmon Δ5-desaturase transgene on Δ6-desaturase and stearoyl-CoA desaturase gene expression and N-3 fatty acid level in common carp (*Cyprinus carpio*). *Transgenic Res.* **2014**, *23*, 729–742. [CrossRef]

72. Bugg, W. The effects of the masou salmon delta5-desaturase transgene on N-3 fatty acid production in F1 transgenic common carp (*Cyprinus carpio*) and channel catfish (*Ictalurus punctatus*). Master's Thesis, Auburn University, Auburn, Alabama, 7 May 2017.

73. Zhang, X.; Pang, S.; Liu, C.; Wang, H.; Ye, D.; Zhu, Z.; Sun, Y. A novel dietary source of EPA and DHA: Metabolic engineering of an important freshwater species—common carp by *fat1*-transgenesis. *Mar. Biotechnol.* **2019**, *21*, 171–185. [CrossRef]

74. Hendersen, R.J.; Tocher, D.R. The lipid composition and biochemistry of freshwater fish. *Prog Lipid Res.* **1987**, *26*, 281–347. [CrossRef]

75. Zhu, G.; Jiang, X.; Ou, Q.; Zhang, T.; Wang, M.; Sun, G.; Wang, Z.; Sun, J.; Ge, T. Enhanced Production of Docosahexaenoic Acid in Mammalian Cells. *PLoS ONE* **2014**, *9*, e96503. [CrossRef]

6

Natural CLA-Enriched Lamb Meat Fat Modifies Tissue Fatty Acid Profile and Increases n-3 HUFA Score in Obese Zucker Rats

Gianfranca Carta [1,*,†], Elisabetta Murru [1,†], Claudia Manca [1], Andrea Serra [2], Marcello Mele [2] and Sebastiano Banni [1]

[1]   Department of Biomedical Sciences, University of Cagliari, 09042 Monserrato, CA, Italy;
      m.elisabetta.murru@gmail.com (E.M.); claumanca@hotmail.com (C.M.); banni@unica.it (S.B.)
[2]   Department of Agriculture, Food and Environment, University of Pisa, 56124 Pisa, Italy;
      andrea.serra@unipi.it (A.S.); marcello.mele@unipi.it (M.M.)
*   Correspondence: giancarta@unica.it
†   These authors contributed equally to this work.

Abstract: Ruminant fats are characterized by different levels of conjugated linoleic acid (CLA) and α-linolenic acid (18:3n-3, ALA), according to animal diet. Tissue fatty acids and their N-acylethanolamides were analyzed in male obese Zucker rats fed diets containing lamb meat fat with different fatty acid profiles: (A) enriched in CLA; (B) enriched in ALA and low in CLA; (C) low in ALA and CLA; and one containing a mixture of olive and corn oils: (D) high in linoleic acid (18:2n-6, LA) and ALA, in order to evaluate early lipid metabolism markers. No changes in body and liver weights were observed. CLA and ALA were incorporated into most tissues, mirroring the dietary content; eicosapentaenoic acid (EPA) and docosahexaenoic acid (DHA) increased according to dietary ALA, which was strongly influenced by CLA. The n-3 highly-unsaturated fatty acid (HUFA) score, biomarker of the n-3/n-6 fatty acid ratio, was increased in tissues of rats fed animal fats high in CLA and/or ALA compared to those fed vegetable fat. DHA and CLA were associated with a significant increase in oleoylethanolamide and decrease in anandamide in subcutaneous fat. The results showed that meat fat nutritional values are strongly influenced by their CLA and ALA contents, modulating the tissue n-3 HUFA score.

Keywords: CLA; conjugated linoleic acid; ALA; α-linolenic acid; n-3 HUFA score; meat fat; vegetable fat

1. Introduction

Dietary fats are often associated with diet-derived health problems such as obesity, coronary heart disease, diabetes, and tumors. Nevertheless, some dietary fatty acids (FAs) have been found to act as preventing factors against cardiovascular disease (CVD) [1] and certain types of tumors [2,3]. The unusual fatty acid, conjugated linoleic acid (CLA) (CD18:2), and n-3 polyunsaturated fatty acids (PUFAs) are considered the major preventing factors that are naturally present in food derived from milk and meat ruminant fats, and from fish, respectively. Consequently, since ruminant products rich in saturated fatty acids (SFAs) are not entirely acceptable to consumers [4], strategies to manipulate the fat content and fatty acid (FA) composition of ruminant meat and milk have been proposed [5,6]. Natural CLA, mainly cis-9, trans-11 CLA (18:2c9,t11), a conjugated dienoic isomer of linoleic acid (18:2n-6, LA), derives from the incomplete biohydrogenation of LA in the rumen, and/or by the action of stearoyl-CoA desaturase (Δ9 desaturase) on vaccenic acid (trans-11 18:1, VA) within the mammary gland [7,8]. Natural CLA and its isomers constitute a special category of trans FAs that have been shown to exert anti-carcinogenic, anti-obesity, and anti-inflammatory effects, among others [2,9,10], by interfering with the metabolism of n-6 PUFAs [11,12]. It is possible to increase the content of CLA in meat and milk from ruminants when animals graze fresh pastures [13], and through supplementation of the diet with oils or seeds [14,15], like safflower oil, which is rich in LA [16], or linseed oil, which is high in α-linolenic acid (18:3n-3, ALA) [17]. CLA is accumulated in a similar fashion as oleic acid (18:1, OA), and desaturated and elongated in tissues to conjugated linolenic acid (CD18:3) and conjugated eicosatrienoic acid (CD20:3), retaining an unaltered conjugated diene structure [18]. CLA is also efficiently β-oxidized in peroxisomes, and acts as avid ligand of peroxisome proliferator-activated receptors type α (PPAR-α) [19], which regulate the expression of genes involved in peroxisomal β-oxidation [20].

The main n-3 long-chain, highly-unsaturated FAs (HUFAs) are represented by eicosapentaenoic acid (20:5n-3, EPA) and docosahexaenoic acid (22:6n-3, DHA), which exert anti-inflammatory and hypolipidemic effects through increased PPAR-α-mediated β-oxidation of FAs [21,22]. EPA and DHA may derive in tissues from seafood products or from the elongation and desaturation process of ALA. However, the conversion rate of ALA to DHA is too low to be considered efficient from a nutritional point of view [23]. Therefore, there is a need to increase the intake of foods with n-3 PUFAs [24].

The balance of LA and ALA, as precursors of the n-6 and n-3 FA families, is critical for the formation of n-3 HUFA [25]. It has been shown that a maximal incorporation of DHA into tissues can be achieved using diets with LA/ALA ratios between 4:1 and 2:1 [26–28]. Lands et al. measured the n-3 HUFAs content of red blood cells and tissues in rats consuming different LA/ALA ratios, and developed the concept that diets high in LA would inhibit the synthesis of n-3 HUFAs by simple competitive inhibition of the Δ6 desaturase (Δ6-D) enzyme and other enzymes [29,30].

The n-3 HUFA score that is obtained as the percentage of n-3 highly-unsaturated fatty acids (HUFA ≥ 20 carbons and ≥ 3 double bonds) in the total HUFAs pool is a potential blood biomarker of n-3 FAs intake and tissue status. HUFAs are mainly incorporated into phospholipids (PLs), and are potential precursors of biologically-active eicosanoids and docosahexaenoids [31]. Because the n-3 HUFA score has been shown to be less variable than n-3 FAs in the blood and tissues of rats, it could serve as a modifiable risk factor for CVD [31].

The ratio between the sum of n-3 HUFAs (EPA, DHA) + dihomo-γ-linolenic acid (20:3n-6, DGLA) and arachidonic acid (20:4n-6, AA), i.e., the anti-inflammatory FA index (AIFAI), provides a marker of the ability to decrease the formation of n-6 eicosanoids [32,33]. In fact, AIFAI has been reported to increase in association with a significant decrease in the formation of PGE2, 6-keto-PGF, prostanoids, and TNFα [34].

Many studies have found that a high fat diet can induce obesity, implying the obesogenic role of dietary fat; however, most of these studies did not take into account the fact that dietary FA composition is crucial in the regulation of body fat deposition and distribution. Some authors have described obesity resistance and reduced hypertrophy of visceral fat pads when employing fish oil-based diets. This might be related to increased lipid oxidation in these animals due to the n-3 HUFA-induced activation of PPAR-α [35,36]. Similarly, dietary CLA has been shown to decrease body fat in animals and humans [37]. Some of the effects of dietary FAs have been shown to be mediated by endocannabinoids (EC), namely anandamide (AEA) and 2-arachidonoyl–glycerol (2-AG), both of which are derived from AA, and by AEA congeners such as N-palmitoylethanolamide (PEA) and N-oleoylethanolamide (OEA), which are avid ligands of PPAR-α. An overactive endocannabinoid system may favor visceral fat deposition and thereby obesity, while the activation of PPAR-α has been shown to reduce body weight [35,36].

In the present study, we aimed to evaluate whether the peculiar nutritional effect of CLA in combination with ALA on increasing n-3 HUFA score, found in humans with dietary CLA-enriched cheese [38,39], would also be confirmed with dietary lamb meat fats which were differentially enriched in CLA and ALA compared to vegetable fats in Zucker rats, a rat model of obesity [40].

2. Materials and Methods

2.1. Reagents

The acetonitrile, methanol, chloroform, n-hexane, ethanol, acetic acid, and fatty acids standards were HPLC grade, and like deferoxamine mesylate, were purchased from Sigma Chemicals Co. (St. Louis, MO, USA). Ascorbic acid, potassium hydroxide, and hydrochloric acid were purchased from Carlo Erba (Milano, Italy). Internal deuterated standards for the AEA, 2-AG, and OEA quantification by isotope dilution ($[^2H]_8$ AEA, $[^2H]_5$ 2-AG, $[^2H]_2$ OEA) were purchased from Cayman Chemicals (MI, USA).

2.2. Animals and Diets

Twenty-four male obese Zucker rats (Harlan) four weeks of age with an initial weight of 200 ± 15 g were randomly assigned to four groups, and fed for four weeks with different diets containing 6% total fat, which provided 14% of the total energy (%en). The diets, based on the AIN-93G formulation with the substitution of soybean oil with experimental fats, differed only for FAs composition: (A) fat enriched in CLA, 1.3 g/kg of total diet, obtained from the meat of suckling milk from grazing ewes; (B) fat enriched in ALA and low in CLA, respectively 0.9 and 0.6 g/kg of the total diet, obtained from the meat of heavy lambs fed a diet based on cereal grains and integrated with rolled linseed + stoned olive cake; (C) fat containing LA, ALA, and CLA, respectively 2.3, 0.3, and 0.4 g/kg of total diet, obtained from the meat of lambs fed a diet based on cereal grains; (D) fat high in LA, and ALA, respectively 13.1 and 0.6 g/kg of total diet, of vegetable origin from a mixture of olive and corn oils, as depicted in Table 1. Diets A and B were characterized by a high content of trans FAs (VA and CLA), and ALA. The SFA content was higher in animal fat, while unsaturated fatty acids were higher in the diet with vegetable fat. Animal fat for the diets was obtained from the carcasses of lambs produced according to the feeding protocols described by Serra et al. [41] for diet A and Mele et al. [42] for diets B and C. The diets were prepared by Harlan.

Table 1. Principal fatty acids in experimental diets. [1]

FA	A [2]	B [2]	C [2]	D [2]
		g/kg diet		
14:0	4.3	2.6	2.4	0.3
16:0	13.1	11.6	12.6	5.9
18:0	8.6	9.5	10.8	1.4
VA	1.4	1.0	0.6	-
OA	22.2	19.7	21.6	33.7
LA	0.8	2.0	2.3	13.1
ALA	0.4	0.9	0.3	0.6
CLA	1.30	0.61	0.41	0.03
EPA	0.09	0.06	0.05	-
DHA	0.13	0.05	0.05	-
SFA	27.9	25.4	27.7	7.6
UFA	30.8	32.9	31.2	47.3
n-6/n-3	1.1	1.1	4.2	22.0
ALA/CLA	0.3	1.5	0.8	22.2
LA/ALA	2.2	2.2	7.0	22.0
total FA	58.7	58.3	58.8	54.9

[1] Diets were AIN-93G by Harlan; standard fat formulation was substituted with experimental fats. [2] Fats in the diets were: (A) enriched in conjugated linoleic acid (CLA); (B) enriched in α-linolenic acid (ALA) and low in CLA; (C) low in ALA and CLA; (D) high in linoleic acid (LA) and trace levels of CLA. Vaccenic acid (VA), oleic acid (OA), eicosapentaenoic acid (EPA), docosahexaenoic acid (DHA), saturated fatty acid (SFA), unsaturated fatty acid (UFA).

Body weight and food intake were measured weekly across the study. Body length, from tip of nose to the base of the tail, was measured at baseline (week 0), and at the study endpoint (week 4).

All experiments were performed according to the guidelines and protocols approved by the European Union (EU Council 86/609; D.L. 27.01.1992, no. 116) and by the Animal Research Ethics Committee of the University of Cagliari, Italy. The authorization number from the Italian Ethical Committee, approved on 28 September 2018, is 733/2018-P.

2.3. Tissues and Blood Sampling

Before sacrifice, rats were fasted for 12 h. After Fentanyl treatment (100 µg/kg of body weight) rats were euthanized without any further anesthesia by decapitation. Immediately after death, liver, heart, hypothalamus, visceral and subcutaneous adipose tissues were taken and stored at −80 °C. Blood was taken and centrifuged at 2000× g for 15 min at room temperature, plasma was stored at −80 °C for future lipid analyses.

2.4. Lipid Analyses

Fatty acid analysis was conducted from the total lipids previously extracted from tissues by the method of Folch [43]. Aliquots of chloroform were dried and mildly saponified as previously described [44] in order to obtain free fatty acids for HPLC analysis. The separation of unsaturated fatty acids was carried out with an Agilent 1100 HPLC system (Palo Alto, CA, USA) equipped with a diode array detector, as previously reported [45]. Since SFAs are transparent to UV detection, they were measured, after methylation, by Agilent 6890 gas chromatography (Palo Alto, CA, USA), as described in [46].

Endocannabinoid and congener quantification is described in [47]. Deuterated EC and congeners were added as internal standards to the samples before extraction. Analyses were carried out by liquid chromatography, atmospheric pressure chemical ionization, and MS (LC–APCI–MS) (Palo Alto, CA, USA), using selected ion monitoring (SIM) at M+1 values for the compounds and their deuterated homologs.

The n-3 HUFA score was calculated as the percentage of the sum of n-3 FAs with 20 or more carbon atoms and three or more double bonds, divided by the sum of total FAs with 20 or more carbon atoms and more than three double bonds [31]:

$$\text{n-3 HUFA score} = (EPA + DHA + \text{docosapentaenoic acid (22:5n-3, DPAn-3))/(EPA} + DHA + DPAn\text{-}3 + DGLA + AA + 22\text{:}4n\text{-}6 + DPAn\text{-}6 + 20\text{:}3n\text{-}9) \times 100 \tag{1}$$

The anti-inflammatory FA index is obtained as [33]:

$$AIFAI = (EPA + DHA + DGLA)/(AA \times 100) \tag{2}$$

2.5. Statistical Analysis

Data are expressed as the mean ± SEM, specifically, fatty acids as nmoles per gram of tissue or ml of plasma, or g/kg diet. EC and congeners are expressed as Mol% compared to total FAs. Multiple unpaired comparison tests were performed by ordinary one-way ANOVA followed by a Tukey's posthoc multiple comparison test in order to check the effect of specific dietary lipids on lipid metabolism in an animal model of obesity. The statistical analyses were performed using the GraphPad Prism 6.01 Software (La Jolla, CA, USA).

3. Results

No variations of food intake, body and liver weights, or liver total lipid concentration were detected in the obese Zucker rats in relation to different dietary fat sources. The BMI of rats fed diet D (0.96 ± 0.022), containing vegetable fat, was slightly, but not significantly increased compared to rats fed diet C (0.89 ± 0.023).

3.1. Tissue FA Profile

An analysis of FAs revealed that the tissue FA profiles, except in the hypothalamus, were strongly influenced by dietary FAs (Table 2). As expected, CLA concentrations mirrored the diet CLA content in the following order A > B > C > D in liver, subcutaneous adipose tissue (SAT), and visceral adipose tissue (VAT) (Table 2). The same pattern was observed in plasma, even though there were not significant differences between diets B and C or C and D; in heart, CLA reached similar concentrations in the A and B groups, i.e., higher than diets C and D, while, as anticipated, no changes were observed in the hypothalamus. The pattern of VA, the other rumen-derived trans FA, did not mirror its dietary concentration, and higher values were observed in all tissues except the hypothalamus in rats fed diet B. Moreover, in liver and both adipose tissues, rats on diet C, displayed higher amount than those fed diet A (Table 2).

Table 2. FA concentrations in obese Zucker rats fed diets A, B, C, or D [1] for 4 wk.

Diet Group	Liver (nmol/g)	Heart (nmol/g)	VAT (nmol/g)	SAT (mol/g)	Plasma (nmol/mL)	Hypothalamus (nmol/g)
			ALA			
A1	898.7 ± 128.3 [ab]	102.5 ± 11.8 [a]	23926.7 ± 1248.4 [ab]	24589.5 ± 1560.4 [b]	97.7 ± 13.5 [a]	ND
B1	1190.2 ± 100.1 [b]	208.6 ± 18.9 [b]	27737.5 ± 978.7 [a]	31367.0 ± 1096.6 [a]	112.1 ± 17.2 [a]	ND
C1	969.6 ± 61.6 [ab]	126.2 ± 10.1 [a]	22779.2 ± 1710.2 [b]	26599.3 ± 1326.9 [b]	103.1 ± 7.1 [a]	ND
D1	596.8 ± 163.3 [a]	103.7 ± 15.9 [a]	20194.1 ± 1017.2 [b]	25711.5 ± 1746.5 [ab]	80.8 ± 5.2 [a]	ND
			EPA			
A1	541.0 ± 92.3 [a]	59.4 ± 3.3 [a]	407.8 ± 35.6 [ab]	506.8 ± 22.6 [ab]	64.5 ± 2.9 [a]	ND
B1	555.9 ± 70.6 [a]	61.9 ± 2.9 [a]	495.9 ± 38.3 [a]	605.9 ± 42.1 [a]	65.5 ± 7.2 [a]	ND
C1	377.1 ± 34.0 [ab]	42.7 ± 1.0 [b]	284.3 ± 17.9 [bc]	386.6 ± 20.6 [bc]	55.1 ± 3.1 [ab]	ND
D1	185.1 ± 59.7 [b]	22.2 ± 1.1 [c]	169.0 ± 21.8 [c]	262.1 ± 28.5 [c]	32.3 ± 2.0 [b]	ND
			DHA			
A1	8242.1 ± 407.5 [a]	4318.2 ± 161.8 [a]	2604.9 ± 372.5 [a]	2403.9 ± 111.3 [a]	480.6 ± 16.9 [a]	11791.5 ± 881.7 [ab]
B1	8101.5 ± 514.4 [a]	3709.7 ± 115.7 [b]	2124.2 ± 187.6 [a]	2014.8 ± 134.8 [ab]	428.3 ± 48.9 [a]	11687.5 ± 349.5 [ab]
C1	7752.8 ± 394.0 [a]	3506.4 ± 95.7 [bc]	2307.0 ± 189.3 [a]	1986.7 ± 106.9 [ab]	439.4 ± 26.1 [d]	11331.8 ± 369.2 [b]
D1	6720.0 ± 671.9 [a]	3068.7 ± 201.6 [c]	1500.0 ± 264.6 [a]	1664.0 ± 23.0 [b]	410.4 ± 45.2 [a]	13612.8 ± 519.5 [a]
			n-3 HUFA score			
A1	32.0 ± 0.8 [a]	34.6 ± 0.3 [a]	22.8 ± 2.4 [ab]	23.7 ± 1.0 [a]	22.8 ± 0.91 [a]	49.8 ± 0.7 [a]
B1	30.3 ± 0.6 [a]	31.9 ± 1.1 [a]	23.5 ± 0.8 [a]	22.2 ± 0.9 [a]	21.2 ± 0.72 [ab]	48.9 ± 0.1 [ab]
C1	28.1 ± 0.2 [b]	32.0 ± 1.1 [a]	19.3 ± 1.0 [b]	18.9 ± 0.7 [b]	20.2 ± 0.33 [b]	48.3 ± 0.2 [b]
D1	24.4 ± 0.23 [c]	25.1 ± 0.7 [b]	14.2 ± 1.1 [c]	12.8 ± 0.7 [c]	16.2 ± 0.02 [c]	47.6 ± 0.2 [b]
			AA			
A1	19396.2 ± 994.3 [a]	9694.9 ± 74.9 [a]	8253.3 ± 277.1 [a]	8558.3 ± 319.4 [b]	2058.7 ± 57.1 [a]	8670.8 ± 648.7 [b]
B1	20722.6 ± 1212.4 [a]	9762.2 ± 117.0 [a]	8138.1 ± 581.0 [a]	8820.2 ± 457.9 [b]	2089.9 ± 152.1 [a]	9066.6 ± 252.8 [b]
C1	21505.6 ± 1048.2 [a]	9638.7 ± 102.4 [a]	9364.8 ± 645.9 [a]	10015.1 ± 418.8 [b]	2237.5 ± 90.7 [a]	9036.5 ± 244.4 [b]
D1	22810.6 ± 2082.8 [a]	9885.1 ± 236.5 [a]	10188.2 ± 889.6 [a]	12291.9 ± 566.0 [a]	2575.2 ± 213.2 [a]	11128.7 ± 436.4 [a]
			CLA			
A1	622.6 ± 66.7 [a]	84.1 ± 10.4 [a]	19892.6 ± 473.5 [a]	17208.7 ± 931.2 [a]	58.9 ± 7.3 [a]	13.4 ± 0.5 [a]
B1	416.2 ± 27.9 [b]	75.8 ± 8.7 [a]	12108.4 ± 396.1 [b]	11697.7 ± 424.9 [b]	35.6 ± 5.3 [b]	17.7 ± 1.3 [a]
C1	280.7 ± 17.4 [c]	43.2 ± 2.9 [b]	8218.0 ± 577.4 [c]	7911.8 ± 251.0 [c]	29.1 ± 1.8 [bc]	11.7 ± 1.5 [a]
D1	58.2 ± 11.1 [d]	12.7 ± 2.8 [c]	1585.7 ± 171.0 [d]	1669.5 ± 181.7 [d]	10.9 ± 1.8 [c]	11.2 ± 4.8 [a]
			VA			
A1	529.7 ± 109.7 [b]	274.1 ± 5.7 [b]	14613.7 ± 200.3 [b]	12199.2 ± 968.5 [b]	76.6 ± 5.79 [a]	656.8 ± 152.5 [a]
B1	1248.4 ± 249.9 [a]	389.7 ± 14.1 [a]	35216.2 ± 384.0 [a]	30441.6 ± 1602.2 [a]	122.5 ± 17.93 [a]	693.5 ± 54.7 [a]
C1	732.1 ± 151.1 [bc]	219.6 ± 11.1 [c]	19793.3 ± 1407.8 [bc]	17789.3 ± 1032.3 [bc]	96.3 ± 6.52 [ab]	707.0 ± 47.3 [a]
D1	214.7 ± 82.3 [bd]	124.8 ± 5.6 [d]	2962.6 ± 342.8 [d]	2844.9 ± 212.1 [d]	47.5 ± 7.87 [b]	744.3 ± 14.3 [a]

Values are means ± SEMs, n = 6/group. Within tissue, labelled means in a variable without a common superscript letter differ, as determined by Tukey's post hoc test after a significant one-way ANOVA, $p < 0.05$; the maximum value is labeled as 'a', the smaller value with difference is marked as 'b', the smaller value than 'b' with difference is marked as 'c', and the smallest value with difference is marked as 'd'. ND not detected. Subcutaneous (SAT), visceral adipose tissue (VAT); arachidonic acid (AA), n-3 highly-unsaturated fatty acids (n-3 HUFA score). [1] Fats in the diets were: (A) enriched in CLA; (B) enriched in ALA and low in CLA; (C) low in ALA and CLA; (D) high in LA and trace level of CLA.

Table 3 shows that CLA was efficiently desaturated to CD18:3 and elongated to CD20:3 in the liver. A similar pattern was observed in other tissues (data not shown), and despite the higher concentration of CLA in diet B compared to diet C, CD20:3 in the B group was not significantly different from the concentrations found in the C group.

Table 3. CLA and its metabolites in liver of obese Zucker rats fed diets A, B, C, or D for 4 wk[1].

Diet Groups	nmol/g Liver		
	CLA	CD18:3	CD20:3
A[1]	622.6 ± 66.7 [a]	34.1 ± 4.5 [a]	59.1 ± 3.1 [a]
B[1]	416.2 ± 27.9 [b]	28.0 ± 4.4 [a]	28.6 ± 1.8 [b]
C[1]	280.7 ± 17.4 [c]	10.3 ± 1.5 [b]	24.2 ± 1.9 [b]
D[1]	58.2 ± 11.1 [d]	6.7 ± 1.3 [b]	7.7 ± 0.6 [c]

Values are means ± SEMs, n = 6/group. Within a variable, labelled means without a common superscript letter differ as determined by Tukey's post hoc test after a significant one-way ANOVA, $p < 0.05$; the maximum value is labeled as 'a', the smaller value with difference is marked as 'b', the smaller value than 'b' with difference is marked as 'c', and the smallest value with difference is marked as 'd'. Conjugated dienes (CD). [1] Fats in the diets were: (A) enriched in CLA; (B) enriched in ALA and low in CLA; (C) low in ALA and CLA; (D) high in LA and trace level of CLA.

As shown in Table 2, ALA concentrations were significantly increased in diet B compared to diet D in the liver, and in diet B compared to all the other groups in heart; in VAT in B compared to C and D; in SAT in B compared to the A and C groups. No significant differences were detected in plasma, while ALA was not detectable in the hypothalamus.

EPA was not detectable in the hypothalamus, and was lower in the tissues of animals fed vegetable fat compared to those fed meat fats. EPA was significantly increased mainly with diets A and B compared to the other groups (Table 2). Among all the dietary groups, DHA significantly increased in SAT compared to D, while in heart, DHA was significantly higher in A compared to all other groups; no changes were observed in liver, plasma, and VAT. In the hypothalamus, DHA levels were significantly reduced by diet C compared to diet D (Table 2). Interestingly, DHA concentrations seem to vary according to the dietary amount of CLA, rather than in relation to the dietary content of its putative precursor, ALA, in liver, heart, and plasma. The maximum yield of DHA was obtained at the lowest ALA/CLA ratio in the diet (0.3 in A). DHA concentrations, except in the hypothalamus, were higher in group A, though significantly, only in heart (Tables 1 and 2).

The n-3 HUFA score was significantly increased in tissues of meat-fat-fed rats compared to those fed vegetable fat, particularly with diets A and B, except in the hypothalamus, in which the level increased significantly only with diet A (Table 2). Interestingly, the n-3 HUFA score was higher in the tissues of rats fed diets with high LA/ALA ratios (diets A and B) (see Tables 1 and 2).

The main SFA concentrations did not change among the groups in liver, heart, SAT, plasma, and hypothalamus, while they were slightly, but significantly, increased in VAT by diets B and C compared to diet D (data not shown).

AIFAI was significantly increased with meat fat diets in liver, heart, and VAT. Specifically, diets A and B induced the highest increase in liver and SAT, while in the heart, A was significantly higher than B, C, and D, while B and C were higher than D. In the hypothalamus, this index was raised only in A compared to the C and D groups, and in plasma in A and B compared to D (Table 4).

Table 4. Anti-inflammatory FA index (AIFAI) obtained as (EPA + DHA + DGLA)/AA in tissues of obese Zucker rats fed diets A, B, C, or D for 4 wk [1].

	Anti-Inflammatory Index				
Diet Groups	Heart	VAT	SAT	Plasma	Hypothalamus
A1	52.8 ± 1.3 [a]	78.8 ± 6.7 [a]	84.3 ± 5.3 [a]	31.3 ± 1.9 [a]	139.9 ± 6.2 [a]
B1	49.1 ± 0.7 [b]	70.3 ± 2.2 [ab]	77.5 ± 4.8 [a]	28 ± 1.7 [a]	131.5 ± 0.6 [ab]
C1	44 ± 0.3 [c]	64.8 ± 1.2 [b]	62.2 ± 2.0 [bc]	26.2 ± 0.7 [ab]	127.6 ± 1.2 [b]
D1	34 ± 0.6 [d]	45.4 ± 1.2 [c]	46.4 ± 5.7 [c]	20.6 ± 0.8 [b]	124.1 ± 1.2 [b]

Values are means ± SEMs, $n = 6$/group. Within a tissue, labelled means without a common superscript letter differ as determined by Tukey's post hoc test after a significant One-way ANOVA, Tukey's multiple comparisons test, $p < 0.05$; the maximum value is labeled as 'a', the smaller value with difference is marked as 'b', and the smallest value with difference is marked as 'c'. Subcutaneous (SAT) and visceral adipose tissue (VAT). [1] Fats in the diets were: (A) enriched in CLA; (B) enriched in ALA and low in CLA; (C) low in ALA and CLA; (D) high in LA and trace level of CLA.

3.2. Tissue Endocannabinoids and Congeners

An analysis of EC and congeners, OEA, and PEA, was performed in all tissues except plasma; these compounds were only marginally influenced by diet. Specifically, AEA in SAT was significantly decreased in rats fed diet A compared to those fed diet C, while OEA was significantly increased in A compared to B (Figure 1). In hypothalamus, 2-AG was significantly increased in A compared to B; AEA in liver was slightly reduced with diet A.

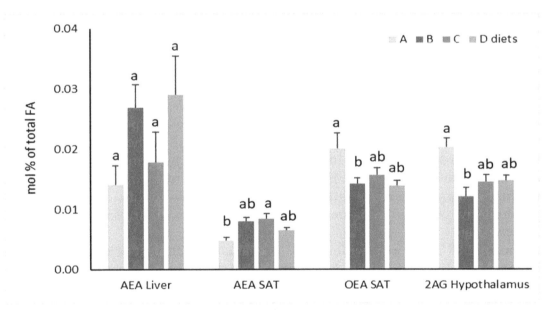

Figure 1. AEA, OEA, and 2-AG in Liver, subcutaneous adipose tissue (SAT), and hypothalamus from rats fed diets A, B, C, or D for 4 wk. Fats in the diets were: (A) enriched in CLA; (B) enriched in ALA and low in CLA; (C) low in ALA and CLA; D) high in LA and trace level of CLA. Values are expressed as mol% of total fatty acids and represent means ± SEMs, $n = 6$/group. Within a tissue, labelled means in a variable without a common superscript letter differ as determined by Tukey's post hoc test after a significant one-way ANOVA, $p < 0.05$; the maximum value is labeled as 'a', the smaller value with difference is marked as 'b'.

4. Discussion

The fatty acid composition of ruminant meat is strongly influenced by the diet of animals. In the present study, we compared three kinds of meat fats obtained from lambs under different feeding regimens: meat from light lambs fed only milk from grazing ewes, meat from heavy lambs maintained under a typical intensive feeding regimen based on cereal grains, integrated or not with rolled linseed as a source of ALA. It is well known that meat from suckling lambs is usually rich in CLA, especially when lactating ewes are fed on pasture [41]. The use of linseed in the diet of intensive rearing heavy lambs has been associated with increasing amounts of ALA in intramuscular fat. At the same time, the ratio n-6/n-3 FA is lower in the meat fat from lambs fed diets integrated with linseed compared to feeding regimens based on cereal grains [42].

The results of the present study suggest that the FA composition of dietary fat does not always anticipate its metabolic impact in tissues. This study confirms that the CLA naturally found in ruminant fat is able to significantly increase n-3 HUFA score. In fact, previously, we found that the intake of CLA naturally incorporated into 90 grams of enriched cheese for four weeks, or 50 grams for two months, significantly increased plasma DHA in humans, suggesting that amount and duration are key aspects of CLA intake to induce DHA biosynthesis [38]. In that study, we observed that enriched cheese intake increased PPAR-α gene expression, which is responsible for the induction of key enzymes of peroxisomal β-oxidation [19,48], which is involved in DHA biosynthesis [49]. Our data indicated that irrespective of the matrix, natural CLA is able to increase the n-3 HUFA score. Most of the effects attributed to the n-3 PUFA family are mainly related to dietary EPA and DHA, while ALA seems to have other beneficial effects which are unconnected to its putative property as precursor of EPA and DHA [50]. In fact, the biosynthesis of DHA requires a crucial step in peroxisome for a partial β-oxidation [51]. Moreover, ALA might act as inhibitor of Δ-6-desaturase, which is essential for DHA synthesis [52]. Therefore, any other event that increases desaturase activities and peroxisomal β-oxidation may also favor DHA biosynthesis.

Our results have shown that, in all tissues except in hypothalamus, ALA and CLA incorporation and metabolization is proportional to their concentration in the diet. On the other hand, DHA concentrations changed mostly depending on the relative amount of CLA and only slightly according to their parent availability in the diet. In our earlier human studies, we found that dietary CLA, in a specific, very low range of ALA/CLA ratio (1:3), was able to significantly increase n-3 HUFA biosynthesis [39], while in a previous study, dietary ALA/CLA in a ratio 11/3.2 failed to enhance DHA biosynthesis [53], suggesting that CLA rather than ALA is crucial to enhance DHA biosynthesis. Therefore, on the basis of the data available in the literature, it seems that CLA products could be an unexpected source of DHA. Diet A, enriched in CLA, contains more DHA and EPA than the other diets; however, the concentration of EPA and DHA are extremely low and, for example, while EPA has similar concentration in diets B and C, and about 1/3 lower than diet A, EPA levels in tissues of rats fed diets A and B are in general significantly higher than in tissues of rats fed the diet C. As a matter of fact, the n-3 UFA score was found to be similarly increased in tissues of rats fed diets A and B, irrespective of EPA and DHA differences in the diets.

The plasma n-3 HUFA score is widely used to evaluate the impact of a nutritional treatment on the balance of n-3/n-6 HUFAs [54]. Our data showed that the highest increase in n-3 HUFA score was induced by A and B meat fat diets which were enriched in ALA and CLA (Table 2).

Interestingly, in group A, characterized by high levels of CLA and an ALA/CLA ratio of 0.3, the increase of n-3 HUFA score was mainly attributed to an increase of DHA, while in diet B, it was characterized by a higher ALA/CLA ratio, i.e., 1.5. The n-3 HUFA score increase was due to a greater tissue concentration of EPA. Since data from human studies are usually limited to plasma analyses, we also evaluated changes in FA metabolism in different tissues.

Interestingly, the higher incorporation in tissues of DHA and CLA induced by diet A was associated with a significant increase of OEA and a concomitant decrease of AEA in some tissues like SAT (Figure 1). Dietary CLA supplementation may increase OEA levels in the livers of obese Zucker rats, possibly by activating PPAR-α [55], which may also contribute to the higher DHA biosynthesis via enhanced peroxisomal β-oxidation. These data are in agreement with what we previously found, i.e., that in obese rats, a diet enriched with n-3 HUFA resulted in the reduction of EC biosynthesis as a result of a decrease in their precursor concentration in membrane PLs, which may account for the reduction of ectopic fat and inflammatory mediators [43], and imply that DHA and CLA may exert a direct effect on EC and the biosynthesis of congeners. Conversely, diet B, with a relatively high dietary ratio ALA/CLA, may result in a lower PPAR-α activation, which may explain the significantly higher accumulation of VA in tissues due to a reduced peroxisomal β-oxidation. In fact, it has been demonstrated that trans FAs are preferentially β-oxidized in peroxisomes [56], which are regulated by PPAR-α [57].

The n-3 HUFA score showed a pattern similar to CLA or VA in liver and plasma, but not in the hypothalamus, as expected, which appeared to be more resistant to FA profile modification by dietary means, with early administration in life and duration of exposure, as well as dietary concentrations, being key factors in the detection of significant alterations [58]. Nevertheless, in hypothalamus, we found a significant increase in the AA concentration with the vegetable oil-based diet D, rich in LA, a precursor of AA; meanwhile, with diet A, enriched in CLA, we found an increase in the n-3 HUFA score. Our data suggest that the n-3 HUFA score change in the hypothalamus was not due to an increased biosynthesis of DHA and EPA in this tissue, but rather, to an increased transportation of these n-3 PUFAs from plasma. One can speculate that changes of FAs in peripheral tissue can directly influence FA concentrations in specific brain areas.

These modest changes in the hypothalamus were in the order of 15–20%, and may not be sufficient to exert significant effects on feeding behavior. Accordingly, the increase of 2-AG found in the hypothalamus of rats fed diet A was not associated with changes in food intake.
We previously demonstrated that CLA passes the blood brain barrier [59]; in the present experimental setting dietary, the CLA level was probably too low to be incorporated into the hypothalamus. Future studies should aim at evaluating whether dietary meat fat higher in CLA and/or longer feeding periods are able to modify hypothalamus CLA levels and influence feeding behavior through PPAR-α activation.

Another remarkable feature which may influence the n-3 HUFA score is the LA/ALA ratio, based on the concept that diets high in LA would inhibit the synthesis of n-3 HUFA by simple competitive inhibition [29,30]. In chickens, it has been found that when the LA/ALA ratio in the diet was above 5, liver PLs were rich in AA and poor in EPA; meanwhile, when the ratio dropped below 5, there was an exchange of AA for EPA [60]. The DHA status increased with a dietary level of ALA of around 1%en in rats, after which DHA accumulation was inhibited and then declined [60]. The LA/ALA ratio in our experimental diets was 2.1 (ALA 0.09%en) in A, 2.2 (ALA 0.21%en) in B, 7.0 (ALA 0.081%en) in C, and 22.0 (ALA 0.1%en) in D (Table 1). We found a significantly reduced n-3 HUFA score with C, and particularly with D in all tissues, while in hypothalamus, as reported, we observed an increase for diet A.
Since obesity is regarded as a low-grade chronic inflammatory condition characterized by increased proinflammatory cytokines in the white adipose tissue [61], it had been suggested that diets that can enhance n-3 PUFA could reduce the synthesis of PGE2 and enhance the production of PG involved in the resolution of inflammatory disorders [34]. We found that AIFAI index was significantly increased in the tissues of obese rats fed meat fat diets compared to obese rats fed vegetable fat diets (Table 4).

Interestingly, as observed in hypothalamus and in the other tissues, our data revealed that AA in SAT was significantly decreased with diets based on meat fat compared to those with vegetable fat. It is possible that increased CLA intake may interfere with the further metabolism of LA. We have previously seen that even though there was no perturbation in tissue LA, LA metabolites (including 18:3n-6, DGLA, and in particular AA) were consistently depressed in tissues by up to 1% CLA in the diet [11]. Consequently, CLA might further enhance the AIFAI index probably by inducing a decrease in AA and an increase in DHA biosynthesis. It is possible that CLA, or its relative metabolites, might differentially modulate the distribution of AA in various PLs [62], competing with AA for incorporation; therefore, this scenario may affect the eicosanoid signaling mechanism.

The consumption of meat fat rich in CLA-ALA resulted in significantly increased accumulation of

DHA and depression of AA synthesis, which may have therapeutic potential to ameliorate clinical symptoms and complications that are secondary to the excessive production of proinflammatory mediators. Our data clearly indicate that metabolic changes by dietary FAs seem to be tissue specific and affected by other factors such as background diet, energy and lipid metabolism. However, in our model of obesity, we didn't find any changes in parameters of metabolic syndrome such as dyslipidemia or fatty liver, probably due to the relative short-term feeding period or to the relatively low CLA concentration.

5. Conclusions

Our data put in evidence that the feeding system of livestock may play an important role in modulating the effect of meat fat on lipid metabolism, as some FAs, like CLA and ALA, improve the tissue FA profile, as shown with the increased n-3 HUFA score. While not being comparable to the direct intake of EPA and DHA through fish products, it seems that meat that is naturally enriched with CLA could be an unexpected source of DHA, provided that a specific ratio of FAs in the pool of total FAs is respected.

These results are promising, especially regarding individuals for whom the intake of fish products is quite low, i.e., far below the recommended daily dose. Future studies are envisaged to evaluate whether dietary fat of different origin and composition is able to modify these parameters in humans.

Author Contributions: S.B., M.M., G.C.: conceived the study, participated in its design; G.C., E.M.: carried out the fatty acid and endocannabinoid and congeners analysis; G.C., S.B., E.M.: performed the statistical analyses, interpreted the data; G.C.: drafted the manuscript; G.C., E.M., C.M., A.S.: contributed to the intellectual content; and all authors: were involved in conducting the research and critically evaluated and revised the final manuscript, read and approved the final manuscript.

References

1. Ulbricht, T.L.; Southgate, D.A. Coronary heart disease: Seven dietary factors. *Lancet* **1991**, *338*, 985–992. [CrossRef]
2. Parodi, P.W. Conjugated linoleic acid and other anticarcinogenic agents of bovine milk fat. *J. Dairy Sci.* **1999**, *82*, 1339–1349. [CrossRef]
3. Ip, C.; Jiang, C.; Thompson, H.J.; Scimeca, A.J. Retention of conjugated linoleic acid in the mammary gland is associated with tumour inhibition during the post initiation phase of carcinogenesis. *Carcinogenesis* **1997**, *18*, 755–759. [CrossRef] [PubMed]
4. Wood, J.D.; Enser, M.; Fisher, A.V.; Nute, G.R.; Sheard, P.R.; Richardson, R.I.; Hughes, S.I.; Whittington, F.M. Fat deposition, fatty acid composition and meat quality: A review. *Meat Sci.* **2008**, *78*, 343–358. [CrossRef] [PubMed]
5. Demirel, G.; Wachira, A.; Sinclair, L.; Wilkinson, R.; Wood, J.; Enser, M. Effects of dietary n-3 polyunsaturated fatty acids, breed and dietary vitamin E on the fatty acids of lamb muscle, liver and adipose tissue. *Br. J. Nutr.* **2004**, *91*, 551–565. [CrossRef] [PubMed]
6. Mele, M. Designing milk fat to improve healthfulness and functional properties of dairy products: From feeding strategies to a genetic approach. *Ital. J. Anim. Sci.* **2009**, *8*, 365–374. [CrossRef]
7. Griinari, J.M.; Corl, B.A.; Lacy, S.H.; Chouinard, P.Y.; Nurmela, K.V.V.; Bauman, D.E. Conjugated linoleic acid is synthesized endogenously in lactating dairy cows by D desaturase. *J. Nutr.* **2000**, *130*, 2285–2291. [CrossRef]
8. Griinari, J.M.; Bauman, D.E. Biosynthesis of conjugated linoleic acid and its incorporation in meat and milk in ruminants. In *Advances in Conjugated Linoleic Acid Research*; Yurawecz, M.P., Mossoba, M.M., Kramer, J.K.G., Pariza, M.W., Nelson, J.G., Eds.; AOCS Press: Champain, IL, USA, 1999; Volume 1, pp. 180–200.
9. Banni, S.; Martin, J.C. Conjugated linoleic acid and metabolites. In *Trans Fatty Acids in Human Nutrition*; Sebedio, J.L., Christie, W.W., Eds.; The Oily Press: Dundee, UK, 1998; Volume 42, pp. 261–302.

10. Kim, J.H.; Kim, Y.; Kim, Y.J.; Park, Y. Conjugated linoleic acid-potential health benefits as a functional food ingredient. *Annu. Rev. Food Sci. T.* **2016**, *7*, 221–244. [CrossRef]

11. Banni, S.; Angioni, E.; Casu, V.; Melis, M.P.; Carta, G.; Corongiu, F.P.; Thompson, H.; Ip, C. Decrease in linoleic acid metabolites as a potential mechanism in cancer risk reduction by conjugated linoleic acid. *Carcinogenesis* **1999**, *20*, 1019–1024. [CrossRef]

12. Banni, S.; Petroni, A.; Blasevich, M.; Carta, G.; Cordeddu, L.; Murru, E.; Melis, M.; Mahon, A.; Belury, M. Conjugated linoleic acids (CLA) as precursors of a distinct family of PUFA. *Lipids* **2004**, *39*, 1143–1146. [CrossRef]

13. Nudda, A.; McGuire, M.A.; Battacone, G.; Pulina, G. Seasonal variation in conjugated linoleic acid and vaccenic acid in milk fat of sheep and its transfer to cheese and ricotta. *J. Dairy Sci.* **2005**, *88*, 1311–1319. [CrossRef]

14. Lawson, R.E.; Moss, A.R.; Givens, D.I. The role of dairy products in supplying conjugated linoleic acid to man's diet: A review. *Nutr. Res. Rev.* **2001**, *14*, 153–172. [CrossRef] [PubMed]

15. Nudda, A.; Palmquist, D.L.; Battacone, G.; Fancellu, S.; Rassu, S.P.G.; Pulina, G. Relationships between the contents of vaccenic acid, CLA and n-3 fatty acids of goat milk and the muscle of their suckling kids. *Livest. Sci.* **2008**, *118*, 195–203. [CrossRef]

16. Boles, J.A.; Kott, R.W.; Hateld, P.G.; Bergman, J.W.; Flynn, C.R. Supplemental safflower oil affects the fatty acid profile including conjugated linoleic acid of lamb. *J. Anim. Sci.* **2005**, *83*, 2175–2181. [CrossRef] [PubMed]

17. Mele, M.; Contarini, G.; Cercaci, L.; Serra, A.; Buccioni, A.; Povolo, M.; Conte, G.; Funaro, A.; Banni, S.; Lercker, G.; et al. Enrichment of Pecorino cheese with conjugated linoleic acid by feeding dairy ewes with extruded linseed: Effect on fatty acid and triglycerides composition and on oxidative stability. *Int. Dairy J.* **2011**, *21*, 365–372. [CrossRef]

18. Banni, S.; Day, B.W.; Evans, R.W.; Corongiu, F.P.; Lombardi, B. Detection of conjugated diene isomers of linoleic acid in liver lipids of rats fed a choline-devoid diet indicates that the diet does not cause lipoperoxidation. *J. Nutr. Biochem.* **1995**, *6*, 281–289. [CrossRef]

19. Moya-Camarena, S.Y.; Vanden Heuvel, J.P.; Blanchard, S.G.; Leesnitzer, L.A.; Belury, M.A. Conjugated linoleic acid is a potent naturally occurring ligand and activator of PPARalpha. *J. Lipid Res.* **1999**, *40*, 1426–1433.

20. Rakhshandehroo, M.; Knoch, B.; Muller, M.; Kersten, S. Peroxisome proliferatoractivated receptor alpha target genes. *PPAR Res.* **2010**, *2010*, 612089. [CrossRef]

21. Calder, P.C. Marine omega-3 fatty acids and inflammatory processes: Effects, mechanisms and clinical relevance. *Biochim. Biophys. Acta* **2015**, *1851*, 469–484. [CrossRef]

22. Flachs, P.; Rossmeisl, M.; Kopecky, J. The effect of n-3 fatty acids on glucose homeostasis and insulin sensitivity. *Physiol. Res.* **2014**, *63*, S93–S118.

23. Burdge, G.C.; Calder, P.C. Conversion of alpha-linolenic acid to longer-chain polyunsaturated fatty acids in human adults. *Reprod. Nutr. Dev.* **2005**, *45*, 581–597. [CrossRef]

24. Watkins, B.A.; Li, Y.; Hennig, B.; Toborek, M. Dietary lipids and health. In *Bailey's industrial oil and fat products. Edible oil and fat products: Chemistry, chemical properties, and health effects*, 6th ed.; Shahidi, F., Ed.; Wiley: Hoboken, NJ, USA, 2005.

25. Albert, C.M.; Hennekens, C.; O'Donnell, C.J.; Ajani, U.A.; Carey, V.J.; Willett, W.C.; Ruskin, J.N.; Manson, J.E. Fish consumption and risk of sudden cardiac death. *JAMA* **1998**, *279*, 23–28. [CrossRef] [PubMed]

26. Blank, C.; Neumann, M.A.; Makrides, M.; Gibson, R.A. Optimizing DHA levels in by lowering the linoleic acid to a-linolenic acid ratio. *J. Lipid Res.* **2002**, *43*, 1537–1543. [CrossRef] [PubMed]

27. Bowen, R.A.; Wierzbicki, A.A.; Clandinin, M.T. Does increasing dietary linolenic acid content increase the docosahexaenoic acid content of phospholipids in neuronal cells of neonatal rats. *Pediatr. Res.* **1999**, *45*, 505–516. [CrossRef] [PubMed]

28. Woods, J.; Ward, G.; Salem, N., Jr. Is docosahexaenoic acid necessary in infant formula? Evaluation of high linolenate diets in the neonatal rat. *Pediatr. Res.* **1996**, *40*, 687–694. [CrossRef]

29. Lands, B. A critique of paradoxes in current advice on dietary lipids. *Prog. Lipid. Res.* **2008**, *47*, 77–106. [CrossRef]

30. Lands, W.E.; Morris, A.; Libelt, B. Quantitative effects of dietary polyunsaturated fats on the composition of fatty acids in rat tissues. *Lipids* **1990**, *25*, 506–516. [CrossRef]

31. Stark, K.D. The percentage of n-3 highly unsaturated fatty acids in total HUFA as a biomarker for omega-3 fatty acid status in tissues. *Lipids* **2008**, *43*, 45–53. [CrossRef]

32. Chavali, S.R.; Zhong, W.W.; Utsunomiya, T.; Forse, R.A. Decreased production of interleukin-1-beta, prostaglandin-E2 and thromboxane-B2, and elevated levels of interleukin-6 and -10 are associated with increased survival during endotoxic shock in mice consuming diets enriched with sesame seed oil supplemented with Quil-A saponin. *Int. Arch. Allergy Immunol.* **1997**, *114*, 153–160. [CrossRef]

33. Grimstad, T.; Berge, R.K.; Bohov, P.; Skorve, J.; Goransson, L.; Omdal, R.; Aasprong, O.G.; Haugen, M.; Meltzer, H.M.; Hausken, T. Salmon diet in patients with active ulcerative colitis reduced the simple clinical colitis activity index and increased the anti-inflammatory fatty acid index–a pilot study. *Scand. J. Clin. Lab. Invest.* **2011**, *71*, 68–73. [CrossRef]

34. Utsunomiya, T.; Chavali, S.R.; Zhong, W.W.; Forse, R.A. Effects of sesamin-supplemented dietary fat emulsions on the ex vivo production of lipopolysaccharide-induced prostanoids and tumor necrosis factor a in rats. *Am. J. Clin. Nutr.* **2000**, *72*, 804–808. [CrossRef] [PubMed]

35. Duplus, E.; Glorian, M.; Forest, C. Fatty acid regulation of gene transcription. *J. Biol. Chem.* **2000**, *275*, 30749–30752. [CrossRef] [PubMed]

36. Jump, D.B. Dietary polyunsaturated fatty acids and regulation of gene transcription. *Curr. Opin. Lipidol.* **2002**, *13*, 155–164. [CrossRef] [PubMed]

37. Belury, M.A.; Mahon, A.; Banni, S. The conjugated linoleic acid (CLA) isomer, t10c12-CLA, is inversely associated with changes in body weight and serum leptin in subjects with type 2 diabetes mellitus. *J. Nutr.* **2003**, *133*, 257S–260S. [CrossRef] [PubMed]

38. Murru, E.; Carta, G.; Cordeddu, L.; Melis, M.P.; Desogus, E.; Ansar, H.; Chilliard, Y.; Ferlay, A.; Stanton, C.; Coakley, M.; et al. Dietary Conjugated Linoleic Acid-Enriched Cheeses Influence the Levels of Circulating n-3 Highly Unsaturated Fatty Acids in Humans. *Int. J. Mol. Sci.* **2018**, *19*, 1730. [CrossRef] [PubMed]

39. Pintus, S.; Murru, E.; Carta, G.; Cordeddu, L.; Batetta, B.; Accossu, S.; Pistis, D.; Uda, S.; Elena Ghiani, M.; Mele, M.; et al. Sheep cheese naturally enriched in alpha-linolenic, conjugated linoleic and vaccenic acids improves the lipid profile and reduces anandamide in the plasma of hypercholesterolaemic subjects. *Br. J. Nutr.* **2013**, *109*, 1453–1462. [CrossRef]

40. Bray, G.A. The Zucker-fatty rat: A review. *Fed. Proc.* **1977**, *36*, 148–153.

41. Serra, A.; Mele, M.; La Comba, F.; Conte, G.; Buccioni, A.; Secchiari, P. Conjugated Linoleic Acid (CLA) content of meat from three muscles of Massese suckling lambs slaughtered at different weights. *Meat Sci.* **2009**, *81*, 396–404. [CrossRef]

42. Mele, M.; Serra, A.; Pauselli, M.; Luciano, G.; Lanza, M.; Pennisi, P.; Conte, G.; Taticchi, A.; Esposto, S.; Morbidini, L. The use of stoned olive cake and rolled linseed in the diet of intensively reared lambs: Effect on the intramuscular fatty-acid composition. *Animal* **2014**, *8*, 152–162. [CrossRef]

43. Folch, J.; Lees, M.; Sloane Stanley, G.H. A simple method for the isolation and purification of total lipides from animal tissues. *J. Biol. Chem.* **1957**, *226*, 497–509.

44. Banni, S.; Carta, G.; Contini, M.S.; Angioni, E.; Deiana, M.; Dessi, M.A.; Melis, M.P.; Corongiu, F.P. Characterization of conjugated diene fatty acids in milk, dairy products, and lamb tissues. *J. Nutr. Biochem.* **1996**, *7*, 150–155. [CrossRef]

45. Melis, M.P.; Angioni, E.; Carta, G.; Murru, E.; Scanu, P.; Spada, S.; Banni, S. Characterization of conjugated linoleic acid and its metabolites by RP-HPLC with diode array detector. *Eur. J. Lipid Sci. Tech.* **2001**, *103*, 617–621. [CrossRef]

46. Batetta, B.; Griinari, M.; Carta, G.; Murru, E.; Ligresti, A.; Cordeddu, L.; Giordano, E.; Sanna, F.; Bisogno, T.; Uda, S.; et al. Endocannabinoids may mediate the ability of (n-3) fatty acids to reduce ectopic fat and inflammatory mediators in obese Zucker rats. *J. Nutr.* **2009**, *139*, 1495–1501. [CrossRef] [PubMed]

47. Piscitelli, F.; Carta, G.; Bisogno, T.; Murru, E.; Cordeddu, L.; Berge, K.; Tandy, S.; Cohn, J.S.; Griinari, M.; Banni, S.; et al. Effect of dietary krill oil supplementation on the endocannabinoidome of metabolically relevant tissues from high-fat-fed mice. *Nutr. Metab. (Lond.)* **2011**, *8*, 51. [CrossRef] [PubMed]

48. Belury, M.A.; Moya-Camarena, S.Y.; Liu, K.L.; Vanden Heuvel, J.P. Dietary conjugated linoleic acid induces peroxisome-specific enzyme accumulation and ornithine decarboxylase activity in mouse liver. *J. Nutr. Biochem.* **1997**, *8*, 579–584. [CrossRef]

49. Ferdinandusse, S.; Denis, S.; Dacremont, G.; Wanders, R.J. Studies on the metabolic fate of n-3 polyunsaturated fatty acids. *J. Lipid Res.* **2003**, *44*, 1992–1997. [CrossRef]

50. Murru, E.; Banni, S.; Carta, G. Nutritional properties of dietary omega-3-enriched phospholipids. *Biomed. Res. Int.* **2013**, *2013*, 965417. [CrossRef]

51. Burdge, G.C. Metabolism of alpha-linolenic acid in humans. *Prostaglandins Leukot. Essent. Fatty Acids* **2006**, *75*, 161–168. [CrossRef]

52. Gibson, R.A.; Neumann, M.A.; Lien, E.L.; Boyd, K.A.; Tu, W.C. Docosahexaenoic acid synthesis from alpha-linolenic acid is inhibited by diets high in polyunsaturated fatty acids. *Prostaglandins Leukot. Essent. Fatty Acids* **2013**, *88*, 139–146. [CrossRef]

53. Attar-Bashi, N.M.; Weisinger, R.S.; Begg, D.P.; Li, D.; Sinclair, A.J. Failure of conjugated linoleic acid supplementation to enhance biosynthesis of docosahexaenoic acid from a-linolenic acid in healthy human volunteers. *Prostaglandins Leukot. Essent. Fatty Acids* **2007**, *76*, 121–130. [CrossRef]

54. Harris, W.S.; Von Schacky, C. The Omega-3 Index: A new risk factor for death from coronary heart disease? *Prev. Med.* **2004**, *39*, 212–220. [CrossRef] [PubMed]

55. Melis, M.; Carta, G.; Pistis, M.; Banni, S. Physiological Role of Peroxisome Proliferator-Activated Receptors Type Alpha on Dopamine Systems. *Cns Neurol. Disord-Dr* **2013**, *12*, 70–77. [CrossRef] [PubMed]

56. Thomassen, M.S.; Christiansen, E.N.; Norum, K.R. Characterization of the stimulatory effect of high-fat diets on peroxisomal beta-oxidation in rat liver. *Biochem. J.* **1982**, *206*, 195–202. [CrossRef] [PubMed]

57. Reddy, J.K.; Hashimoto, T. Peroxisomal beta-oxidation and peroxisome proliferator-activated receptor alpha: An adaptive metabolic system. *Annu. Rev. Nutr.* **2001**, *21*, 193–230. [CrossRef]

58. Rapoport, S.I.; Rao, J.S.; Igarashi, M. Brain metabolism of nutritionally essential polyunsaturated fatty acids depends on both the diet and the liver. *Prostaglandins Leukot. Essent. Fatty Acids* **2007**, *77*, 251–261. [CrossRef]

59. Fa, M.; Diana, A.; Carta, G.; Cordeddu, L.; Melis, M.P.; Murru, E.; Sogos, V.; Banni, S. Incorporation and metabolism of c9,t11 and t10,c12 conjugated linoleic acid (CLA) isomers in rat brain. *Biochim. Biophys. Acta* **2005**, *1736*, 61–66. [CrossRef]

60. Gibson, R.A.; Muhlhausler, B.; Makrides, M. Conversion of linoleic acid and alpha-linolenic acid to long-chain polyunsaturated fatty acids (LCPUFAs), with a focus on pregnancy, lactation and the first 2 years of life. *Matern. Child Nutr.* **2011**, *7* (Suppl. 2), 17–26. [CrossRef]

61. Balistreri, C.R.; Caruso, C.; Candore, G. The role of adipose tissue and adipokines in obesity-related inflammatory diseases. *Mediators Inflamm.* **2010**, *2010*, 802078. [CrossRef]

62. Ip, C.; Thompson, H.J.; Ganther, H.E. Selenium modulation of cell proliferation and cell cycle biomarkers in normal and premalignant cells of the rat mammary gland. *Cancer Epidem. Biomar.* **2000**, *9*, 49–54.

7

Long-Chain Omega-3 Polyunsaturated Fatty Acids in Natural Ecosystems and the Human Diet

Michail I. Gladyshev [1,2,*] **and Nadezhda N. Sushchik** [1,2]

[1]   Institute of Biophysics of Siberian Branch of Russian Academy of Sciences, Akademgorodok, 50/50, Krasnoyarsk 660036, Russia; labehe@ibp.ru
[2]   Siberian Federal University, Svobodny av. 79, Krasnoyarsk 660041, Russia
*   Correspondence: glad@ibp.ru

Abstract: Over the past three decades, studies of essential biomolecules, long-chain polyunsaturated fatty acids of the omega-3 family (LC-PUFAs), namely eicosapentaenoic acid (20:5n-3, EPA) and docosahexaenoic acid (22:6n-3, DHA), have made considerable progress, resulting in several important assumptions. However, new data, which continue to appear, challenge these assumptions. Based on the current literature, an attempt is made to reconsider the following assumptions: 1. There are algal classes of high and low nutritive quality. 2. EPA and DHA decrease with increasing eutrophication in aquatic ecosystems. 3. Animals need EPA and DHA. 4. Fish are the main food source of EPA and DHA for humans. 5. Culinary treatment decreases EPA and DHA in products. As demonstrated, some of the above assumptions need to be substantially specified and changed.

Keywords: eicosapentaenoic acid; docosahexaenoic acid; nutritive quality; eutrophication; fish; culinary treatments

---

1. Introduction

Polyunsaturated fatty acids in the omega-3 family (PUFAs) are a focus of interest in many fields of science: biochemistry, physiology, dietetics, pharmacology, agriculture, aquaculture, ecology, etc. [1–7]. For many animals and humans, long-chain polyunsaturated fatty acids (LC-PUFAs), namely eicosapentaenoic acid (20:5n-3, EPA) and docosahexaenoic acid (22:6n-3, DHA), are precursors of signaling molecules (bioactive lipid mediators) and essential components of cell membranes in neural and muscle tissues [8–11]. The number of publications on EPA and DHA in scientific journals has substantially increased since the 1970s (Figure 1). It is impossible to equally review LC-PUFA studies in all the fields of science; therefore, we will primarily take into consideration environmental issues, because natural and agricultural ecosystems are the source of EPA and DHA for human nutrition. There are many papers that report the fatty acid (FA) profiles of diverse microorganisms, plants, and animals, and address EPA and DHA as components of these profiles. However, we will consider only those studies that emphasize the role of these LC-PUFAs as well as their precursors.

In the field of ecology, even when studies have focused on primary producers (microalgae), the importance of EPA and DHA for human health was the rationale for the study of the content and composition of LC-PUFAs [12]. Indeed, only some taxa of microalgae can synthesize large amounts of EPA and DHA, while animals, including humans, have a comparatively low ability for such synthesis via conversion of the precursor, a short-chain PUFA, alpha-linolenic acid (18:3n-3, ALA) [13–15]. ALA, which is synthesized by plants, can be obtained by most animals only from food [16,17]. It should be noted that in contrast to EPA and DHA, which are synthesized by some algae, ALA is synthesized by terrestrial vascular plants and is the main component of the photosynthetic membranes of chloroplasts [18–21]. Since EPA and DHA can be efficiently synthesized *de novo* only by some taxa of algae, aquatic ecosystems are recognized as the main source of these LC-PUFAs in the biosphere [22]. The algae-synthesized EPA and DHA are transferred through trophic chains to organisms at higher trophic levels, invertebrates, and fish, and then to terrestrial consumers, including humans.

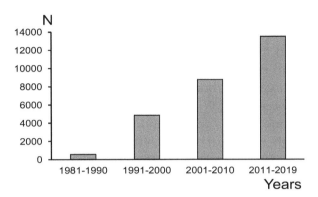

Figure 1. Number of publications containing (*N*) the terms 'eicosapentaenoic' or 'docosahexaenoic' in the Web of Science Core Collection during the last four decades.

In studies of EPA and DHA transfer from microalgae to humans, which inevitably included the culinary treatment of aquatic products for human nutrition, considerable progress has been made in recent decades. Many important findings have been summarized in several keynote statements [1,2,6,12,22], and a number of assumptions have appeared. New data appear continuously, and new questions arise, which naturally challenge some parts of the former assumptions. The aim of this paper is to consider and specify, if necessary, some important assumptions in the field of LC-PUFA production and regarding LC-PUFA transfer from natural ecosystems to the human diet.

2. Assumption 1: There Are Algal Classes of High and Low Nutritive Quality

In the nutrition ecology of zooplankton and other microalgaevorous organisms, microalgae have been subdivided into groups representing good and poor nutritive value according to their size and shape, and later to their carbon:nitrogen:phosphorus (C:N:P) ratio [23]. Starting with the milestone work of Ahlgren et al. [24], the PUFA content of microalgae became an important indicator of the nutritive quality of algae for consumers. In this work, it was found that Cryptophyceae and Dinophyceae (Peridinea) had high levels of EPA and DHA, and were the best food for zooplankton [24]. In turn, Chlorophyceae (green algae) and cyanobacteria contained no LC-PUFAs but certain levels of PUFAs, namely ALA, and had comparatively lower nutritive value [24]. Then, in another milestone paper, Muller-Navarra [25] demonstrated that Bacillariophyceae (diatoms), which had a high content of EPA, were a higher-quality food source for *Daphnia* than chlorophytes.

Since then, depending on their content of LC-PUFAs, microalgae have been subdivided into classes representing high and low nutritive quality, Cryptophyceae, Dinophyceae, and Bacillariophyceae vs. Chlorophyceae and cyanobacteria, respectively [26]. Some other classes of microalgae, Eustigmatophyceae, Prymnesiophyceae, Prasinophyceae, Chrysophyceae (golden algae), and Euglenophyceae, have also been shown to have a high content (mg g^{-1} C) of EPA and especially DHA [27–30]. Based on FA percentages, Trebouxiophyceae and Raphidophyceae were classified as intermediate and excellent food resources for zooplankton, respectively [31]. Recently, the division of phytoplankton classes into four categories was suggested based on the content of several biomolecules that included EPA and DHA: poor, medium, high, and superior quality food [30].

On the one hand, the division of microalgae into classes of high and low nutritive value in terms of their EPA and DHA content is reasonable. Indeed, Chlorophyceae and cyanobacteria (blue-green algae) do not produce these LC-PUFAs [32,33]. On the other hand, species from the classes lacking LC-PUFAs can have a high content of ALA, as mentioned above, and thereby provide a high growth rate for consumers that can efficiently convert ALA to EPA and DHA (e.g., [34–37] see also Assumption 3 below). For instance, there were no statistically significant differences among the growth rates of *Daphnia magna* when feeding on chlorophytes, chrysophytes, and diatoms [30]. The cited authors emphasized that the reproduction of *D. magna* in feeding experiments was dependent on total n-3 FAs rather than only on EPA.

Moreover, within algal classes that can synthesize EPA and/or DHA, there are species with low contents of these LC-PUFAs and thereby with low nutritive value. Indeed, EPA levels (as a percent of total FA) in 17 marine diatom species used in aquaculture ranged from 5 to 30% [38]. The EPA content in diatoms per gram of organic carbon varied from 1.7 mg g^{-1} C in *Cyclotella meneghiniana* [30] to 45.9 mg g^{-1} C in *Thalassiosira oceanica* [39]. Moreover, in addition to inter-species variability, high variation in EPA content within one species occurs. Indeed, despite the lowest value for *C. meneghiniana* given above, a considerably higher content of EPA, 40.8 mg g^{-1} C, was reported for this species [25]. In natural ecosystems, it has been shown that some marine plankton diatoms have little EPA [40]. In a freshwater reservoir, *Cyclotella* was not associated with the EPA content in seston, while there was a significant correlation between *Stephanodiscus* and the content of EPA [41]. Similar results demonstrating contrasting levels of EPA in different diatom taxa were obtained for river littoral epilithic microalgae [42]. Evidently, diatom species can differ strongly in EPA content, and the common point of view that all diatoms are the superior quality food should be revised.

Consequently, dividing microalgae on the basis of their LC-PUFA content into classes appears to be too coarse for assessing nutritional value for consumers. It is worth noting that 'nutritional value' is not characteristic of a food item only, but indicates the demands of the consumer as well. Indeed, if a consumer does not need considerable amounts of EPA and DHA (see also Assumption 3 below), the above subdivision of microalgae into classes of high and low nutritive value is irrelevant. As found in a study of a freshwater plankton community, "there were no phytoplankton species of clearly high or low nutritive value. All phytoplankters, or at least detritus, that originated from them may meet the specific elemental and biochemical requirements of specific groups of zooplankton" [43]. Thus, Assumption 1 should be improved in future studies.

3. Assumption 2: EPA and DHA Decrease with the Increasing Eutrophication of Aquatic Ecosystems

The primary producers of EPA and DHA, Bacillariophyceae, Cryptophyceae and Dinophyceae are known to mainly inhabit oligotrophic aquatic ecosystems with low concentrations of total phosphorus (TP), while eutrophic (high TP) aquatic ecosystems are dominated by green algae and cyanobacteria, which do not produce LC-PUFAs (e.g., [44,45]). Thus, a high TP concentration decreases the contents of EPA and DHA in seston due to an increase of cyanobacteria, and thereby decreases the transfer of these LC-PUFAs to higher trophic levels [45,46]. Indeed, the EPA(DHA)-to-carbon content ratio in lake seston had a statistically significant negative relationship with TP concentration in lake water [46], and the EPA + DHA content in perch (per unit muscle mass) also had a statistically significant negative relationship with lake TP [45]. Thus, the nutritive quality of fish for humans becomes lower with eutrophication [45]. An increase in lake eutrophication, measured via the chlorophyll *a* concentration, also resulted in a significant decrease in the EPA content in bighead carp [47].

However, in the above publications, the relative content of LC-PUFAs per unit of organic carbon, C, or fish mass was estimated. These measures of relative content definitely indicate food quality, namely, the nutritive value of seston for zooplankton [46] and fish for humans [45,47]. In addition, if the quantification of the LC-PUFA supply for human nutrition is to be regarded as the paramount aim of relevant ecological studies, it is necessary to quantify this supply as the EPA and DHA yield, $Y_{\text{LC-PUFA}}$ value, which has the units µg or mg of LC-PUFA per m^2 or m^3 in an aquatic ecosystem per day or year. Thus, to convert food quality, i.e., the relative content of LC-PUFAs (µg mg C^{-1}), to the quantity, $Y_{LC-PUFA}$ ($\mu g\ m^{-3}\ day^{-1}$), it is necessary to take into account the production of organic carbon in an aquatic ecosystem, V ($mg\ C\ m^{-3}\ day^{-1}$):

$$Y_{LC-PUFA} = LC\text{-}PUFA \cdot V. \qquad (1)$$

For instance, Muller-Navarra et al. [46] provided the following relation of the content of EPA ($\mu g\ mg^{-1}\ C$) in lake seston to the concentration of total phosphorus (TP; $\mu g\ L^{-1}$) in lake water:

$$\ln EPA = -0.69 \cdot \ln TP + 2.78. \qquad (2)$$

From Equation (2), the following dependence can be obtained:

$$EPA = e^{(-0.69 \cdot \ln TP + 2.78)} = 16.12 \cdot TP^{-0.69}. \qquad (3)$$

The dependence of the rate of photosynthesis (primary production), V ($mg\ C\ m^{-3}\ day^{-1}$), in lake phytoplankton on TP was given in [48]:

$$V = 10.4 \cdot TP - 79. \qquad (4)$$

To determine the value of the yield of EPA, Y_{EPA} ($\mu g\ m^{-3}\ day^{-1}$) (Equation (1)), i.e., its amount produced in a lake, it is necessary to multiply Equation (3) by Equation (4):

$$Y_{EPA} = 16.12 \cdot TP^{-0.69} \cdot (10.4\ TP - 79) = 167.6 \cdot TP^{0.31} - 1273.5 \cdot TP^{-0.69}. \qquad (5)$$

Graphs of Equations (3)–(5) are given in Figure 2. Evidently, with a decrease in the relative content of EPA in seston, there is an increase in EPA yield with increasing TP, i.e., with increasing lake eutrophication. The same is true for DHA if the analogue for Equation (1) is taken from Muller-Navarra et al. [46].

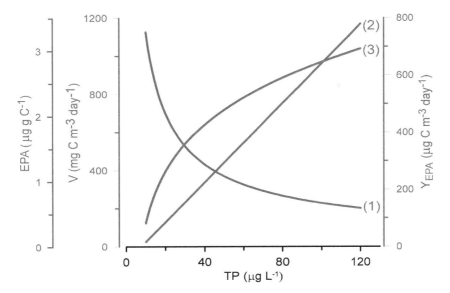

Figure 2. Dependence of the content of eicosapentaenoic acid in lake seston, EPA (1), rate of photosynthesis (primary production), V (2), and the yield of EPA, Y_{EPA} (3), on the total phosphorus concentration in water, TP.

It is worth noting that according to Equation (4), the LC-PUFA yield produced by microalgae (lake phytoplankton) increases gradually with increasing TP (Figure 2). This relationship contradicts taxonomic changes in phytoplankton that usually occur during eutrophication. If primary production in a water body increases only at the expense of cyanobacterial or green algal growth, there would be no increase in LC-PUFA yield. In contrast, the EPA + DHA yield obtained from lakes where perch were caught increased up to a TP concentration of ~40 μg L^{-1} and then decreased with increasing eutrophication [49]. Thus, the maximum LC-PUFA yield (g km^{-2} year^{-1}) associated with fish catches occurred in mesotrophic rather than oligotrophic or eutrophic aquatic ecosystems [49].

Moreover, in eutrophic temperate lakes, cyanobacteria dominate in the phytoplankton community only in summer, while in spring, at low water temperatures, "blooms" of psychrophilic diatoms with high contents of EPA often occur. The above equation describing the relation between EPA and TP (Equation (2)) was obtained by Muller-Navarra et al. [46] for summer phytoplankton only. However, if a whole year is taken into consideration, in a eutrophic reservoir, a spring "bloom" of psychrophilic diatoms can produce a large pool of EPA, which can then be transferred through the trophic chain and peak in zooplankton and fish biomass with a time lag during summer [50]. It should also be noted that with decreasing phosphate availability, the proportion of EPA in some freshwater algae significantly decreased [51].

Thus, the common assumption that EPA and DHA decrease with increasing eutrophication of aquatic ecosystems should be improved. Such a decrease occurs in the relative contents of these LC-PUFAs per unit of sestonic organic carbon or fish biomass. However, with regard to EPA and DHA yield, including that available for human nutrition, measured as LC-PUFA quantity per unit area or volume of an aquatic ecosystem per unit time, such a decrease may not occur. At present, mesotrophic aquatic ecosystems are believed to provide a maximum supply of EPA + DHA for humans via fish catches. Ranking aquatic ecosystems on the basis of their ability to produce LC-PUFAs for human nutrition is an important challenge for future research.

4. Assumption 3: Animals Need EPA and DHA

As mentioned above, EPA and DHA play important biochemical and physiological roles in humans and many animals, and the common point of view is that all vertebrates and most invertebrate groups require these LC-PUFAs [52]. The low ability of animals to synthesize EPA and DHA from short-chain ALA necessitates them to obtain these LC-PUFAs from food. However, the commonly used terms of "many animals" and "most invertebrates" as well as "low ability" and "necessity" have not been specified or quantified yet. Furthermore, the hypothesis that in natural ecosystems "many" animals can be limited by a low EPA and DHA supply is the important premise of a number of studies [22,52]. Thus, the specification and quantification of the above terms seems to be an important challenge for relevant ecological studies [22,53].

First, it should be noted that a large group of invertebrates, terrestrial insects, practically do not have EPA and DHA [54–59]. Indeed, terrestrial insects use EPA only as the precursor of lipid mediators, eicosanoids, and thereby synthesize it from consumed ALA in small quantities, at the level of vitamins [60]. In contrast to terrestrial insects, aquatic (amphibiotic) insects have high levels of EPA in their biomass, but contain very low, if any, DHA [61–64]. As mentioned above, DHA is the main component of the phospholipids of the cell membranes of vertebrate neural tissues, including retinal photoreceptors [3,4,65,66] (Figure 3). However, instead of DHA, there are 18C PUFAs in the eyes of terrestrial insects [3,67] (Figure 3), and EPAs in the eyes of amphibiotic insects [68] (Figure 3). Thus, terrestrial insects evidently do not need EPA or DHA, and aquatic insects do not need DHA in their food or in their biomass in considerable amounts.

Other terrestrial invertebrates, earthworms (*Lumbricus terrestris*), likely need EPA since they have a comparatively high content of this LC-PUFA in their biomass [69]. However, earthworms do not need to obtain these biomolecules from their food, since they likely obtain EPA from their gut microflora [69]. According to our unpublished data obtained using GC-MS and internal standards as in [61,70], Californian worms (*Eisenia foetida*) from a laboratory culture [71] contain 0.37 ± 0.02 and 0.02 ± 0.02 (n = 3) mg g^{-1} WW EPA and DHA, respectively. Moreover, some species of soil nematodes can *de novo* synthesize omega-3 PUFAs, namely ALA and EPA [72].

Herbivorous terrestrial vertebrates that consume green parts of plants can satisfy their physiological needs for EPA and DHA through the conversion of ALA [5,15,73]. Some omnivorous terrestrial vertebrates with high metabolic rates, such as the rattlesnake (*Crotalus atrox*), hummingbird (*Archilochus colubris*), white-throated sparrow (*Zonotrichia albicollis*), deer mouse (*Peromyscus maniculatus*), and bank vole (*Myodes glareolus*), have high proportions of DHA in their muscle phospholipids, accounting for up to 33% of the total fatty acids [74–77]. However, these animals evidently have no dietary source of this LC-PUFA, because even aquatic insects or earthworms, if presented in the diet, could provide them with only EPA rather than DHA.

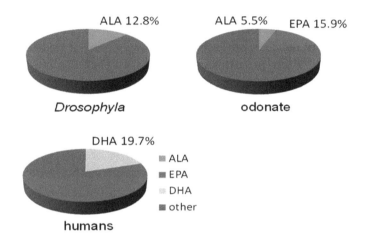

Figure 3. Levels (% of total fatty acids) of alpha-linolenic (ALA), eicosapentaenoic (EPA) and docosahexaenoic (DHA) acids in the heads of *Drosophila* [67], the eyes of odonates [68] and human retinas [3].

The high levels of n-3 LC-PUFAs in the functional lipids and organs of some consumers, such as those mentioned above, certainly indicate the physiological significance of these compounds. Recently, new promising approaches have been used to confirm that a consumer has a dietary need for EPA and DHA. One of them combines the elucidation of a dietary source of LC-PUFAs for a consumer and possible physiological consequences of deprivation of this dietary source. For instance, wolf spiders (*Tigrosa georgicola*) that inhabited wetlands and consumed aquatic insects had higher tissue levels of aquatically derived LC-PUFAs and elevated immune function in comparison to upland spiders [78]. Another way to obtain evidence of a dietary need for EPA and DHA is to measure tissue LC-PUFA pools formed due to direct incorporation from an aquatic diet versus conversion from dietary ALA of terrestrial origin. For instance, Twining and colleagues [79] showed that the ALA content of terrestrial insects, and the ALA-to-EPA conversion efficiency, are insufficient to supply insectivorous tree swallow chicks with the n-3 LC-PUFAs that they require. The authors concluded that EPA-rich aquatic insects are ecologically essential resources during a critical ontogenetic period in this bird.

Thus, the statement that "many" animals need LC-PUFAs and must consume them can be challenged by the statement that "many" animals do not need EPA and DHA or do not need to obtain them from food. Evidently, more work is needed to specify species of aquatic and terrestrial animals that truly depend on dietary sources of LC-PUFAs.

5. Assumption 4: Fish Are the Main Food Source of EPA and DHA for Humans

Fish and aquatic invertebrates (shellfish, crayfish, etc.) are known to be the main source of LC-PUFAs for humans [7,80–84]. In general, i.e., on a global scale, this statement is absolutely true. However, in some cases, it should be used with caution. First, the contents of the sum of EPA + DHA in the edible parts (muscle tissue) of fish species varied ~400-fold, from 32.78 mg g^{-1} wet weight (WW) in the boganid char (*Salvelinus boganidae*) [85] to 0.08 mg g^{-1} WW in red hybrid tilapia (*Oreochromis* sp.) [86].

To obtain health benefits, the World Health Organization and some national nutrition and health organizations recommend daily personal consumption of 0.5–1.0 g of EPA + DHA [6,7,81,87,88] and even >1.0 g for a Western-type diet [7,88]. Thus, the recommended personal daily dose of EPA + DHA is contained in 15–30 g of meat of boganid char or in 6250–12500 g of meat of the tilapia. Naturally, a question arises: is the fish red hybrid tilapia a real source of LC-PUFAs for humans if 6–12 kg of its meat should be consumed daily to obtain the recommended healthy dose? If 1.0 g of EPA and DHA is considered as the daily dose for the reliable prevention of heart diseases and 1 kg as the maximum portion of fish per serving, the lower threshold value of EPA+DHA content in edible fish biomass is 1.0 mg g^{-1} [85]. All fish species with LC-PUFA contents below this threshold cannot be regarded as a "main source" of EPA and DHA for humans. Nevertheless, these "low LC-PUFA fish" are valuable sources of protein, vitamins, and other nutrients.

Second, some people do not eat fish, and so fish naturally cannot be the "main source" of EPA and DHA for them. For such people, there are alternative dietary LC-PUFA sources, for instance, the livers of terrestrial production animals, which contain EPA + DHA at a level close to the above threshold value of 1.0 mg g^{-1} [83]. However, as mentioned above, on a global scale, the livers of production animals, such as cattle, pigs, and chickens, cannot be alternatives to fish but represent an additional source only. Indeed, the global EPA + DHA supply through the wild fish catch is ~180 10^6 kg year^{-1}, while the global production of both of these LC-PUFAs together in animal livers is ~4 10^6 kg year^{-1} [83]. In addition, some game birds from the order Passeriformes, which have EPA + DHA contents in their meat above the threshold value, from 1.8 to 3.7 mg g^{-1} [89], may also be an alternative source of these LC-PUFAs for humans who do not eat fish.

Human populations have different diets, e.g., the vegetarian diet in South and Eastern Asia and omnivorous (Western-type) diet in Europe and North America. As a rule, the LC-PUFA status of individuals from populations with a vegetarian diet, i.e., the content of EPA and DHA in various tissues and organs, is significantly lower than that of their omnivorous counterparts [90]. They commonly have a very low intake of preformed LC-PUFAs, ca. 10–74 mg per day, and most EPA and DHA is obtained as a result of the conversion of the dietary short-chain PUFA, ALA [90,91]. The following question arises: how does this conversion meet the demands for LC-PUFAs in vegetarian humans, since its rate is known to be low? Recent studies have found marked global polymorphism in the FADS (fatty acid desaturase) gene cluster, which is strongly associated with the efficiency of the conversion of linoleic and linolenic acids to LC-PUFAs of the corresponding family [92–94]. Human populations that have moved to more vegetarian diets are adapted to a low intake of LC-PUFAs, and these alleles provide the more efficient conversion of ALA to EPA and DHA. The selective patterns in FADS genes have been suggested to be driven by a change in the dietary composition of fatty acids following the transition to agriculture. Overall, there is a premise that vegetarian humans can function adequately with the found LC-PUFA status. More studies are necessary to assess the physiological and pathological outcomes of a vegetarian diet in terms of individual- and population-based genetic differences in the metabolism of dietary 18C-PUFAs [91,94].

When discussing fish as the main source of EPA and DHA for humans, the following paradox should be mentioned. Since the growing human population requires an increasing supply of essential LC-PUFAs, and wild catch fisheries are at exploitable limits, a greater proportion of food fish are obtained from aquaculture [95]. However, farmed fish, such as one of the most popular and valuable species, Atlantic salmon (*Salmo salar*), need high levels of EPA and DHA in their diet, which are obtained from the limited wild catch fisheries in fishmeal and fish oil [95]. Thus, in aquaculture oilseed, plant sources are increasingly used in feed to substitute the finite fishmeal and oil. This replacement of fish oil with the sustainable alternative, vegetable oils, has no detrimental effect on fish growth but results in a dramatic decrease in EPA and DHA in their flesh. For instance, the contents of EPA+DHA in the flesh of farmed Atlantic salmon in Scotland decreased from 27.4 mg g^{-1} in 2006 to 13.6 mg g^{-1} in 2015 [95]. Evidently, sustainable sources of EPA and DHA, in addition to fish, must inevitably satisfy the growing human population.

6. Assumption 5: Culinary Treatments Decrease EPA and DHA in Products

Polyunsaturated fatty acids are known to be preferentially affected by oxidation during heating [96–98]. Thus, the degradation of LC-PUFAs in food during cooking and other culinary treatments has been reported by many authors [99–111]. However, other authors reported no decrease in EPA and DHA during cooking [70,83,112–123]. Indeed, products such as raw fish and production animals do not contain EPA and DHA in pure chemical form, but as components of phospholipids integrated into cell membranes, which have comparatively low susceptibility to degradation [82,124].

It is important to note that the above data on the degradation of LC-PUFAs during culinary treatments are based on measurements of their relative levels as a percentage of total FA. Furthermore, it has been demonstrated that the use of relative values (%) instead of the absolute content of EPA and DHA (mg g^{-1} WW) for the estimation of nutritive value gives erroneous conclusions regarding the nutritive values of fish and other products for humans (e.g., [83,95,117,125–128]). For instance, there are many fish species with high EPA+DHA contents, >8 mg g^{-1}, but a low percentage, <20%, e.g., chum salmon (*Oncorhynchus keta*), coho salmon (*Oncorhynchus kisutch*), and lake trout (*Salvelinus namaycush*), while there are species with a high percentage, >40%, and low content, <3 mg g^{-1}, e.g., Atlantic cod (*Gadus morhua*) and whiting (*Merlangius merlangus*) [129].

Another striking example is related to edible macroalgae (seaweeds): red algae (Rhodophyta), (*Palmaria palmata*), have extremely high levels of EPA, ca. 50% of the total fatty acids, but because of their very low total lipid contents, at realistic daily consumption levels, they cannot satisfy the LC-PUFA requirements of humans [126].

There is no correlation, or even a negative correlation, between EPA and DHA levels (%) and contents (mg g^{-1} WW) in fish [129,130]. The explanation for the above phenomenon is believed to be as follows: EPA and DHA are mostly contained in phospholipids (PLs), i.e., in the structural lipids of cell membranes, which should remain nearly constant in proportion to functional muscle tissues, while many other fatty acids are contained in reserve neutral lipids, triacylglycerols (TAGs), whose composition is highly variable in fish biomass [125,130–132]. For this reason, in fish which are considered to be fatty, i.e., accumulating comparatively more TAG [133], EPA and DHA are "diluted", and their relative levels decrease.

Thus, the nutritive value of products for human nutrition should be estimated on the basis of the contents of EPA and DHA, mg per g of product, which can be obtained using internal standards during chromatography, rather than levels, or the % of total FAs [83,117,127,134,135]. Furthermore, content estimates based on internal standards are scarce, and most data are published as the level, or %, of total FAs [100–102,105–111]. Here, we provide data obtained using internal standards, which allows us to compare the real nutritive value of fish and production animal products (Table 1). As noted in Assumption 4, some products prepared from terrestrial animals are comparable in their EPA and DHA contents with those of some fish (Table 1). In general, according to the quantitative data, mg per g of product, there is no decrease in LC-PUFA contents following most culinary treatments, and cooked products prepared from relevant raw biomass are good sources of EPA and DHA for humans.

Table 1. Contents of the sum of eicosapentaenoic and docosahexaenoic fatty acids (EPA+DHA, mg g^{-1} of product) in cooked fish and the daily portion of products (DP, g) that need to be consumed to obtain the recommended intake of EPA+DHA for humans, 1 g day^{-1}.

Product	EPA + DHA	DP	Reference
Atlantic salmon *Salmo salar* (fried)	40.1	25	[121]
Pacific saury *Cololabis saira* (canned, brand H)	37.9	26	[136]
Atlantic salmon *Salmo salar* (fried)	19.6	51	[114]
Pacific herring *Clupea harengus* (canned)	17.9	56	[118]
Atlantic salmon *Salmo salar* (fried)	17.0	59	[112]
Baltic sprat *Sprattus sprattus* (canned)	14.3	70	[118]
Pacific saury *Cololabis saira* (canned, brand N)	13.1	76	[136]
King salmon *Oncorhynchus tshawytscha* (baked)	12.4	81	[137]
Lake trout *Salvelinus namaycush* (baked)	12.4	81	[122]
Lake trout *Salvelinus namaycush* (fried)	12.4	81	[122]
Lake trout *Salvelinus namaycush* (broiled)	12.3	81	[122]
King salmon *Oncorhynchus tshawytscha* (steamed)	11.9	84	[137]
King salmon *Oncorhynchus tshawytscha* (fried)	11.5	87	[137]
King salmon *Oncorhynchus tshawytscha* (microwaved)	10.4	96	[137]
King salmon *Oncorhynchus tshawytscha* (poached)	10.0	100	[137]
Sardine *Sardina pilchardus* (fried)	8.8	114	[112]
Humpback salmon *Oncorhynchus gorbuscha* (boiled)	6.0	167	[116]
Brown trout *Salmo trutta* (boiled)	5.7	175	[117]
Humpback salmon *Oncorhynchus gorbuscha* (stewed)	5.3	189	[116]
Humpback salmon *Oncorhynchus gorbuscha* (roasted)	5.0	200	[116]
Humpback salmon *Oncorhynchus gorbuscha* (fried)	4.3	233	[116]
Brown trout *Salmo trutta* (fried)	4.1	244	[117]
Cod *Gadus morhua* (fried)	4.1	244	[114]
Spanish mackerel *Scomberomorus commerson* (fried)	3.9	256	[112]
Pacific herring *Clupea harengus* (boiled)	3.9	256	[117]
Pacific herring *Clupea harengus* (fried)	3.8	263	[117]
Rock sole *Lepidopsetta bilineata* (boiled)	3.6	278	[117]
Chinook salmon *Oncorhynchus tshawytscha* (fried)	3.2	313	[122]
Rock sole *Lepidopsetta bilineata* (fried)	3.1	323	[117]
Chinook salmon *Oncorhynchus tshawytscha* (baked)	3.1	323	[122]
White sucker *Catostomus commersonii* (baked)	2.3	435	[122]
Cod *Gadus morhua* (boiled)	2.4	417	[117]
Chinook salmon *Oncorhynchus tshawytscha* (fried)	2.8	357	[122]
Cod *Gadus morhua* (fried)	2.2	455	[121]
Walleye (*Sander vitreus*) (baked)	2.1	476	[122]
White sucker *Catostomus commersonii* (broiled)	2.1	476	[122]
White sucker *Catostomus commersonii* (fried)	2.0	500	[122]
Walleye (*Sander vitreus*) (broiled)	1.9	526	[122]
Walleye (*Sander vitreus*) (fried)	1.9	526	[122]
Prawn *Macrobrachium acanthurus* (fried)	1.8	556	[138]
Beef liver (boiled)	1.3	769	[83]
Zander *Sander lucioperca* (boiled)	1.1	909	[70]
Pork liver (boiled)	1.0	1000	[83]
Zander *Sander lucioperca* (stewed)	1.0	1000	[70]
Zander *Sander lucioperca* (fried)	1.0	1000	[70]
Common carp *Cyprinus carpio* (fried)	1.0	1000	[122]
Chicken liver (boiled)	0.7	1429	[83]
Common carp *Cyprinus carpio* (baked)	0.7	1429	[122]
Gilthead sea bream *Sparus aurata* (fried)	0.6	1667	[139]
Common carp *Cyprinus carpio* (broiled)	0.5	2000	[122]
Pork (fried)	0.3	3333	[119]
White rabbit (baked)	0.1	10000	[140]

7. Conclusions

A number of assumptions important for developing LC-PUFA studies in the ecological and food sciences should be improved:

1. Dividing microalgae on the basis of their LC-PUFA content into classes of high and low nutritive value appeared to be too coarse. Although there are no Chlorophyceae (green algae) that contain EPA and DHA, there are Bacillariophyceae (diatoms) with low contents of LC-PUFAs.

2. The maximum LC-PUFA yield (g km^{-2} $year^{-1}$) that can be ultimately obtained by humans occurs in mesotrophic rather than oligotrophic aquatic ecosystems.

3. Many animals and terrestrial insects do not need EPA, and aquatic insects do not need DHA in any considerable quantity. Many other animals do not need LC-PUFAs in their food: some worms can obtain these biomolecules from their intestine microflora, and strictly herbivorous terrestrial mammals can synthesize required quantities of EPA and DHA from ALA obtained from the green parts of consumed plants.

4. There are many fish species that are not adequate sources of EPA and DHA for humans, especially for those with a Western-type diet. In turn, there are products of terrestrial animals that can be a source of LC-PUFAs for persons who do not eat fish. In human populations with a vegetarian diet, the conversion of dietary C18-PUFAs is considered to be sufficient to meet the demands for LC-PUFAs based on the found genetic patterns; however, this statement requires further study.

5. Most common culinary treatments do not decrease the EPA and DHA contents in fish and other animal products.

Author Contributions: Conceptualization, calculations, writing—original draft preparation, M.I.G.; writing—review and editing, N.N.S.

Acknowledgments: We are grateful to anonymous Reviewers for their valuable recommendations to improve the manuscript.

References

1. Simopoulos, A.P. Human requirement for n-3 polyunsaturated fatty acids. *Poult. Sci.* **2000**, *79*, 961–970. [CrossRef] [PubMed]

2. Arts, M.T.; Ackman, R.G.; Holub, B.J. "Essential fatty acids" in aquatic ecosystems: A crucial link between diet and human health and evolution. *Can. J. Fish. Aquat. Sci.* **2001**, *58*, 122–137. [CrossRef]

3. Lauritzen, L.; Hansen, H.S.; Jorgensen, M.H.; Michaelsen, K.F. The essentiality of long chain n-3 fatty acids in relation to development and function of the brain and retina. *Prog. Lipid Res.* **2001**, *40*, 1–94. [CrossRef]

4. SanGiovanni, J.P.; Chew, E.Y. The role of omega-3 long-chain polyunsaturated fatty acids in health and disease of the retina. *Prog. Retin. Eye Res.* **2005**, *24*, 87–138. [CrossRef] [PubMed]

5. Wood, J.D.; Enser, M.; Fisher, A.V.; Nute, G.R.; Sheard, P.R.; Richardson, R.I.; Hughes, S.I.; Whittington, F.M. Fat deposition, fatty acid composition and meat quality: A review. *Meat Sci.* **2008**, *78*, 343–358. [CrossRef]

6. Kris-Etherton, P.M.; Grieger, J.A.; Etherton, T.D. Dietary reference intakes for DHA and EPA. *Prostaglandins Leukot. Essent. Fat. Acids* **2009**, *81*, 99–104. [CrossRef]

7. Tocher, D.R.; Betancor, M.B.; Sprague, M.; Olsen, R.E.; Napier, J.A. Omega-3 long-chain polyunsaturated fatty acids, EPA and DHA: Bridging the gap between supply and demand. *Nutrients* **2019**, *11*, 89. [CrossRef]

8. Casula, M.; Soranna, D.; Catapano, A.L.; Corrao, G. Longterm effect of high dose omega-3 fatty acid supplementation for secondary prevention of cardiovascular outcomes: A meta-analysis of randomized, double blind, placebo controlled trials. *Atheroscler. Suppl.* **2013**, *14*, 243–251. [CrossRef]

9. Dyall, S.C. Long-chain omega-3 fatty acids and the brain: A review of the independent and shared effects of EPA, DPA and DHA. *Front. Aging Neurosci.* **2015**, *7*, 52. [CrossRef]

10. Weiser, M.J.; Butt, C.M.; Mohajeri, M.H. Docosahexaenoic acid and cognition throughout the lifespan. *Nutrients* **2016**, *8*, 99. [CrossRef]

11. Calder, P.C. Very long-chain n-3 fatty acids and human health: Fact, fiction and the future. *Proc. Nutr. Soc.* **2018**, *77*, 52–72. [CrossRef] [PubMed]

12. Ahlgren, G.; Gustafsson, I.B.; Boberg, M. Fatty acid content and chemical composition of freshwater microalgae. *J. Phycol.* **1992**, *28*, 37–50. [CrossRef]

13. Davis, B.C.; Kris-Etherton, P.M. Achieving optimal essential fatty acid status in vegetarians: Current knowledge and practical implications. *Am. J. Clin. Nutr.* **2003**, *78*, 640S–646S. [CrossRef] [PubMed]

14. Stark, A.H.; Crawford, M.A.; Reifen, R. Update on alpha-linolenic acid. *Nutr. Rev.* **2008**, *66*, 326–332. [CrossRef] [PubMed]

15. Kang, X.; Bai, Y.; Sun, G.; Huang, Y.; Chen, Q.; Han, R.; Li, G.; Li, F. Molecular cloning, characterization, and expression analysis of chicken Δ-6 desaturase. *Asian-Australas. J. Anim. Sci.* **2010**, *23*, 116–121. [CrossRef]

16. Uttaro, A.D. Biosynthesis of polyunsaturated fatty acids in lower eukaryotes. *IUBMB Life* **2006**, *58*, 563–571. [CrossRef] [PubMed]

17. Zhou, X.-R.; Green, A.G.; Singh, S.P. *Caenorhabditis elegans* Δ12-desaturase FAT-2 is a bifunctional desaturase able to desaturate a diverse range of fatty acid substrates at the Δ12 and Δ15 positions. *J. Biol. Chem.* **2011**, *286*, 43644–43650. [CrossRef] [PubMed]

18. Harwood, J.L. Recent advances in the biosynthesis of plant fatty acids. *Biochim. Biophys. Acta* **1996**, *1301*, 7–56. [CrossRef]

19. Sayanova, O.V.; Napier, J.A. Eicosapentaenoic acid: Biosynthetic routes and the potential for synthesis in transgenic plants. *Phytochemistry* **2004**, *65*, 147–158. [CrossRef]

20. Ward, O.P.; Singh, A. Omega-3/6 fatty acids: Alternative sources of production. *Process Biochem.* **2005**, *40*, 3627–5652. [CrossRef]

21. Ruiz-Lopez, N.; Sayanova, O.; Napier, J.A.; Haslam, R.P. Metabolic engineering of the omega-3 long chain polyunsaturated fatty acid biosynthetic pathway into transgenic plants. *J. Exp. Bot.* **2012**, *63*, 2397–2410. [CrossRef] [PubMed]

22. Gladyshev, M.I.; Arts, M.T.; Sushchik, N.N. Preliminary estimates of the export of omega-3 highly unsaturated fatty acids (EPA+DHA) from aquatic to terrestrial ecosystems. In *Lipids in Aquatic Ecosystems*; Arts, M.T., Kainz, M., Brett, M.T., Eds.; Springer: New York, NY, USA, 2009; pp. 179–209.

23. Sterner, R.W.; Schulz, K.L. Zooplankton nutrition: Recent progress and a reality check. *Aquat. Ecol.* **1998**, *32*, 261–279. [CrossRef]

24. Ahlgren, G.; Lundstedt, L.; Brett, M.; Forsberg, C. Lipid composition and food quality of some freshwater phytoplankton for cladoceran zooplankters. *J. Plankton Res.* **1990**, *12*, 809–818. [CrossRef]

25. Muller-Navarra, D.C. Biochemical versus mineral limitation in Daphnia. *Limnol. Oceanogr.* **1995**, *40*, 1209–1214. [CrossRef]

26. Gulati, R.D.; DeMott, W.R. The role of food quality for zooplankton: Remarks on the state-of-the-art, perspectives and priorities. *Freshw. Biol.* **1997**, *38*, 753–768. [CrossRef]

27. Wacker, A.; Becher, P.; Von Elert, E. Food quality effects of unsaturated fatty acids on larvae of the zebra mussel *Dreissena polymorpha*. *Limnol. Oceanogr.* **2002**, *47*, 1242–1248. [CrossRef]

28. Chu, F.L.E.; Lund, E.D.; Podbesek, J.A. Quantitative significance of n-3 essential fatty acid contribution by heterotrophic protists in marine pelagic food webs. *Mar. Ecol. Prog. Ser.* **2008**, *354*, 85–95. [CrossRef]

29. Martin-Creuzburg, D.; Wacker, A.; Basena, T. Interactions between limiting nutrients: Consequences for somatic and population growth of *Daphnia magna*. *Limnol. Oceanogr.* **2010**, *55*, 2597–2607. [CrossRef]

30. Peltomaa, E.T.; Aalto, S.L.; Vuorio, K.M.; Taipale, S.J. The Importance of phytoplankton biomolecule availability for secondary production. *Front. Ecol. Evol.* **2017**, *5*, 128. [CrossRef]

31. Taipale, S.; Strandberg, U.; Peltomaa, E.; Galloway, A.W.E.; Ojala, A.; Brett, M.T. Fatty acid composition as biomarkers of freshwater microalgae: Analysis of 37 strains of microalgae in 22 genera and in seven classes. *Aquat. Microb. Ecol.* **2013**, *71*, 165–178. [CrossRef]

32. Petkov, G.; Garcia, G. Which are fatty acids of the green alga *Chlorella*. *Biochem. Syst. Ecol.* **2007**, *35*, 281–285. [CrossRef]

33. Iliev, I.; Petkov, G.; Lukavsky, J.; Furnadzhieva, S.; Andreeva, R. Do cyanobacterial lipids contain fatty acids longer than 18 carbon atoms. *Z. Naturforsch. C* **2011**, *66*, 267–276. [CrossRef] [PubMed]

34. Weers, P.M.M.; Gulati, R.D. Growth and reproduction of *Daphnia galeata* response to changes in fatty acids, phosphorus, and nitrogen in *Chlamydomonas reinhardtii*. *Limnol. Oceanogr.* **1997**, *42*, 1584–1589. [CrossRef]

35. Nanton, D.A.; Castell, J.D. The effects of dietary fatty acids on the fatty acid composition of the harpacticoid copepod, *Tisbe* sp., for use a live food for marine fish larvae. *Aquaculture* **1998**, *163*, 251–261. [CrossRef]

36. Wacker, A.; Von Elert, E. Polyunsaturated fatty acids: Evidence for non-substitutable biochemical resources in *Daphnia galeata*. *Ecology* **2001**, *82*, 2507–2520. [CrossRef]

37. Sushchik, N.N.; Gladyshev, M.I.; Kalachova, G.S.; Kravchuk, E.S.; Dubovskaya, O.P.; Ivanova, E.A. Particulate fatty acids in two small Siberian reservoirs dominated by different groups of phytoplankton. *Freshw. Biol.* **2003**, *48*, 394–403. [CrossRef]

38. Brown, M.R.; Jeffrey, S.W.; Volkman, J.K.; Dunstan, G.A. Nutritional properties of microalgae for mariculture. *Aquaculture* **1997**, *151*, 315–331. [CrossRef]

39. Chen, X.; Wakeham, S.G.; Fisher, N.S. Influence of iron on fatty acid and sterol composition of marine phytoplankton and copepod consumers. *Limnol. Oceanogr.* **2011**, *56*, 716–724. [CrossRef]

40. Claustre, H.; Marty, J.C.; Cassiani, L.; Dagaut, J. Fatty acid dynamics in phytoplankton and microzooplankton communities during a spring bloom in the coastal Ligurian Sea: Ecological implications. *Mar. Microb. Food Webs* **1989**, *3*, 51–66.

41. Sushchik, N.N.; Gladyshev, M.I.; Makhutova, O.N.; Kalachova, G.S.; Kravchuk, E.S.; Ivanova, E.A. Associating particulate essential fatty acids of the ω3 family with phytoplankton species composition in a Siberian reservoir. *Freshw. Biol.* **2004**, *49*, 1206–1219. [CrossRef]

42. Sushchik, N.N.; Gladyshev, M.I.; Ivanova, E.A.; Kravchuk, E.S. Seasonal distribution and fatty acid composition of littoral microalgae in the Yenisei River. *J. Appl. Phycol.* **2010**, *22*, 11–24. [CrossRef]

43. Gladyshev, M.I.; Sushchik, N.N.; Kolmakova, A.A.; Kalachova, G.S.; Kravchuk, E.S.; Ivanova, E.A.; Makhutova, O.N. Seasonal correlations of elemental and ω3 PUFA composition of seston and dominant phytoplankton species in a eutrophic Siberian Reservoir. *Aquat. Ecol.* **2007**, *41*, 9–23. [CrossRef]

44. Ahlgren, G.; Sonesten, L.; Boberg, M.; Gustafsson, I.-B. Fatty acid content of some freshwater fish in lakes of different trophic levels—A bottom-up effect? *Ecol. Freshw. Fish* **1996**, *5*, 15–27. [CrossRef]

45. Taipale, S.J.; Vuorioc, K.; Strandberg, U.; Kahilainen, K.K.; Jarvinen, M.; Hiltunen, M.; Peltomaa, E.; Kankaala, P. Lake eutrophication and brownification downgrade availability and transfer of essential fatty acids for human consumption. *Environ. Int.* **2016**, *96*, 156–166. [CrossRef] [PubMed]

46. Muller-Navarra, D.C.; Brett, M.T.; Park, S.; Chandra, S.; Ballantyne, A.P.; Zorita, E.; Goldman, C.R. Unsaturated fatty acid content in seston and tropho-dynamic coupling in lakes. *Nature* **2004**, *427*, 69–72. [CrossRef] [PubMed]

47. Razavi, N.R.; Arts, M.T.; Qua, M.; Jin, B.; Rend, W.; Wang, Y.; Campbell, L.M. Effect of eutrophication on mercury, selenium, and essential fatty acids in Bighead Carp (*Hypophthalmichthys nobilis*) from reservoirs of eastern China. *Sci. Total Environ.* **2014**, *499*, 36–46. [CrossRef]

48. Smith, V.H. Nutrient dependence of primary productivity in lakes. *Limnol. Oceanogr.* **1979**, *24*, 1051–1064. [CrossRef]

49. Gladyshev, M.I. Quality and quantity of biological production in water bodies with different concentration of phosphorus: Case study of Eurasian perch. *Dokl. Biochem. Biophys.* **2018**, *478*, 1–3. [CrossRef]

50. Sushchik, N.N.; Gladyshev, M.I.; Makhutova, O.N.; Kravchuk, E.S.; Dubovskaya, O.P.; Kalacheva, G.S. Seasonal transfer of the pool of the essential eicosapentaenoic acid along the pelagic trophic chain of a eutrophic reservoir. *Dokl. Biol. Sci.* **2008**, *422*, 355–356. [CrossRef]

51. Khozin-Goldberg, I.; Cohen, Z. The effect of phosphate starvation on the lipid and fatty acid composition of the fresh water eustigmatophyte *Monodus subterraneus*. *Phytochemistry* **2006**, *67*, 696–701. [CrossRef]

52. Twining, C.W.; Brenna, J.T.; Hairston, N.G., Jr.; Flecker, A.S. Highly unsaturated fatty acids in nature: What we know and what we need to learn. *Oikos* **2015**, *125*, 749–760. [CrossRef]

53. Hixson, S.M.; Sharma, B.; Kainz, M.J.; Wacker, A.; Arts, M.T. Production, distribution, and abundance of long-chain omega-3 polyunsaturated fatty acids: A fundamental dichotomy between freshwater and terrestrial ecosystems. *Environ. Rev.* **2015**, *23*, 414–424. [CrossRef]

54. Stanley-Samuelson, D.W.; Jurenka, R.A.; Cripps, C.; Blomquist, G.J.; de Renobales, M. Fatty acids in insects: Composition, metabolism, and biological significance. *Arch. Insect Biochem. Physiol.* **1988**, *9*, 1–33. [CrossRef]

55. Buckner, J.S.; Hagen, M.M. Triacylglycerol and phospholipid fatty acids of the silverleaf whitefly: Composition and biosynthesis. *Arch. Insect Biochem. Physiol.* **2003**, *53*, 66–79. [CrossRef]

56. Wang, Y.; Lin, D.S.; Bolewicz, L.; Connor, W.E. The predominance of polyunsaturated fatty acids in the butterfly *Morpho peleides* before and after metamorphosis. *J. Lipid Res.* **2006**, *47*, 530–536. [CrossRef]

57.   Rumpold, B.A.; Schluter, O.K. Nutritional composition and safety aspects of edible insects. *Mol. Nutr. Food Res.* **2013**, *57*, 802–823. [CrossRef]

58.   Barroso, F.G.; de Haro, C.; Sanchez-Muros, M.J.; Venegas, E.; Martinez-Sanchez, A.; Perez-Ban, C. The potential of various insect species for use as food for fish. *Aquaculture* **2014**, *422*, 193–201. [CrossRef]

59.   Sanchez-Muros, M.-J.; Barroso, F.G.; Manzano-Agugliaro, F. Insect meal as renewable source of food for animal feeding: A review. *J. Clean. Prod.* **2014**, *65*, 16–27. [CrossRef]

60.   Stanley, D.; Kim, Y. Eicosanoid signaling in insects: From discovery to plant protection. *Crit. Rev. Plant Sci.* **2014**, *33*, 20–63. [CrossRef]

61.   Sushchik, N.N.; Yurchenko, Y.A.; Gladyshev, M.I.; Belevich, O.E.; Kalachova, G.S.; Kolmakova, A.A. Comparison of fatty acid contents and composition in major lipid classes of larvae and adults of mosquitoes (*Diptera: Culicidae*) from a steppe region. *Insect Sci.* **2013**, *20*, 585–600. [CrossRef]

62.   Borisova, E.V.; Makhutova, O.N.; Gladyshev, M.I.; Sushchik, N.N. Fluxes of biomass and essential polyunsaturated fatty acids from water to land via chironomid emergence from a mountain lake. *Contemp. Probl. Ecol.* **2016**, *9*, 446–457. [CrossRef]

63.   Makhutova, O.N.; Borisova, E.V.; Shulepina, S.P.; Kolmakova, A.A.; Sushchik, N.N. Fatty acid composition and content in chironomid species at various life stages dominating in a saline Siberian lake. *Contemp. Probl. Ecol.* **2017**, *10*, 230–239. [CrossRef]

64.   Popova, O.N.; Haritonov, A.Y.; Sushchik, N.N.; Makhutova, O.N.; Kalachova, G.S.; Kolmakova, A.A.; Gladyshev, M.I. Export of aquatic productivity, including highly unsaturated fatty acids, to terrestrial ecosystems via Odonata. *Sci. Total Environ.* **2017**, *581*, 40–48. [CrossRef] [PubMed]

65.   Politi, L.; Rotstein, N.; Carri, N. Effects of docosahexaenoic acid on retinal development: Cellular and molecular aspects. *Lipids* **2001**, *36*, 927–935. [CrossRef] [PubMed]

66.   Bazan, N.G. Cellular and molecular events mediated by docosahexaenoic acid-derived neuroprotectin D1 signaling in photoreceptor cell survival and brain protection. *Prostaglandins Leukot. Essent. Fatty Acids* **2009**, *81*, 205–211. [CrossRef]

67.   Ziegler, A.B.; Ménagé, C.; Grégoire, S.; Garcia, T.; Ferveur, J.-F.; Bretillon, L.; Grosjean, Y. Lack of dietary polyunsaturated fatty acids causes synapse dysfunction in the *Drosophila* visual system. *PLoS ONE* **2015**, *10*, e0135353. [CrossRef] [PubMed]

68.   Sushchik, N.N.; Popova, O.N.; Makhutova, O.N.; Gladyshev, M.I. Fatty acid composition of odonate's eyes. *Dokl. Biochem. Biophys.* **2017**, *475*, 280–282. [CrossRef] [PubMed]

69.   Sampedro, L.; Jeannotte, R.; Whalen, J.K. Trophic transfer of fatty acids from gut microbiota to the earthworm *Lumbricus terrestris* L. *Soil Biol. Biochem.* **2006**, *38*, 2188–2198. [CrossRef]

70.   Gladyshev, M.I.; Sushchik, N.N.; Gubanenko, G.A.; Makhutova, O.N.; Kalachova, G.S.; Rechkina, E.A.; Malyshevskaya, K.K. Effect of the way of cooking on contents of essential polyunsaturated fatty acids in filets of zander. *Czech. J. Food Sci.* **2014**, *32*, 226–231. [CrossRef]

71.   Manukovsky, N.S.; Kovalev, V.S.; Gribovskaya, I.V. Two-stage biohumus production from inedible potato biomass. *Bioresour. Technol.* **2001**, *78*, 273–275. [CrossRef]

72.   Menzel, R.; von Chrzanowski, H.; Tonat, T.; van Riswyck, K.; Schliesser, P.; Ruess, L. Presence or absence? Primary structure, regioselectivity and evolution of Delta 12/omega 3 fatty acid desaturases in nematodes. *Biochim. Biophys. Acta Mol. Cell Biol. Lipids* **2019**, *1864*, 1194–1205. [CrossRef] [PubMed]

73.   Kouba, M.; Mourot, J. A review of nutritional effects on fat composition of animal products with special emphasis on n-3 polyunsaturated fatty acids. *Biochimie* **2011**, *93*, 13–17. [CrossRef]

74.   Infante, J.P.; Kirwan, R.C.; Brenna, J.T. High levels of docosahexaenoic acid (22:6*n*-3)-containing phospholipids in high-frequency contraction muscles of hummingbirds and rattlesnakes. *Comp. Biochem. Physiol. Part B* **2001**, *130*, 291–298. [CrossRef]

75.   Geiser, F.; McAllan, B.M.; Kenagy, G.J.; Hiebert, S.M. Photoperiod affects daily torpor and tissue fatty acid composition in deer mice. *Naturwissenschaften* **2007**, *94*, 319–325. [CrossRef] [PubMed]

76.   Klaiman, J.M.; Price, E.R.; Guglielmo, C.G. Fatty acid composition of pectoralis muscle membrane, intramuscular fat stores and adipose tissue of migrant and wintering white-throated sparrows (*Zonotrichia albicollis*). *J. Exp. Biol.* **2009**, *212*, 3865–3872. [CrossRef]

77.   Stawski, C.; Valencak, T.G.; Ruf, T.; Sadowska, E.T.; Dheyongera, G.; Rudolf, A.; Maiti, U.; Koteja, P. Effect of selection for high activity-related metabolism on membrane phospholipid fatty acid composition in bank voles. *Physiol. Biochem. Zool.* **2015**, *88*, 668–679. [CrossRef]

78. Fritz, K.A.; Kirschman, L.J.; McCay, S.D.; Trushenski, J.T.; Warne, R.W.; Whiles, M.R. Subsidies of essential nutrients from aquatic environments correlate with immune function in terrestrial consumers. *Freshw. Sci.* **2017**, *36*, 893–900. [CrossRef]

79. Twining, C.W.; Lawrence, P.; Winkler, D.W.; Flecker, A.S.; Brenna, J.T. Conversion efficiency of α-linolenic acid to omega-3 highly unsaturated fatty acids in aerial insectivore chicks. *J. Exp. Biol.* **2018**, *221*, jeb165373. [CrossRef] [PubMed]

80. Robert, S.S. Production of eicosapentaenoic and docosahexaenoic acid-containing oils in transgenic land plants for human and aquaculture nutrition. *Mar. Biotechnol.* **2006**, *8*, 103–109. [CrossRef]

81. Adkins, Y.; Kelley, D.S. Mechanisms underlying the cardioprotective effects of omega-3 polyunsaturated fatty acids. *J. Nutr. Biochem.* **2010**, *2*, 781–792. [CrossRef]

82. Gladyshev, M.I.; Sushchik, N.N.; Makhutova, O.N. Production of EPA and DHA in aquatic ecosystems and their transfer to the land. *Prostaglandins Other Lipid Mediat.* **2013**, *107*, 117–126. [CrossRef] [PubMed]

83. Gladyshev, M.I.; Makhutova, O.N.; Gubanenko, G.A.; Rechkina, E.A.; Kalachova, G.S.; Sushchik, N.N. Livers of terrestrial production animals as a source of long-chain polyunsaturated fatty acids for humans: An alternative to fish? *Eur. J. Lipid Sci. Technol.* **2015**, *117*, 1417–1421. [CrossRef]

84. Cladis, D.P.; Kleiner, A.C.; Freiser, H.H.; Santerre, C.R. Fatty acid profiles of commercially available finfish fillets in the United States. *Lipids* **2014**, *49*, 1005–1018. [CrossRef] [PubMed]

85. Gladyshev, M.I.; Glushchenko, L.A.; Makhutova, O.N.; Rudchenko, A.E.; Shulepina, S.P.; Dubovskaya, O.P.; Zuev, I.V.; Kolmakov, V.I.; Sushchik, N.N. Comparative analysis of content of omega-3 polyunsaturated fatty acids in food and muscle tissue of fish from aquaculture and natural habitats. *Contemp. Probl. Ecol.* **2018**, *11*, 297–308. [CrossRef]

86. Teoh, C.Y.; Ng, W.K. The implications of substituting dietary fish oilwith vegetable oils on the growth performance, fillet fatty acid profile and modulation of the fatty acid elongase, desaturase and oxidation activities of red hybrid tilapia, *Oreochromis* sp. *Aquaculture* **2016**, *465*, 311–322. [CrossRef]

87. Harris, W.S.; Mozaffarian, D.; Lefevre, M.; Toner, C.D.; Colombo, J.; Cunnane, S.C.; Holden, J.M.; Klurfeld, D.M.; Morris, M.C.; Whelan, J. Towards establishing dietary reference intakes for eicosapentaenoic and docosahexaenoic acids. *J. Nutr.* **2009**, *139*, 804S–819S. [CrossRef] [PubMed]

88. Nagasaka, R.; Gagnon, C.; Swist, E.; Rondeau, I.; Massarelli, I.; Cheung, W.; Ratnayake, W.M.N. EPA and DHA status of South Asian and white Canadians living in the National Capital Region of Canada. *Lipids* **2014**, *49*, 1057–1069. [CrossRef]

89. Gladyshev, M.I.; Popova, O.N.; Makhutova, O.N.; Zinchenko, T.D.; Golovatyuk, L.V.; Yurchenko, Y.A.; Kalachova, G.S.; Krylov, A.V.; Sushchik, N.N. Comparison of fatty acid compositions in birds feeding in aquatic and terrestrial ecosystems. *Contemp. Probl. Ecol.* **2016**, *9*, 503–513. [CrossRef]

90. Burdge, G.C.; Henry, C.J. Omega-3 Polyunsaturated Fatty Acid Metabolism in Vegetarians. In *Polyunsaturated Fatty Acid Metabolism*; Burdge, G.C., Ed.; Elsevier Inc.: Amsterdam, The Netherland; AOCS Press: Urbana, IL, USA, 2018; pp. 193–204.

91. Burdge, G.C. Is essential fatty acid interconversion an important source of PUFA in humans? *Br. J. Nutr.* **2019**, *121*, 615–624. [CrossRef]

92. Kothapalli, K.S.; Ye, K.; Gadgil, M.S.; Carlson, S.E.; O'Brien, K.O.; Zhang, J.Y.; Park, H.G.; Ojukwu, K.; Zou, J.; Hyon, S.S.; et al. Positive selection on a regulatory insertion-deletion polymorphism in FADS2 influences apparent endogenous synthesis of arachidonic acid. *Mol. Biol. Evol.* **2016**, *33*, 1726–1739. [CrossRef]

93. Buckley, M.T.; Racimo, F.; Allentoft, M.E.; Jensen, M.K.; Jonsson, A.; Huang, H.; Hormozdiari, F.; Sikora, M.; Marnetto, D.; Eskin, E.; et al. Selection in Europeans on fatty acid desaturases associated with dietary changes. *Mol. Biol. Evol.* **2017**, *34*, 1307–1318. [CrossRef] [PubMed]

94. Chilton, F.H.; Dutta, R.; Reynolds, L.M.; Sergeant, S.; Mathias, R.A.; Seeds, M.C. Precision nutrition and omega-3 polyunsaturated fatty acids: A case for personalized supplementation approaches for the prevention and management of human diseases. *Nutrients* **2017**, *9*, 1165. [CrossRef] [PubMed]

95. Sprague, M.; Dick, J.R.; Tocher, D.R. Impact of sustainable feeds on omega-3 long-chain fatty acid levels in farmed Atlantic salmon, 2006–2015. *Sci. Rep.* **2016**, *6*, 21892. [CrossRef] [PubMed]

96. Estevez, M.; Ventanas, S.; Cava, R. Oxidation of lipids and proteins in frankfurters with different fatty acid compositions and tocopherol and phenolic contents. *Food Chem.* **2007**, *100*, 55–63. [CrossRef]

97. Ruiz-Rodriguez, A.; Marin, F.R.; Ocana, A.; Soler-Rivas, C. Effect of domestic processing on bioactive compounds. *Phytochem. Rev.* **2008**, *7*, 345–384. [CrossRef]

98. Ganhão, R.; Estévez, M.; Armenteros, M.; Morcuende, D. Mediterranean berries as inhibitors of lipid oxidation in porcine burger patties subjected to cooking and chilled storage. *J. Integr. Agric.* **2013**, *12*, 1982–1992. [CrossRef]

99. Ohshima, T.; Shozen, K.; Usio, H.; Koizumi, C. Effects of grilling on formation of cholesterol oxides in seafood products rich in polyunsaturated fatty acids. *LWT-Food Sci. Technol.* **1996**, *29*, 94–99. [CrossRef]

100. Tarley, C.R.T.; Visentainer, J.V.; Matsushita, M.; de Souza, N.E. Proximate composition, cholesterol and fatty acids profile of canned sardines (*Sardinella brasiliensis*) in soybean oil and tomato sauce. *Food Chem.* **2004**, *88*, 1–6. [CrossRef]

101. Sampaio, G.R.; Bastos, D.H.M.; Soares, R.A.M.; Queiroz, Y.S.; Torres, E.A.F.S. Fatty acids and cholesterol oxidation in salted and dried shrimp. *Food Chem.* **2006**, *95*, 344–351. [CrossRef]

102. De Castro, F.A.F.; Sant'Ana, H.M.P.; Campos, F.M.; Costa, N.M.B.; Silva, M.T.C.; Salaro, A.L.; Franceschini, S.D.C.C. Fatty acid composition of three freshwater fishes under different storage and cooking processes. *Food Chem.* **2007**, *103*, 1080–1090. [CrossRef]

103. Saldanha, T.; Bragagnolo, N. Relation between types of packaging, frozen storage and grilling on cholesterol and fatty acids oxidation in Atlantic hake fillets (*Merluccius hubbsi*). *Food Chem.* **2008**, *106*, 619–627. [CrossRef]

104. Weber, J.; Bochi, V.C.; Ribeiro, C.P.; Victorio, A.D.M.; Emanuelli, T. Effect of different cooking methods on the oxidation, proximate and fatty acid composition of silver catfish (*Rhamdia quelen*) fillets. *Food Chem.* **2008**, *106*, 140–146. [CrossRef]

105. Mnari, A.B.; Jrah, H.H.; Dhibi, M.; Bouhlel, I.; Hammami, M.; Chaouch, A. Nutritional fatty acid quality of raw and cooked farmed and wild sea bream (*Sparus aurata*). *J. Agric. Food Chem.* **2010**, *58*, 507–512. [CrossRef] [PubMed]

106. Zotos, A.; Kotaras, A.; Mikras, E. Effect of baking of sardine (*Sardina pilchardus*) and frying of anchovy (*Engraulis encrasicholus*) in olive and sunflower oil on their quality. *Food Sci. Technol. Int.* **2013**, *19*, 11–23. [CrossRef] [PubMed]

107. Sampels, S.; Zajíc, T.; Mráz, J. Effects of frying fat and preparation on carp (*Cyprinus carpio*) fillet lipid composition and oxidation. *Czech J. Food Sci.* **2014**, *32*, 493–502. [CrossRef]

108. Ferreira, F.S.; Sampaio, G.R.; Keller, L.M.; Sawaya, A.C.H.F.; Chavez, D.W.H.; Torres, E.A.F.S.; Saldanha, T. Impact of air frying on cholesterol and fatty acids oxidation in sardines: Protective effects of aromatic herbs. *J. Food Sci.* **2017**, *82*, 2823–2831. [CrossRef] [PubMed]

109. Shalini, R.; Jeya Shakila, R.; Palani Kumar, M.; Jeyasekaran, G. Changes in the pattern of health beneficial omega 3 fatty acids during processing of sardine fish curry. *Indian J. Fish.* **2017**, *64*, 260–263. [CrossRef]

110. Dong, X.P.; Li, D.Y.; Huang, Y.; Wu, Q.; Liu, W.T.; Qin, L.; Zhou, D.Y.; Prakash, S.; Yu, C.X. Nutritional value and flavor of turbot (*Scophthalmus maximus*) muscle as affected by cooking methods. *Int. J. Food Prop.* **2018**, *21*, 1972–1985. [CrossRef]

111. Chaula, D.; Laswai, H.; Chove, B.; Dalsgaard, A.; Mdegela, R.; Hyldig, G. Fatty acid profiles and lipid oxidation status of sun dried, deep fried, and smoked sardine (*Rastrineobola argentea*) from Lake Victoria, Tanzania. *J. Aquat. Food Prod. Technol.* **2019**, *2*, 165–176. [CrossRef]

112. Candela, M.; Astiasaran, I.; Bello, J. Deep-fat frying modifies high-fat fish lipid fraction. *J. Agric. Food Chem.* **1998**, *46*, 2793–2796. [CrossRef]

113. Echarte, M.; Zulet, M.A.; Astiasaran, I. Oxidation process affecting fatty acids and cholesterol in fried and roasted salmon. *J. Agric. Food Chem.* **2001**, *49*, 5662–5667. [CrossRef] [PubMed]

114. Sioen, I.; Haak, L.; Raes, K.; Hermans, C.; De Henauw, S.; De Smet, S.; Van Camp, J. Effects of pan-frying in margarine and olive oil on the fatty acid composition of cod and salmon. *Food Chem.* **2006**, *98*, 609–617. [CrossRef]

115. Stolyhwo, A.; Kolodziejska, I.; Sikorski, Z.E. Long chain polyunsaturated fatty acids in smoked Atlantic mackerel and Baltic sprats. *Food Chem.* **2006**, *94*, 589–595. [CrossRef]

116. Gladyshev, M.I.; Sushchik, N.N.; Gubanenko, G.A.; Demirchieva, S.M.; Kalachova, G.S. Effect of way of cooking on content of essential polyunsaturated fatty acids in muscle tissue of humpback salmon (*Oncorhynchus gorbuscha*). *Food Chem.* **2006**, *96*, 446–451. [CrossRef]

117. Gladyshev, M.I.; Sushchik, N.N.; Gubanenko, G.A.; Demirchieva, S.M.; Kalachova, G.S. Effect of boiling and frying on the content of essential polyunsaturated fatty acids in muscle tissue of four fish species. *Food Chem.* **2007**, *101*, 1694–1700. [CrossRef]

118. Gladyshev, M.I.; Sushchik, N.N.; Makhutova, O.N.; Kalachova, G.S. Content of essential polyunsaturated fatty acids in three canned fish species. *Int. J. Food Sci. Nutr.* **2009**, *60*, 224–230. [CrossRef] [PubMed]

119. Haak, L.; Sioen, I.; Raes, K.; Van Camp, J.; De Smet, S. Effect of pan-frying in different culinary fats on the fatty acid profile of pork. *Food Chem.* **2007**, *102*, 857–864. [CrossRef]

120. Yanar, Y.; Büyükçapar, H.M.; Yanar, M.; Göcer, M. Effect of carotenoids from red pepper and marigold flower on pigmentation, sensory properties and fatty acid composition of rainbow trout. *Food Chem.* **2007**, *100*, 326–330. [CrossRef]

121. Ansorena, D.; Guembe, A.; Mendizabal, T.; Astiasaran, I. Effect of fish and oil nature on frying process and nutritional product quality. *J. Food Sci.* **2010**, *75*, H62–H67. [CrossRef] [PubMed]

122. Neff, M.R.; Bhavsar, S.P.; Braekevelt, E.; Arts, M.T. Effects of different cooking methods on fatty acid profiles in four freshwater fishes from the Laurentian Great Lakes region. *Food Chem.* **2014**, *164*, 544–550. [CrossRef] [PubMed]

123. Cheung, L.K.Y.; Tomita, H.; Takemori, T. Mechanisms of docosahexaenoic and eicosapentaenoic acid loss from Pacific saury and comparison of their retention rates after various cooking methods. *J. Food Sci.* **2016**, *81*, C1899–C1907. [CrossRef] [PubMed]

124. Kronberg, S.L.; Scholljegerdes, E.J.; Maddock, R.J.; Barcelo-Coblijn, G.; Murphy, E.J. Rump and shoulder muscles from grass and linseed fed cattle as important sources of n-3 fatty acids for beef consumers. *Eur. J. Lipid Sci. Technol.* **2017**, *119*, 1600390. [CrossRef]

125. Litzow, M.A.; Bailey, K.M.; Prahl, F.G.; Heintz, R. Climate regime shifts and reorganization of fish communities: The essential fatty acid limitation hypothesis. *Mar. Ecol. Prog. Ser.* **2006**, *315*, 1–11. [CrossRef]

126. Wells, M.L.; Potin, P.; Craigie, J.S.; Raven, J.A.; Merchant, S.S.; Helliwell, K.E.; Smith, A.G.; Camire, M.E.; Brawley, S.H. Algae as nutritional and functional food sources: Revisiting our understanding. *J. Appl. Phycol.* **2017**, *29*, 949–982. [CrossRef] [PubMed]

127. Woods, V.B.; Fearon, A.M. Dietary sources of unsaturated fatty acids for animals and their transfer into meat, milk and eggs: A review. *Livest. Sci.* **2009**, *126*, 1–20. [CrossRef]

128. Turchini, G.M.; Hermon, K.M.; Francis, D.S. Fatty acids and beyond: Fillet nutritional characterisation of rainbow trout (*Oncorhynchus mykiss*) fed different dietary oil sources. *Aquaculture* **2018**, *491*, 391–397. [CrossRef]

129. Gladyshev, M.I.; Sushchik, N.N.; Tolomeev, A.P.; Dgebuadze, Y.Y. Meta-analysis of factors associated with omega-3 fatty acid contents of wild fish. *Rev. Fish. Biol. Fish.* **2018**, *28*, 277–299. [CrossRef]

130. Mairesse, G.; Thomas, M.; Gardeur, J.-N.; Brun-Bellut, J. Effects of geographic source rearing system, and season on the nutritional quality of wild and farmed *Perca fluviatilis*. *Lipids* **2006**, *41*, 221–229. [CrossRef]

131. Kiessling, A.; Pickova, J.; Johansson, L.; Esgerd, T.; Storebakken, T.; Kiessling, K.-H. Changes in fatty acid composition in muscle and adipose tissue of farmed rainbow trout (*Oncorhynchus mykiss*) in relation to ration and age. *Food Chem.* **2001**, *73*, 271–284. [CrossRef]

132. Benedito-Palos, L.; Calduch-Giner, J.A.; Ballester-Lozano, G.F.; Pérez-Sánchez, J. Effect of ration size on fillet fatty acid composition, phospholipid allostasis and mRNA expression patterns of lipid regulatory genes in gilthead sea bream (*Sparus aurata*). *Br. J. Nutr.* **2013**, *109*, 1175–1187. [CrossRef]

133. Moths, M.D.; Dellinger, J.A.; Holub, B.; Ripley, M.P.; McGraw, J.E.; Kinnunen, R.E. Omega-3 fatty acids in fish from the Laurentian Great Lakes tribal fisheries. *Hum. Ecol. Risk Assess.* **2013**, *19*, 1628–1643. [CrossRef]

134. Huynh, M.D.; Kitts, D.D. Evaluating nutritional quality of pacific fish species from fatty acid signatures. *Food Chem.* **2009**, *114*, 912–918. [CrossRef]

135. Joordens, J.C.A.; Kuipers, R.S.; Wanink, J.H.; Muskiet, F.A.J. A fish is not a fish: Patterns in fatty acid composition of aquatic food may have had implications for hominin evolution. *J. Hum. Evol.* **2014**, *77*, 107–116. [CrossRef] [PubMed]

136. Anishchenko, O.V.; Sushchik, N.N.; Makhutova, O.N.; Kalachova, G.S.; Gribovskaya, I.V.; Morgun, V.N.; Gladyshev, M.I. Benefit-risk ratio of canned pacific saury (*Cololabis saira*) intake: Essential fatty acids vs. heavy metals. *Food Chem. Toxicol.* **2017**, *101*, 8–14. [CrossRef] [PubMed]

137. Larsen, D.; Quek, S.Y.; Eyres, L. Effect of cooking method on the fatty acid profile of New Zealand King Salmon (*Oncorhynchus tshawytscha*). *Food Chem.* **2010**, *119*, 785–790. [CrossRef]

138. Simon, S.J.G.B.; Sancho, R.A.S.; Lima, F.A.; Cabral, C.C.V.Q.; Souza, T.M.; Bragagnolo, N.; Lira, G.M. Interaction between soybean oil and the lipid fraction of fried pitu prawn. *LWT-Food Sci. Technol.* **2012**, *48*, 120–126. [CrossRef]

139. Amira, M.B.; Hanene, J.H.; Madiha, D.; Imen, B.; Mohamed, H.; Abdelhamid, C. Effects of frying on the fatty acid composition in farmed and wild gilthead sea bream (*Sparus aurata*). *Int. J. Food Sci. Technol.* **2010**, *45*, 113–123. [CrossRef]

140. Kouba, M.; Benatmane, F.; Blochet, J.E.; Mourot, J. Effect of a linseed diet on lipid oxidation, fatty acid composition of muscle, perirenal fat, and raw and cooked rabbit meat. *Meat Sci.* **2008**, *80*, 829–834. [CrossRef]

8

Comparison of Fatty Acid Contents in Major Lipid Classes of Seven Salmonid Species from Siberian Arctic Lakes

Nadezhda N. Sushchik [1,2,*], Olesia N. Makhutova [1,2], Anastasia E. Rudchenko [1,2], Larisa A. Glushchenko [2], Svetlana P. Shulepina [2], Anzhelika A. Kolmakova [1] and Michail I. Gladyshev [1,2]

[1] Institute of Biophysics of Federal Research Center "Krasnoyarsk Science Center" of Siberian Branch of Russian Academy of Sciences, Akademgorodok, 50/50, Krasnoyarsk 660036, Russia; makhutova@ibp.krasn.ru (O.N.M.); rudchenko.a.e@gmail.com (A.E.R.); angelika_@inbox.ru (A.A.K.); glad@ibp.ru (M.I.G.)

[2] Siberian Federal University, Svobodny av., 79, Krasnoyarsk 660041, Russia; loraglushchenko@gmail.com (L.A.G.); shulepina@mail.ru (S.P.S.)

* Correspondence: labehe@ibp.ru

Abstract: Long-chain omega-3 polyunsaturated fatty acids (LC-PUFA) essential for human nutrition are mostly obtained from wild-caught fish. To sustain the LC-PUFA supply from natural populations, one needs to know how environmental and intrinsic factors affect fish fatty acid (FA) profiles and contents. We studied seven Salmoniformes species from two arctic lakes. We aimed to estimate differences in the FA composition of total lipids and two major lipid classes, polar lipids (PL) and triacylglycerols (TAG), among the species and to evaluate LC-PUFA contents corresponding to PL and TAG in muscles. Fatty acid profiles of PL and TAG in all species were characterized by the prevalence of omega-3 LC-PUFA and C16-C18 monoenoic FA, respectively. Fish with similar feeding spectra were identified similarly in multivariate analyses of total lipids, TAG and PL, due to differences in levels of mostly the same FA. Thus, the suitability of both TAG and total lipids for the identification of the feeding spectra of fish was confirmed. All species had similar content of LC-PUFA esterified as PL, 1.9–3.5 mg g^{-1}, while the content of the TAG form strongly varied, from 0.9 to 9.8 mg g^{-1}. The LC-PUFA-rich fish species accumulated these valuable compounds predominately in the TAG form.

Keywords: arctic; Salmoniformes; long-chain polyunsaturated fatty acids; polar lipids; triacylglycerols; eicosapentaenoic acid; docosahexaenoic acid

1. Introduction

Long-chain omega-3 polyunsaturated fatty acids (LC-PUFA), such as eicosapentaenoic acid (EPA) and docosahexaenoic acid (DHA), are known to be essential compounds for human nutrition, since they can modulate the functioning of cardiovascular and neural systems and general metabolism, being the precursors for the synthesis of diverse lipid mediators and directly affecting membrane properties [1–5]. Most international and national health agencies and foundations recommended personal consumption of 0.5–1.0 g of EPA+DHA per day for reducing the risk of cardiovascular diseases and other metabolic disorders [6–8]. Although a lot of potential sources of LC-PUFA are now being considered, natural fish populations are still the major source of these compounds for human nutrition [9,10]. Recent reviews showed the deficiency of the LC-PUFA supply with fish caught from natural populations and emphasized the potential negative influence of some global threats, like climate change, pollution, eutrophication, etc. [10–13]. To challenge above threats and to sustain the LC-PUFA supply from natural populations, one needs to know how environmental and intrinsic factors affect fish fatty acid profiles and content, including those of EPA and DHA. Causes of variations of fatty acid composition and content in wild fish are still incompletely understood [10,14].

The ability of fish to deposit fat (lipids) in muscles varies from species to species and may be a crucial intrinsic factor [15]. According to their functions, lipids in fish, like in other animals, could roughly be divided into energy-reserve and membrane-structural groups [16,17]. Fish reserve lipids are primarily represented by triacylglycerols (TAG) and include mostly fatty acids that come from food sources. Fatty acid profiles of TAG are generally considered as valuable trophic markers due to their resemblance with fatty acid profiles of particular food sources [18]. In addition, the TAG fraction in fish can also contain high levels of monoenoic C16-C18 fatty acids that are intensively synthesized in so called "fatty" fish species to provide energy reserves. TAG molecules are accumulated either directly in muscle cells as droplets or in specific adipocytes which may be integrated in muscle tissues or form separate layers of adipose tissue.

The structural polar lipids (PL) that form fish cellular and intracellular membranes mostly comprise phospholipids [19,20]. As known, fatty acid composition of PL affects physico-chemical properties of cellular membranes. Hence, PL are considered to have conservative fatty acid profiles which slightly reflect that of diet. The essential omega-3 LC-PUFA are preferentially accumulated in PL fraction of muscle tissues due to their strong membrane-modulating properties. Thus, fatty acid profiles of the major lipid classes, TAG and PL, in fish muscles are different in general [19].

TAG content per mass unit of fish muscles is highly variable due to influence of many factors [15]. In contrast, PL content per mass unit of muscles is fairly constant [21]. Thereby, we hypothesize that PL specific content has a putative upper threshold, because amounts of PL molecules in tissue are likely determined by a volume of membranes.

Contents of omega-3 LC-PUFA in muscle tissue of diverse fish species greatly vary, approximately ~400-fold [22]. The question arises what part of this variation in total EPA and DHA contents is provided by TAG or PL variability? There is a basic assumption in the current literature that a major part of omega-3 LC-PUFA presents as acyl groups of membrane phospholipid molecules [11].

To evaluate contribution of the two major lipid fractions in total content of LC-PUFA in edible muscle tissue (filets) we studied seven commercial species of the order Salmoniformes that inhabit oligotrophic non-polluted lakes in Arctic Siberia. The fish species vary in their feeding habits and habitats and have different fat content in filets. Using data on these fish we aimed to compare distribution of fatty acids, including omega-3 LC-PUFA in total lipids and two major lipid classes: TAG and PL. Specifically, we aimed (i) to check if the fish species with various feeding spectra can be differentiated basing on FA profiles of total lipids, TAG or PL, (ii) to evaluate LC-PUFA content corresponded to TAG and PL classes in muscles, (iii) to range species according to their nutritive value for humans in respect of LC-PUFA content.

2. Materials and Methods

2.1. Sampling

Fish specimen of commercial sizes were collected during July 2017 from catches of local authorized fishers. Following sampling was done in accordance with the BioEthics Protocol on Animal Care approved by Siberian Federal University. The catches were from two oligotrophic arctic lakes, Sobachye and Pyasino. Sobachye Lake was previously characterized elsewhere [23]. Briefly, it is located at 69°01′ N 91°05′ E and has maximum depth of 162 m and area equal to 99 km². Pyasino Lake situates at 69°40′ N 87°51′ E, has average depth of 4 m and an area equal to 735 km² [24].

Whitefish *Coregonus lavaretus* (Linnaeus, 1758), non-identified form of whitefish *C. lavaretus*, round whitefish *Prosopium cylindraceum* (Pennant, 1784) and charr *Salvelinus drjagini* Logashev, 1940 were caught in Sobachye Lake; broad whitefish *Coregonus nasus* (Pallas, 1776), muksun *Coregonus muksun*

(Pallas, 1814) and inconnu *Stenodus leucichthys nelma* (Guldenstadt, 1772) were caught in Pyasino Lake. All studied fish species are of the Salmoniformes order. Numbers of samples, average individual sizes and weights, and main food sources for the studied fish species are given in Table 1. Stomach contents of all specimen were studied under a light microscope, and main food items were identified to a possible taxon level.

Table 1. The basic biological and sampling information on fish species (Salmoniformes order) from Siberian arctic lakes, 2017: *n*—number of sampled individuals; L—total length, cm (mean ± SE); W—total weight, g (mean ± SE); Food—items found in stomachs.

Common and Species Name	Lake	*n*	L	W	Food
Charr *Salvelinus drjagini*	Sobachye	9	608 ± 17	2371 ± 271	Fish (salmonids)
Whitefish *Coregonus lavaretus*	Sobachye	7	480 ± 23	1153 ± 167	Amphipods, mollusks, chironomid larvae
Muksun *Coregonus muksun*	Pyasino	8	492 ± 14	1271 ± 160	Ostracods, mollusks, chironomid larvae, detritus
Inconnu *Stenodus leucichthys nelma*	Pyasino	5	675 ± 86	3239 ± 1581	Fish
Broad whitefish *Coregonus nasus*	Pyasino	10	563 ± 17	1916 ± 183	Gastropods, detritus
Round whitefish *Prosopium cylindraceum*	Sobachye	7	409 ± 7	488 ± 27	Caddisfly and chironomid larvae, filamentous algae
Whitefish *Coregonus lavaretus* non-identified form	Sobachye	7	402 ± 12	568 ± 80	Chironomid and other insect pupa and adults

For biochemical analyses, we cut slices of fish white muscles of approximately 2–3 g, 2–3 cm below the dorsal fin. The samples were subdivided into two parts: for FA and moisture analyses. For FA analyses, ca. 1 g of muscle tissues was immediately placed into a volume of 3 mL of chloroform/methanol (2:1, by vol.) and kept until further analysis at −20 °C. Another subsample of ca. 1–2 g of wet weight was weighed, dried at 105 °C until constant weight, and weighed dry for moisture calculation.

2.2. Lipid and Fatty Acid Analyses

In laboratory, lipids were extracted with chloroform/methanol (2:1, by vol.) in triplicate, simultaneously with homogenizing tissues with glass beads in a mortar. Prior to the extraction, an aliquot of 19:0-fatty acid methyl ester (FAME) chloroform solution, as the internal standard, was added to samples for quantification of chromatographic peaks. The extracts were combined and dried with anhydrous Na_2SO_4 and the solvents were roto-evaporated under vacuum at 35 °C. The extracted lipids were redissolved in a 1 mL portion of chloroform and separated in two equal parts. To analyze the fatty acid composition of total lipids, one part of the lipid extract was methylated in the following way. The lipids were hydrolysed under reflux at 90° C for 10 min in 0.8 mL of methanolic sodium hydroxide solution (8 g/L). Then the mixture was cooled for 5 min at room temperature. Next, 1 mL of methanol/sulphuric acid (97:3, by vol.) was added and the mixture was heated under reflux at 90 °C for 10 min to methylate free fatty acids. At the end, 5 mL of saturated solution of NaCl and 3 mL of hexane were added. The FAMEs were extracted for 1 min, the mixture was transferred to a separatory funnel, and the lower aquatic layer was discarded. The hexane layer was additionally washed once with an aliquot of the NaCl solution and twice with 5 mL of distilled water. Then the hexane solution of FAMEs was dried with anhydrous Na_2SO_4, and hexane was removed by roto-evaporating at 30 °C.

We fractionated another part of the lipid extracts by thin layer chromatography (TLC) on silica gel G with hexane-diethyl ether-acetic acid mixture (85:15:1, by vol.). We prepared a reference mixture containing triolein, oleic acid, cholesterol, and phosphatidylcholine (Sigma, St. Louis, MO, USA), which was applied on the side lanes of the silica gel plates. To identify the lipid composition of fish species, we applied aliquots of samples to the plates, and then developed them as described above. After developing, the plates were sprayed with mixture of ethanol/sulphuric acid (90:10, by vol.) and gently heated until grey spots of lipid classes appeared. To evaluate fatty acid profiles and quantify dominant lipid classes, we separated a main portion of lipid extracts on the silica gel plates, and then visualized only the side lanes corresponding to the reference–compound mixture by a reaction with phosphomolybdic acid ethanolic solution. Lipid spots of the fish samples on the plates were blind detected according to positions of the reference compounds. The lipid spots containing TAG and PL fractions were scraped off from the silica gel plates and placed in flasks. Aliquots of 19:0-FAME hexane solution (t internal standard) were added into the flasks containing silica gel powder with the lipid fractions. To prepare FAMEs, 1ml of hexane and 0.2 mL of fresh 3 M sodium methoxide methanolic solution was added, and the mixture was shaken vigorously for 1 min. Subsequently, the mixture was kept calm at room temperature for 5 min, and finally 3 mL of hexane and 5 mL of a saturated solution of NaCl were added. The next procedure of FAME extraction and washing was the same as for those prepared from total lipids.

Analyses of all FAMEs were done with a gas chromatograph equipped with a mass spectrometer detector (model 6890/5975C; Agilent Technologies, Santa Clara, CA, USA) and with a 30-m long, 0.25-mm internal diameter capillary HP-FFAP column. Detailed descriptions of the chromatographic and mass-spectrometric conditions are given in [23].

2.3. Statistical Analysis

The Kolmogorov–Smirnov one-sample test for normality D_{K-S}, standard errors (SE), Student's *t*-tests, one-way ANOVA with post hoc Tukey HSD test, Kruskal–Wallis test (in the absence of normal distribution) and canonical correspondence analysis (CCA) were calculated conventionally, using STATISTICA software, version 9.0 (StatSoft Inc., Tulsa, OK, USA).

3. Results

According to gut content analysis, *C. lavaretus* in Sobachye Lake was benthivorous (Table 1). Whitefish of a non-identified form in this lake consumed mostly pupa and adult insects, i.e., foraged near the water surface. Round whitefish fed on benthic invertebrates and algae (Table 1). Both broad whitefish and muksun in Pyasino Lake consumed benthic food items, including detritus. Charr in Sobachye Lake and inconnu in Pyasino Lake were piscivorous (Table 1).

Average values of moisture content, lipid content and sum of fatty acid content for total lipids in the studied fish are given in Table 2. Lower moisture values were characteristic of the species with higher values of lipid and sum FA content, charr and whitefish (Table 2). In contrast, round whitefish and whitefish of the non-identified form had the maximum moisture content and the minimal contents of lipids and sum FA. Sum FA content of total lipids significantly varied ~7-fold among the studied species (Table 2). Based on the averages of lipid and sum FA contents, charr and whitefish are further considered as "fatty" fish, muksun, inconnu and broad whitefish as "medium fat" fish, and round whitefish and the non-identified form of whitefish as "lean" fish (Table 2).

Table 2. Average (± SE—standard errors) moisture content (% wet weight), lipid content (mg g^{-1} wet weight) and sum fatty acid content for total lipids (mg g^{-1} wet weight) in muscle tissues of fish species caught in Siberian arctic lakes, 2017. Means of total fatty acids labeled with the same letter are not significantly different at $P < 0.05$ after ANOVA post hoc Tukey HSD test.

Common and Species Name	Moisture	Lipids	Total Fatty Acids
Charr *Salvelinus drjagini*	69.8 ± 1.3	155.8 ± 7.4	78.8 ± 5.1 [D]
Whitefish *Coregonus lavaretus*	69.8 ± 1.4	82.0 ± 1.9	62.7 ± 7.2 [CD]
Muksun *Coregonus muksun*	74.4 ± 1.0	n.d.	45.4 ± 8.7 [BC]
Inconnu *Stenodus leucichthys nelma*	72.1 ± 2.6	68.4 ± 11.4	36.5 ± 10.0 [ABC]
Broad whitefish *Coregonus nasus*	73.8 ± 1.4	n.d.	31.9 ± 5.1 [AB]
Round whitefish *Prosopium cylindraceum*	76.1 ± 0.6	39.1 ± 3.3	13.8 ± 1.3 [A]
Whitefish *Coregonus lavaretus* non-identified form	76.0 ± 0.9	41.1 ± 2.1	11.5 ± 1.9 [A]

n.d.—no data.

Levels of 25 prominent individual FA and their structural groups in total lipids are showed in Table 3. Charr had the highest levels of 20:2n-6, 20:3n-3, 20:4n-3, 22:4n-3 and C24 PUFA among the studied species (Table 3). Whitefish had the significantly highest levels of 16:1n-7 and C16 PUFA, and tended to be higher in levels of 18:1n-9 and 20:5n-3. Muksun tended to have higher levels of 14:0, 20:1 and 22:5n-6 (Table 3). Broad whitefish had higher levels of 16:1n-9, C15-17 BFA (branched-chain fatty acids), 18:0, 18:1n-7, 18:2n-6, 18:3n-3 (Table 3). Whitefish of the non-identified form had higher levels of 16:0 and 22:6n-3 compared to those of the other fish. Inconnu and round whitefish had intermediate levels of all FA in total lipids (Table 3).

Table 3. Mean levels of fatty acids (% of the total) in total lipids of species of Salmoniformes order: charr—*S. drjagini* from Sobachye Lake; whitefish—*C. lavaretus* from Sobachye Lake; muksun—*C. muksun* from Pyasino Lake; inconnu—*S. leucichthys nelma* from Pyasino Lake; broad whitefish—*C. nasus* from Pyasino Lake; round whitefish—*P. cylindraceum* from Sobachye Lake; whitefish nd—the non-identified form of *C. lavaretus* from Sobachye Lake. Cases (fatty acids) with normal distribution are given in bold. Means labeled with the same letter are not significantly different at $P < 0.05$ after ANOVA post hoc Tukey HSD test (cases with normal distribution) or Kruskal–Wallis test. If ANOVA is insignificant, letters are absent.

Fatty Acid	Charr	Whitefish	Muksun	Inconnu	Broad Whitefish	Round Whitefish	Whitefish nd
14:0	4.0 ± 0.1^{A}	2.9 ± 0.1^{B}	4.2 ± 0.2^{A}	3.4 ± 0.1^{AC}	2.5 ± 0.2^{B}	2.2 ± 0.2^{B}	2.3 ± 0.2^{B}
15:0	0.3 ± 0.0^{A}	0.2 ± 0.0^{CD}	0.5 ± 0.0^{B}	0.4 ± 0.0^{A}	0.5 ± 0.0^{B}	0.2 ± 0.0^{D}	0.3 ± 0.0^{C}
16:0	15.4 ± 0.2^{C}	14.9 ± 0.2^{C}	15.2 ± 0.2^{C}	16.3 ± 0.7^{AC}	18.2 ± 0.4^{AB}	16.6 ± 0.2^{AC}	18.9 ± 0.8^{B}
16:1n-9	0.4 ± 0.0^{ABD}	0.2 ± 0.0^{B}	0.3 ± 0.0^{CD}	0.4 ± 0.0^{AC}	0.5 ± 0.1^{B}	0.2 ± 0.0^{C}	0.4 ± 0.1^{ABD}
16:1n-7	6.6 ± 0.2^{B}	17.4 ± 0.5^{D}	10.1 ± 0.3^{AC}	13.4 ± 1.0^{ACE}	11.8 ± 0.8^{C}	15.2 ± 0.4^{DE}	9.7 ± 1.4^{BC}
15-17BFA	$1.8 + 0.0^{A}$	1.0 ± 0.0^{CD}	1.8 ± 0.1^{A}	1.3 ± 0.0^{AC}	2.5 ± 0.2^{B}	0.6 ± 0.0^{D}	1.0 ± 0.1^{CD}
16PUFA	0.2 ± 0.0^{B}	4.6 ± 0.2^{C}	2.3 ± 0.2^{DE}	1.5 ± 0.2^{AD}	1.3 ± 0.1^{A}	2.9 ± 0.1^{E}	1.5 ± 0.4^{AD}
17:0	0.2 ± 0.0^{A}	0.3 ± 0.0^{D}	0.4 ± 0.0^{B}	0.2 ± 0.0^{AD}	0.4 ± 0.0^{B}	0.1 ± 0.0^{C}	0.2 ± 0.0^{A}
18:0	3.2 ± 0.1^{C}	1.9 ± 0.0^{D}	2.4 ± 0.1^{A}	2.5 ± 0.2^{AB}	3.0 ± 0.1^{BC}	2.8 ± 0.0^{AB}	2.5 ± 0.2^{A}
18:1n-9	17.9 ± 0.1^{AC}	19.8 ± 0.2^{C}	16.1 ± 0.4^{A}	16.8 ± 0.8^{AC}	16.3 ± 0.7^{A}	11.7 ± 0.6^{B}	12.3 ± 1.0^{B}
18:1n-7	3.1 ± 0.0^{C}	3.9 ± 0.1^{AC}	4.0 ± 0.3^{AC}	4.3 ± 0.2^{AB}	5.1 ± 0.1^{B}	4.6 ± 0.1^{AB}	3.7 ± 0.5^{AC}
18:2n-6	3.0 ± 0.1^{AC}	2.1 ± 0.1^{A}	2.8 ± 0.1^{AC}	2.7 ± 0.2^{AC}	4.7 ± 0.6^{B}	3.7 ± 0.1^{BC}	2.7 ± 0.3^{AC}
18:3n-3	2.6 ± 0.1^{AC}	1.3 ± 0.0^{A}	2.9 ± 0.2^{BC}	2.2 ± 0.2^{AC}	3.8 ± 0.6^{B}	2.7 ± 0.2^{AB}	1.6 ± 0.1^{AC}
18:4n-3	1.7 ± 0.0	1.8 ± 0.1	1.8 ± 0.1	1.6 ± 0.1	1.5 ± 0.2	1.4 ± 0.2	1.3 ± 0.2
Σ20:1 *	1.6 ± 0.0^{AD}	1.2 ± 0.1^{BCD}	2.3 ± 0.1^{A}	2.1 ± 0.9^{ABC}	1.7 ± 0.3^{AB}	0.6 ± 0.0^{C}	0.8 ± 0.1^{BC}
20:2n-6	1.0 ± 0.0^{C}	0.3 ± 0.0^{A}	0.6 ± 0.0^{B}	0.4 ± 0.0^{A}	0.6 ± 0.0^{B}	0.3 ± 0.0^{A}	0.4 ± 0.1^{A}
20:4n-6	1.9 ± 0.0^{AC}	1.6 ± 0.0^{CD}	2.4 ± 0.1^{B}	2.3 ± 0.3^{AB}	2.8 ± 0.1^{B}	1.4 ± 0.1^{D}	2.6 ± 0.2^{B}
20:3n-3	1.5 ± 0.0^{B}	0.2 ± 0.0^{C}	0.6 ± 0.0^{AB}	0.4 ± 0.0^{ABC}	0.3 ± 0.0^{AC}	0.2 ± 0.0^{C}	0.4 ± 0.1^{AC}
20:4n-3	2.9 ± 0.0^{A}	0.7 ± 0.0^{BC}	1.2 ± 0.1^{AC}	1.1 ± 0.1^{AC}	0.6 ± 0.0^{B}	0.8 ± 0.0^{BC}	0.8 ± 0.1^{BC}
20:5n-3	4.8 ± 0.1^{C}	10.4 ± 0.2^{A}	9.6 ± 0.2^{AB}	7.4 ± 0.7^{ABC}	6.5 ± 0.3^{BC}	10.2 ± 0.1^{A}	10.3 ± 0.3^{A}
22:5n-6	1.2 ± 0.0^{B}	0.3 ± 0.0^{C}	1.3 ± 0.1^{B}	1.0 ± 0.1^{AB}	0.8 ± 0.1^{A}	0.3 ± 0.0^{C}	0.8 ± 0.1^{A}
22:4n-3	1.5 ± 0.1^{C}	0.0 ± 0.0^{AB}	0.1 ± 0.0^{AC}	0.2 ± 0.1^{AC}	0.0 ± 0.0^{B}	0.0 ± 0.0^{AB}	0.1 ± 0.0^{ABC}
22:5n-3	3.0 ± 0.1^{CD}	2.5 ± 0.1^{AC}	2.5 ± 0.1^{AC}	2.3 ± 0.1^{A}	1.7 ± 0.1^{B}	3.0 ± 0.0^{D}	2.4 ± 0.1^{A}
22:6n-3	12.1 ± 0.1^{AC}	6.3 ± 0.2^{D}	9.6 ± 0.6^{ADE}	11.3 ± 1.1^{ACD}	7.7 ± 0.9^{AD}	14.4 ± 1.5^{BCE}	20.1 ± 3.0^{B}
24PUFA	4.3 ± 0.3^{B}	1.0 ± 0.1^{AB}	1.1 ± 0.1^{BC}	1.0 ± 0.2^{AC}	0.7 ± 0.0^{A}	0.8 ± 0.0^{A}	0.6 ± 0.1^{AC}

* sum of 20:1n-11, 20:1n-9 and 20:1n-7, here and in other Tables.

We performed CCA of FA profiles of total lipids of the fish species to find out their overall differences (Figure 1). Along Dimension 1, most difference in FA composition was observed between charr, on the one hand, and whitefish, on the other hand. The differences were primarily caused by higher percentages of 22:4n-3, 20:3n-3 and C24 PUFA in charr, and higher percentages of C16 PUFA and 16:1n-7 in whitefish. In Dimension 2, whitefish of a non-identified form and round whitefish located at the one end and broad whitefish was at the other end (Figure 1). Whitefish of the non-identified form and round whitefish were separated due to higher levels of 22:6n-3 and 20:5n-3, and partial separation of broad whitefish was due to higher levels of C15-17 BFA (Table 3). Samples of the non-identified whitefish were markedly scattered (Figure 1) due to the high variability in percentages of 22:6n-3 (Table 3).

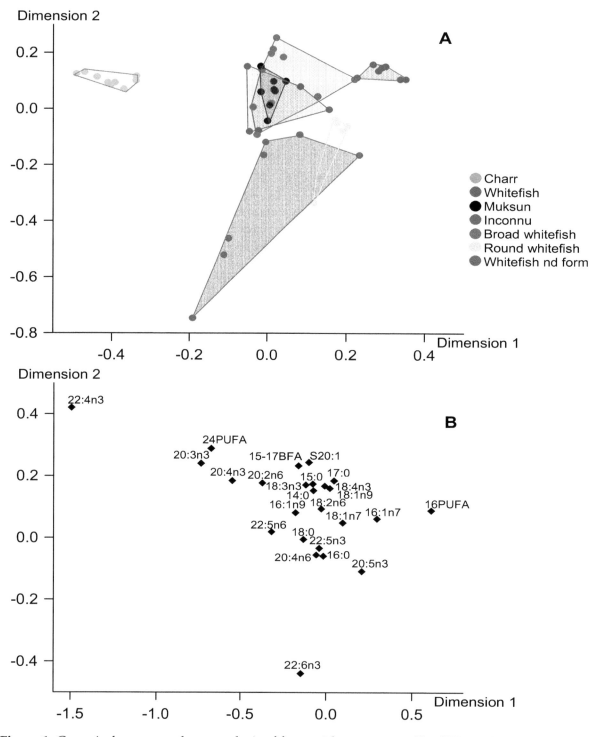

Figure 1. Canonical correspondence analysis of fatty acid percentages (% of FA sum) in total lipids of muscles of seven fish species from arctic lakes (Siberia, Russia). **A**—individual variables, **B**—factor structure coefficients for fatty acids. Dimension 1 and Dimension 2 represented 33.9% and 21.2% of inertia, respectively.

In PL of all studied species, 22:6n-3, 16:0 and 20:5n-3 were dominant fatty acids (Table 4). Charr had the highest levels of 22:6n-3, 22:4n-3, 22:5n-6, 20:4n-3 among the studied species (Table 4). Whitefish had a significantly higher level of 20:5n-3 compared to that of the other fish. Inconnu tended to be higher in 18:1n-9 level. Broad whitefish had the highest levels of 20:4n-6 and 18:2n-6 (Table 4). Round whitefish had significantly higher levels of 16:1n-7, 18:1n-7 and C16 PUFA compared to those of the other fish. Muksun and whitefish of the non-identified form had intermediate levels of most FA in PL (Table 4).

Table 4. Mean levels of fatty acids (% of the total) and total content of fatty acids (Sum FA, mg g^{-1} wet weight) in polar lipids of Salmoniformes species: charr—*S. drjagini* from Sobachye Lake; whitefish—*C. lavaretus* from Sobachye Lake; muksun—*C. muksun* from Pyasino Lake; inconnu—*S. leucichthys nelma* from Pyasino Lake; broad whitefish—*C. nasus* from Pyasino Lake; round whitefish—*P. cylindraceum* from Sobachye Lake; whitefish nd—the non-identified form of *C. lavaretus* from Sobachye Lake. Cases (fatty acids) with normal distribution are given in bold. Means labeled with the same letter are not significantly different at $P < 0.05$ after ANOVA post hoc Tukey HSD test (cases with normal distribution) or Kruskal–Wallis test. If ANOVA is insignificant, letters are absent.

Fatty Acid	Charr	Whitefish	Muksun	Inconnu	Broad Whitefish	Round Whitefish	Whitefish nd
14:0	1.2 ± 0.1 AB	0.9 ± 0.0 A	1.1 ± 0.1 AB	1.5 ± 0.2 B	1.2 ± 0.1 AB	0.9 ± 0.1 A	1.4 ± 0.1 B
15:0	0.2 ± 0.0 A	0.2 ± 0.0 A	0.4 ± 0.0 CD	0.3 ± 0.0 BC	0.5 ± 0.0 D	0.2 ± 0.0 A	0.3 ± 0.0 AB
16:0	24.2 ± 0.4 A	29.4 ± 0.5 BC	26.3 ± 0.6 AB	25.9 ± 0.5 AB	27.3 ± 0.8 AB	29.1 ± 1.3 BC	31.7 ± 1.1 C
16:1n-9	0.3 ± 0.0 AB	0.1 ± 0.0 A	0.3 ± 0.0 BC	0.2 ± 0.1 AB	0.5 ± 0.1 C	0.2 ± 0.0 AB	0.3 ± 0.1 BC
16:1n-7	1.3 ± 0.1 A	2.2 ± 0.1 BC	1.9 ± 0.2 AB	2.3 ± 0.1 B	2.9 ± 0.2 B	4.1 ± 0.2 D	3.0 ± 0.2 C
15-17BFA	0.6 ± 0.0 C	0.1 ± 0.0 A	0.4 ± 0.0 B	0.3 ± 0.0 AB	0.6 ± 0.1 C	0.2 ± 0.0 AB	0.3 ± 0.1 AB
16PUFA	0.0 ± 0.0 A	0.2 ± 0.0 B	0.1 ± 0.1 AB	0.1 ± 0.0 AB	0.1 ± 0.0 AB	0.3 ± 0.0 C	0.1 ± 0.0 AB
17:0	0.2 ± 0.0 A	0.2 ± 0.0 AB	0.3 ± 0.0 B	0.2 ± 0.0 AB	0.3 ± 0.0 B	0.1 ± 0.0 A	0.2 ± 0.0 A
18:0	2.8 ± 0.1 AB	2.1 ± 0.1 A	3.0 ± 0.1 B	2.9 ± 0.3 AB	2.2 ± 0.2 A	3.4 ± 0.2 B	2.8 ± 0.3 AB
18:1n-9	6.2 ± 0.3 AB	6.5 ± 0.2 AB	6.6 ± 0.4 AB	7.8 ± 0.4 B	6.1 ± 0.5 A	5.6 ± 0.3 A	6.9 ± 0.3 AB
18:1n-7	1.6 ± 0.1 A	1.7 ± 0.1 AB	2.2 ± 0.2 BC	2.4 ± 0.1 C	2.5 ± 0.1 C	3.3 ± 0.2 D	2.4 ± 0.3 C
18:2n-6	0.7 ± 0.0 A	0.8 ± 0.0 A	1.0 ± 0.0 A	1.1 ± 0.1 A	2.5 ± 0.4 B	1.6 ± 0.0 A	1.4 ± 0.3 A
18:3n-3	0.6 ± 0.0 AB	0.4 ± 0.0 A	1.0 ± 0.1 ABC	1.0 ± 0.2 ABC	2.1 ± 0.4 C	1.3 ± 0.1 BC	0.7 ± 0.1 ABC
18:4n-3	0.1 ± 0.0 AB	0.0 ± 0.0 A	0.2 ± 0.0 B	0.2 ± 0.0 AB	0.1 ± 0.0 AB	0.3 ± 0.0 B	0.2 ± 0.1 AB
Σ20:1	0.2 ± 0.0 BC	0.1 ± 0.0 A	0.3 ± 0.0 C	0.1 ± 0.0 ABC	0.1 ± 0.0 AB	0.2 ± 0.0 ABC	0.2 ± 0.1 AC
20:2n-6	0.2 ± 0.0 BC	0.0 ± 0.0 A	0.1 ± 0.0 AB	0.1 ± 0.0 A	0.2 ± 0.0 C	0.1 ± 0.0 AC	0.1 ± 0.0 AB
20:4n-6	3.9 ± 0.1 BC	3.0 ± 0.1 AB	4.1 ± 0.2 C	3.8 ± 0.3 BC	5.6 ± 0.3 D	2.4 ± 0.1 A	3.2 ± 0.2 ABC
20:3n-3	0.5 ± 0.0 C	0.1 ± 0.0 A	0.2 ± 0.0 B	0.1 ± 0.0 AB	0.2 ± 0.0 B	0.1 ± 0.0 AB	0.1 ± 0.0 AB
20:4n-3	1.1 ± 0.0 B	0.3 ± 0.0 A	0.5 ± 0.0 AB	0.5 ± 0.1 AB	0.4 ± 0.1 A	0.5 ± 0.0 AB	0.4 ± 0.0 A
20:5n-3	7.9 ± 0.2 A	16.6 ± 0.6 D	12.5 ± 0.3 BC	11.2 ± 1.1 BC	10.9 ± 0.4 B	13.6 ± 1.0 C	10.0 ± 0.5 AB
22:5n-6	2.4 ± 0.1 C	0.6 ± 0.0 A	2.0 ± 0.3 C	1.7 ± 0.2 BC	2.2 ± 0.2 C	0.4 ± 0.0 A	0.9 ± 0.1 AB
22:4n-3	0.2 ± 0.0 B	0.0 ± 0.0 A	0.0 ± 0.0 A	0.0 ± 0.0 A	0.0 ± 0.0 A	0.0 ± 0.0 A	0.0 ± 0.0 A
22:5n-3	2.5 ± 0.0 A	2.7 ± 0.2 AB	2.5 ± 0.1 A	2.4 ± 0.3 A	2.6 ± 0.1 AB	3.2 ± 0.3 B	2.1 ± 0.1 A
22:6n-3	39.9 ± 0.5 D	31.1 ± 0.7 BC	31.8 ± 0.5 C	32.4 ± 0.9 C	26.4 ± 0.6 A	27.9 ± 0.7 AB	30.1 ± 1.1 BC
24PUFA	0.3 ± 0.1	0.0 ± 0.0	0.0 ± 0.0	0.4 ± 0.4	0.1 ± 0.0	0.0 ± 0.0	0.0 ± 0.0
Sum FA	*3.0 ± 0.2 AB*	*3.3 ± 0.5 AB*	*3.2 ± 0.9 AB*	*2.6 ± 0.2 AB*	*3.4 ± 0.2 B*	*3.2 ± 0.7 AB*	*2.1 ± 0.2 A*

To reveal overall differences in PL FA, CCA was performed (Figure 2). In the first dimension, a conspicuous difference of round whitefish versus charr was found. This difference was provided mostly by the greater levels of C16 PUFA, 16:1n-7 and 18:2n-6 in PL of round whitefish and by greater levels of C24 PUFA and 22:4n-3 in that of charr (Figure 2). The variation in the second dimension of CCA was related to differences between whitefish and broad whitefish due to levels of 15:0 and 14:0 versus levels of C16 PUFA, 18:2n-6 and 18:3n-3.

In FA composition of triacylglycerols of all the studied arctic fish, 18:1n-9, 16:1n-7, 16:0 and 20:5n-3 dominated (Table 5). Charr had the highest levels of 22:6n-3, C24 PUFA, 22:5n-3, 20:4n-3 and 20:3n-3 compared to that of the other studied species. Whitefish had the maximum levels of 18:1n-9 and 20:5n-3 (Table 5). Muksun had the significantly higher level of 14:0 than the other fish. Inconnu and whitefish of the non-identified form had intermediate FA levels in TAG (Table 5). Broad whitefish had the maximum levels of 16:0, C15-17 BFA, 17:0, 18:2n-6, 18:3n-3, and 20:4n-6 among the studied fish. Round whitefish had the significantly higher percentages of 16:1n-7 and C16 PUFA in TAG (Table 5).

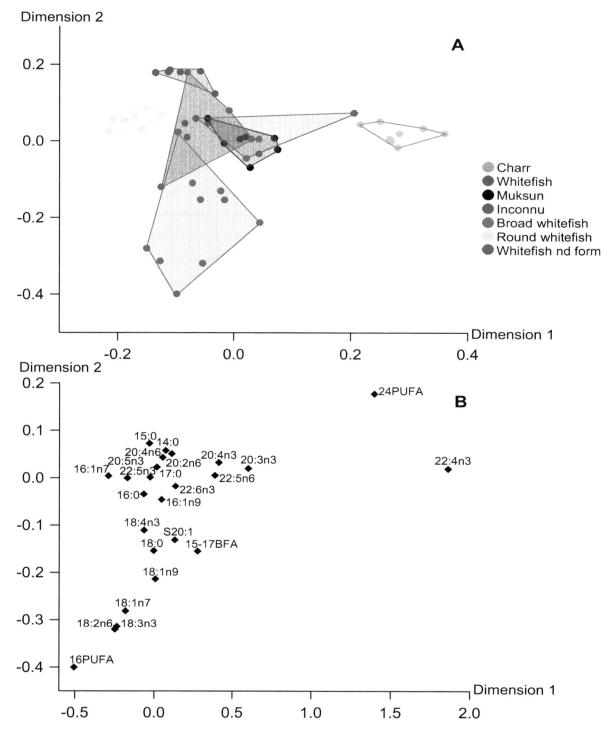

Figure 2. Canonical correspondence analysis of fatty acid percentages (% of FA sum) in polar lipids of muscles of seven fish species from arctic lakes (Siberia, Russia). **A**—individual variables, **B**—factor structure coefficients for fatty acids. Dimension 1 and Dimension 2 represented 30.9% and 22.5% of inertia, respectively.

Table 5. Mean levels of fatty acids (% of the total) and total content of fatty acids (Sum FA, mg g^{-1} wet weight) in triacylglycerols of Salmoniformes species: charr—*S. drjagini* from Sobachye Lake; whitefish—*C. lavaretus* from Sobachye Lake; muksun—*C. muksun* from Pyasino Lake; inconnu—*S. leucichthys nelma* from Pyasino Lake; broad whitefish—*C. nasus* from Pyasino Lake; round whitefish—*P. cylindraceum* from Sobachye Lake; whitefish nd—the non-identified form of *C. lavaretus* from Sobachye Lake. Cases (fatty acids) with normal distribution are given in bold. Means labeled with the same letter are not significantly different at $P < 0.05$ after ANOVA post hoc Tukey HSD test (cases with normal distribution) or Kruskal–Wallis test. If ANOVA is insignificant, letters are absent.

Fatty Acid	Charr	Whitefish	Muksun	Inconnu	Broad Whitefish	Round Whitefish	Whitefish nd
14:0	4.4 ± 0.1 CD	3.3 ± 0.1 B	4.8 ± 0.3 D	3.9 ± 0.0 BC	2.7 ± 0.2 A	3.1 ± 0.2 B	3.5 ± 0.4 ABC
15:0	0.3 ± 0.0 B	0.2 ± 0.0 AB	0.5 ± 0.0 C	0.4 ± 0.0 BC	0.5 ± 0.1 C	0.2 ± 0.0 A	0.3 ± 0.0 AB
16:0	16.7 ± 0.4 C	14.6 ± 0.2 AB	14.6 ± 0.8 AB	15.9 ± 0.7 B	17.4 ± 0.3 C	12.9 ± 0.4 A	15.7 ± 0.5 B
16:1n-9	0.8 ± 0.3 B	0.3 ± 0.0 A	0.4 ± 0.0 AB	0.4 ± 0.1 AB	0.8 ± 0.1 B	0.3 ± 0.0 A	0.5 ± 0.1 AB
16:1n-7	7.9 ± 0.4 A	19.4 ± 0.8 C	12.6 ± 1.3 B	15.9 ± 1.4 BC	13.2 ± 0.7 B	24.5 ± 1.2 D	15.7 ± 0.9 BC
15-17BFA	2.0 ± 0.2 BCD	1.0 ± 0.0 AB	1.9 ± 0.1 CD	1.1 ± 0.1 ABC	2.8 ± 0.4 D	0.7 ± 0.1 A	1.4 ± 0.1 AB
16PUFA	0.4 ± 0.1 A	4.8 ± 0.2 C	3.0 ± 0.6 B	1.6 ± 0.4 AB	1.3 ± 0.2 AB	5.4 ± 0.7 C	2.2 ± 0.5 B
17:0	0.2 ± 0.0 AB	0.3 ± 0.0 BC	0.3 ± 0.0 C	0.2 ± 0.0 AB	0.5 ± 0.0 D	0.1 ± 0.0 A	0.3 ± 0.0 BC
18:0	3.4 ± 0.1 C	1.9 ± 0.0 A	2.4 ± 0.1 A	2.5 ± 0.3 AB	3.2 ± 0.1 BC	2.5 ± 0.1 A	2.6 ± 0.3 A
18:1n-9	19.2 ± 0.6 BC	21.5 ± 0.4 C	16.8 ± 1.0 AB	18.9 ± 1.4 B	17.8 ± 0.8 AB	14.6 ± 0.8 A	18.3 ± 0.6 B
18:1n-7	2.9 ± 0.4 A	4.2 ± 0.1 AB	4.6 ± 0.4 BC	4.9 ± 0.2 BC	5.9 ± 0.3 C	5.9 ± 0.4 C	4.8 ± 0.5 BC
18:2n-6	3.2 ± 0.1 AB	2.2 ± 0.1 A	3.0 ± 0.1 AB	3.0 ± 0.4 AB	5.4 ± 0.6 D	5.1 ± 0.1 CD	3.9 ± 0.3 BC
18:3n-3	2.6 ± 0.1 AB	1.4 ± 0.1 A	3.0 ± 0.2 BC	2.4 ± 0.3 A	4.0 ± 0.5 C	3.6 ± 0.4 BC	2.2 ± 0.2 AB
18:4n-3	1.6 ± 0.1	1.8 ± 0.1	1.8 ± 0.1	1.7 ± 0.2	1.5 ± 0.2	1.9 ± 0.3	2.1 ± 0.4
$\sum$20:1	1.3 ± 0.0 BC	1.0 ± 0.1 AB	1.9 ± 0.2 C	2.0 ± 0.9 ABC	1.7 ± 0.3 BC	0.7 ± 0.1 A	1.3 ± 0.1 ABC
20:2n-6	0.9 ± 0.0 C	0.2 ± 0.0 A	0.6 ± 0.0 B	0.4 ± 0.0 AB	0.4 ± 0.1 AB	0.3 ± 0.0 AB	0.5 ± 0.1 B
20:4n-6	1.7 ± 0.0 BC	1.4 ± 0.0 B	2.1 ± 0.1 D	1.9 ± 0.2 CD	2.3 ± 0.1 D	0.6 ± 0.0 A	1.7 ± 0.1 BCD
20:3n-3	1.4 ± 0.0 C	0.1 ± 0.0 A	0.5 ± 0.1 BC	0.4 ± 0.0 ABC	0.3 + 0.0 AB	0.2 ± 0.0 AB	0.4 ± 0.1 AB
20:4n-3	2.7 ± 0.1 C	0.7 ± 0.1 AB	1.2 ± 0.1 BC	1.2 ± 0.1 ABC	0.6 ± 0.0 A	0.9 ± 0.1 AB	0.9 ± 0.1 AB
20:5n-3	4.4 ± 0.1 A	9.4 ± 0.2 D	9.0 ± 0.3 D	6.5 ± 1.0 BC	5.3 ± 0.4 AB	7.0 ± 0.5 BC	8.6 ± 0.5 CD
22:5n-6	0.9 ± 0.0 D	0.2 ± 0.0 AB	1.0 ± 0.1 D	0.7 ± 0.1 CD	0.5 ± 0.1 BC	0.0 ± 0.0 A	0.4 ± 0.1 AB
22:4n-3	1.3 ± 0.0 B	0.0 ± 0.0 A	0.1 ± 0.0 A	0.2 ± 0.1 A	0.0 ± 0.0 A	0.0 ± 0.0 A	0.1 ± 0.0 A
22:5n-3	2.7 ± 0.1 B	2.2 ± 0.1 B	2.4 ± 0.2 B	2.1 ± 0.2 B	1.4 ± 0.1 A	2.3 ± 0.1 B	2.1 ± 0.2 B
22:6n-3	9.9 ± 0.2 C	4.1 ± 0.2 B	6.4 ± 0.3 C	7.8 ± 0.8 C	4.1 ± 0.3 B	2.4 ± 0.2 A	6.6 ± 0.5 C
24PUFA	2.9 ± 0.1 B	0.7 ± 0.1 A	0.9 ± 0.1 AB	0.6 ± 0.3 A	0.7 ± 0.0 A	0.6 ± 0.1 A	0.7 ± 0.1 A
Sum FA	*30.3 ± 4.7 AB*	*33.2 ± 6.9 AB*	*41.6 ± 15.0 B*	*18.7 ± 7.0 AB*	*19.5 ± 7.0 AB*	*4.9 ± 1.2 A*	*2.0 ± 0.5 A*

To study differences in fish reserve lipids, we performed CCA of FA in TAG (Figure 3). Like the multidimensional analysis for PL, Dimension 1 showed a marked difference of round whitefish versus charr. This difference was provided mostly by the greater levels of C16 PUFA and 16:1n-7 in TAG of round whitefish and by greater levels of 22:4n-3 and 20:3n-3 in that of charr (Figure 3). The second dimension of CCA for TAG also showed a similar trend to that observed in CCA of PL (Figures 2 and 3). In this dimension, most prominent difference was found between whitefish and broad whitefish due to variability in levels of C15-17 BFA, 17:0 and C16 PUFA (Figure 3).

In general, positioning of fish species in the biplot for reserve TAG well corresponded to that in biplot for structural PL (Figures 2 and 3). It should be also noted that physiologically significant EPA and DHA were not found among the FA markers responsible for separation of fish species in CCA for TAG and PL (Figures 2 and 3). Positioning of fish species in the biplot for total lipids generally corresponded to that in biplots for the lipid classes, with exception of whitefish of the non-identified form (Figures 1–3). The fatty acid markers responsible for separation of the fish samples in CCA were generally similar for total lipids, TAG and PL with exception of DHA and EPA in CCA of total lipids (Figures 1–3).

A visual analysis of all thin-layer chromatograms showed a marked dominance of TAG, PL and sterols as major lipid fractions. Spots that corresponded to other lipid classes were negligible. Therefore, we considered TAG and PL as major acyl-containing fractions, summarized their FA contents per mass unit (Tables 4 and 5) and calculated their parts in the sum of FA in the fish muscles (Figure 4A). Polar lipids constituted from 10.3 to 57.0% of the acyl-containing lipid sum, being the highest in whitefish of the non-identified form (Figure 4A). Triacylglycerols were the dominant acyl-containing lipid fraction for majority of the studied fish and exceeded 85% in charr, whitefish and muksun (Figure 4A). Note that increase in lipid content and total fatty acids for the studied species well corresponded with the increase in the TAG proportion of the acyl-containing lipids (Table 2, Figure 4A).

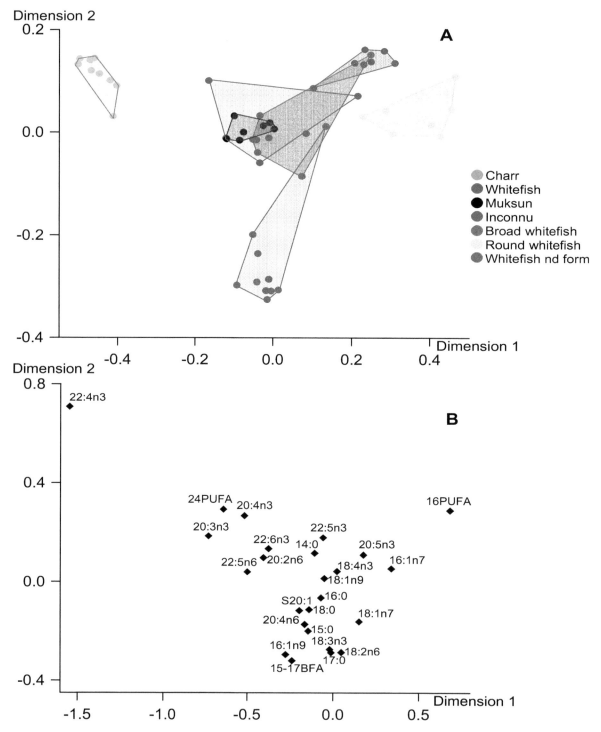

Figure 3. Canonical correspondence analysis of fatty acid percentages (% of FA sum) in triacylglycerols of muscles of seven fish species from arctic lakes (Siberia, Russia). **A**—individual variables, **B**—factor structure coefficients for fatty acids. Dimension 1 and Dimension 2 represented 51.1% and 15.6% of inertia, respectively.

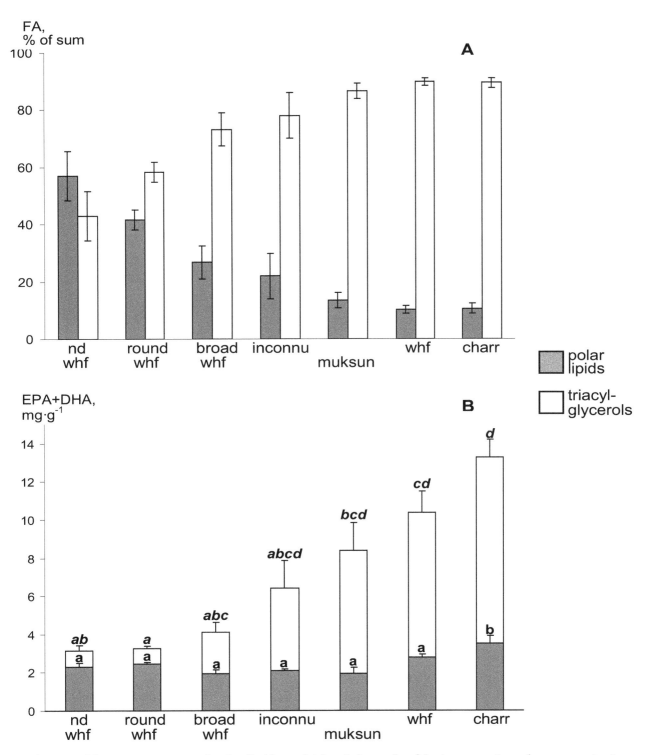

Figure 4. Mean percentages of polar lipids and triacylglycerols of their sum—**A**, and mean contents (mg g^{-1} wet weight) of sum of eicosapentaenoic and docosahexaenoic acids that corresponded to polar lipids and triacylglycerols—**B**, in muscles of seven fish species from arctic lakes (Siberia, Russia): nd whf—the non-identified form of *C. lavaretus* from Sobachye Lake, round whf—*P. cylindraceum* from Sobachye Lake, broad whf—*C. nasus* from Pyasino Lake, inconnu—*S. leucichthys nelma* from Pyasino Lake, muksun—*C. muksun* from Pyasino Lake, whf—*C. lavaretus* from Sobachye Lake, charr—*S. drjagini* from Sobachye Lake. Bars represent standard errors. Means labelled with the same letter are not significantly different at $P < 0.05$ after Tukey HSD post hoc test.

Using the PL and TAG percentages of the acyl-containing lipid sum and content of EPA and DHA of total lipids per mass unit, we calculated parts of EPA + DHA that provided by polar lipids versus triacylglycerols and expressed them as mg g^{-1} wet weight of muscle tissue (Figure 4B). Contents of EPA + DHA provided by PL fraction varied from 1.9 to 3.5 mg g^{-1} (Figure 4B). The average value for the seven fish species was 2.4 ± 0.2 mg g^{-1}, and coefficient of variation, CV, was 8.7%. Contents of EPA + DHA in TAG were of a greater range, from 0.9 to 9.8 mg g^{-1}; the average value was to 4.4 ± 0.2 mg g^{-1}, and CV was 28.9% (Figure 4B).

4. Discussion

4.1. Main Finding

All the taxonomically related species of order Salmoniformes had nearly similar content of EPA+DHA in PL, 2.4 ± 0.2 mg g^{-1}, in average. In contrast, content of EPA + DHA esterified as TAG varied ~10-fold among the studied salmonids. Thus, all variations of nutritive value, i.e., EPA + DHA content per mass unit of filet, were caused by TAG fraction, while PL had a constant species (taxon—specific physiologically optimum content. As a result, fatty fish, charr, whitefish and muksun, contained most amounts of nutritionally valuable EPA and DHA in TAG fraction of muscles. Conversely, the lean species, round whitefish and whitefish of the non-identified form, had the omega-3 LC-PUFA contained mostly in polar lipids. Regarding nutritive value, the fatty species with higher proportion of TAG in muscles, charr, whitefish and muksun, appeared to be most valuable and had 13.3 ± 0.8, 10.4 ± 1.1 and 8.4 ± 1.4 mg g^{-1}, respectively.

4.2. Fatty Acid Markers in Fish Total Lipids, TAG and PL

In our snap-shot field study of salmonids from arctic lakes we found significant differences between fatty acid profiles of the main lipid classes, with prevalence of n-3 LC-PUFA and monoenoic 16-18 FA in structural PL and storage TAG, respectively. In addition, both FA profiles of TAG and PL, as well as profiles of total lipids, had distinct peculiarities among the studied fish that allowed separating the most species in the CCA biplots (Figures 1–3). It should be emphasized that the separations of the fish in the multivariate analyses of all three lipid fractions were provided mainly by the same marker FA: C24 PUFA, 22:4n-3, 20:3n-3, C16 PUFA, 16:1n-7 and C15-17 BFA. The conspicuous exception was 22:6n-3 and 20:5n-3 in total lipid CCA that separated whitefish of the non-identified form and round whitefish from the other species (Figure 1). PL had a predominant contribution to total lipids of these lean fish (Figure 4A). Therefore, FA profiles of total lipids of whitefish of the non-identified form and round whitefish mostly reflected FA composition of PL which were considerably rich in DHA and EPA (Table 4). As a result, DHA and EPA, dominant FA of PL, played as markers for the lean fish in CCA of total lipids.

Similar patterns of biomarker FA, characteristic of zoobenthic, algal, terrestrial and other food sources, within TAG and PL fractions of the studied species allowed to use both these fractions, as well as their sum, total lipids, for identification of food sources of wild fish (Figures 1–3). In ecological studies, analysis of FA trophic markers of various consumers is often performed for TAG assuming that they generally deposit fatty acid molecules coming from food assimilation [18,25,26]. Meanwhile, many studies used FA composition of total lipids to elucidate trophic relations of various fish species, e.g., [27–31]. In overall, both approaches base on a premise that biochemical composition of food sources is reflected by FA profiles of TAG or total lipids.

Another lipid class of high concern is polar lipids comprised mostly phospholipids and glycolipids that are main constituents of cell membranes. As known, the specific FA composition of PL provides proper membrane structure and functions. As a result, FA composition of PL are considered to be highly conserved relative to diet and tended to reflect FA biosynthetic capacities of an organism [18].

In contrast to the common opinion on conserved PL composition, in some studies FA profiles of PL were successfully used as trophic markers. For instance, fatty acid profiles of both polar and neutral lipids conspicuously differed among three species, benthivorous whitefish *Coregonus clupeaformis*, and piscivorous walleye *Sander vitreus* and northern pike *Esox lucius*, due to different feeding habits of the fish [32]. We also confirmed that PL FA profiles of muscles allow to identify feeding spectra of fish similarly that TAG profiles do (Figures 2 and 3). For instance, round whitefish was one of the most separated species in the both multivariate analyses of TAG and PL due to the greatest levels of C16 PUFA and 16:1n-7 which originated from diatoms and green algae [33]. Indeed, the algae were one of the dominant items in stomach content of this species (Table 1).

In CCA biplots of total lipids, PL and TAG, charr had a particular position due to higher levels of minor n-3 PUFA, like 22:4-3, C24 PUFA and 20:3n-3. These FA were not assigned as trophic markers, whereas some of them, C24 PUFA, were considered as intermediate compounds indicative for conversion of C20 to C22 PUFA [34,35]. In TAG of charr the percentage of 22:4n-3 accounted for 1.3% of FA sum, being absent or found in traces in other studied fish. The presence of this PUFA was previously reported for least cisco *Coregonus sardinella*, small-sized pelagic fish inhabited Sobachye Lake [23]. The studied charr from Sobachye Lake was piscivorous (Table 1), thus, it could obtain this PUFA from the consumed least cisco. Like in our study, species of the same genus and its prey, lake trout *Salvelinus namaycush* and cisco from Great Bear Lake, were together separated from other hydrobionts in a multivariate analysis due to higher levels of 22:4-3 and 20:3n-3 [31]. Alternative explanation based on coincidence between 22:4n-3 and C24 PUFA levels is that the fatty acid 22:4n-3 may be a marker of LC-PUFA conversion in fish. Anyway, we suppose that considerable levels of 22:4n-3, 20:3n-3 and C24 PUFA might be a characteristic feature of FA profiles of *Salvelinus* genus.

In both CCA analyses of TAG and PL, broad whitefish well separated from the other species due to higher levels of C15-17 BFA, 17:0, 18:2n-6 and 18:3n-3. The two former FA are known to be markers of bacterial organic matter, while the two latter are considered as markers of terrestrial organic matter [31,33]. Broad whitefish is a typical benthivorous species that likely got these marker fatty acids from detritus enriched with bacterial and terrestrial organic matter. The species had the highest levels of 18:2n-6 in TAG, and 20:4n-6 in PL, relatively. This finding likely indicates for the initial storage of dietary 18:2n-6 in TAG and its consequent conversion to 20:4n-6 with further transfer to PL.

The similarity of FA sets that are markers for food sources between TAG and PL classes likely indicates that the studied wild fish are able to directly incorporate dietary biochemical components, i.e., fatty acyl groups, into membrane PL. Besides, FA originated from food assimilation, fish are able to include in lipid molecules fatty acyl groups obtained due to biosynthesis or conversion from precursors. Freshwater fish are known to have capacity to synthesize LC-PUFA from the shorter chain precursors [35,36]. Indeed, some studied fish, e.g., charr, contained in TAG and PL certain amounts of C24 PUFA and 20:4n-3 that likely were intermediates of DHA and EPA synthesis.

4.3. Content of Essential LC-PUFA in Fish PL and TAG

The studied seven salmonid species varied ~ 7-fold in total lipid and fatty acid contents per a mass unit of muscle tissues. Most of this variation was related with different TAG content in muscles (Table 5, Figure 4), whereas PL content evaluated as their FA sum varied slightly (Table 4). The observed variation in lipid class contents in the studied fish is in agreement with well-known notion that polar lipids comprise cellular membranes and, as a result, have a relatively constant content in muscle cells, in contrast to that of triacylglycerols [19,21,37]. For instance, an absence of relation between total lipid and phospholipid contents and a strong relation between total lipid and triacylglycerol contents expressed as percentages of muscle mass were previously shown for a number of marine myctophid species [38].

Fish polar lipids are commonly considered as a physiologically crucial lipid class that are rich in LC-PUFA, mostly in DHA and, to a lesser extent, in EPA [19,37,39]. Indeed, percentages of DHA and EPA in PL of various wild marine and freshwater species ranged as 11.5–55.7% and 2.6–14.6%, with average values of 31.4% and 7.4%, respectively [40–46]. The average levels of EPA and DHA in the fish species from our study coincided with the above ranges, except the EPA value of whitefish, 16.6%, which was a bit higher than the known values.

Triacylglycerols are considered to be relatively poor in LC-PUFA and preferably accumulate monoenoic C16-22 FA [18,19,37]. According to the available data, levels of DHA and EPA in fish TAG varied in intervals of 2.3%-23.3% and 1.1%–14.1%, with average values of 8.8% and 5.3%, respectively [40–46]. The percentages of both EPA and DHA of TAG in the fish species from our study well coincided with the above ranges (Table 5).

Triacylglycerols commonly comprise a large part of total lipids in muscles of medium-fat and fatty fish species. For instance, TAG achieved 80%, 90% and 51.5% in marine species arrow-tooth flounder (*Atheresthes stomias)* and golden pompano (*Trachinotus blochii*) and freshwater whitefish (*Coregonus lavaretus*), respectively [43,47,48]. In the studied freshwater salmonids, TAG percentages varied from 43.4% to 89.7% of the sum of two acyl-containing lipid classes (Figure 4A). Such high TAG levels may be explained by adaptation of the fish species to low-temperature conditions in the studied arctic lakes [49,50].

Regarding the relatively high contents of TAG per mass unit and percentages of EPA and DHA in TAG, we hypothesized that content of EPA and DHA in TAG would appreciably contribute to total content of EPA and DHA and would increase along with lipid content in muscles of the studied fish. Hence, we compared content of EPA and DHA esterified as TAG versus that esterified as PL. Content of EPA+DHA in PL per mass unit of muscle tissues were similar among the studied salmonids, moderately varying in the interval of 1.9–3.5 mg g^{-1} (Figure 4B). In contrast, values of EPA+DHA in TAG of the fish species greatly varied, ~10-fold. Lean fish, i.e., whitefish of the non-identified form, round whitefish and broad whitefish, contained only 25%–47% of EPA + DHA of total content of these PUFA in the muscles esterified as TAG molecules. In contrast, the medium-fat and fatty fish, inconnu, muksun, whitefish and charr, had more than half of their muscle EPA and DHA content as TAG molecules, up to 72%. Thus, the wild salmonids that had relatively high content of n-3 LC-PUFA in muscles (~ > 5 mg g^{-1}) contained the major portion of these nutritionally valuable compounds in the storage lipids. Our finding evidently contradicts a common notion that lean and medium-fat fish that have PL as a main lipid class in the muscles are the best dietary sources of n-3 LC-PUFA for humans [19,51]. Wild fatty fish which are able to deposit large amounts of storage lipids in their muscles appear to be the most valuable sources of n-3 LC-PUFA in human diet.

Further, our results are in a good accordance with many studies showed strong relationship between total lipid and EPA, DHA or their sum contents in fish muscles. Such relation was found for marine species, sprat *Sprattus sprattus* and herring *Clupea harengus* from Baltic Sea [52], for five marine species from the northeast Pacific [38] and for several freshwater species from a subalpine lake [53]. The significant relation was also found across farmed families of Atlantic salmon *Salmo salar* [54].

It is interesting to note that percentages of the n-3 LC-PUFA and lipid (or total FA as its proxy) content were negatively correlated in aforementioned and other studies [10]. The reported negative correlation was explained by the fact that total lipids increase preferably at the expense of TAG, whereas content of the membrane phospholipids, which are rich in n-3 PUFA remains fairly constant [21,37]. As a result, the proportions of EPA and DHA in muscle total lipids become diluted due to the accumulation of neutral lipids, which have high levels of monounsaturated FA. Although increase of total lipid content at the expense of TAG in fish muscles, as a rule, leads to decrease of n-3 LC-PUFA percentage, this does not mean that a concomitant decrease of nutritional quality of a fish occurs. Nutritional quality of fish products must be estimated on quantitative base expressed as mg FA per gram of tissue rather than percentage base [10].

Quantitative (mg per gram of tissue) measurements of TAG versus PL contribution in n-3 LC-PUFA of fish muscles are very scarce. Some studies gave indirect evidence of significant TAG contribution. For instance, among four fish species commercially harvested in Alaskan waters, arrow-tooth flounder *A. stomias* had maximum contents of EPA and DHA, 7.0 mg g^{-1}, as well as maximum levels of TAG,

80% of total lipids in edible muscles [47]. Myctophid fish species with higher total lipid content (proxy for TAG content) also had higher contents of EPA and DHA esterified as TAG [38].

Some direct measurements showed that lean fish contained less than half of EPA and DHA in their muscles esterified as TAG, e.g., wild white seabream *Diplodus sargus* [40], whitefish *Coregonus lavaretus* [43], six commercial Chilean marine species [46]. In contrast, the only studied medium-fat fish (2–4% lipid content of wet mass), Pacific sandperch *Prolatilus jugularis*, had approximately 60% of EPA+DHA esterified as TAG [46]. Similar to latter finding, farmed *C. lavaretus* which had one of the highest known values of EPA and DHA in muscles, 18.6 mg g^{-1} wet weight, had 61% of that value in TAG [43]. In our study, the fish species were strongly variable in lipid and total FA content and, as a result, in n-3 LC-PUFA content esterified as TAG. Like in the abovementioned studies, the fatty fish, muksun, whitefish and charr, had relatively higher content of EPA + DHA per mass unit that were mostly esterified as TAG (Figure 4B).

If we take the threshold of the recommended personal daily dose of EPA + DHA as 1 g and the average per serve portion of fish as 200 g [55,56], a fish of proper nutritional value should contain EPA + DHA nearly or more than 5 mg g^{-1} of filet [57]. The obtained data on lipid class composition and content mean that when such fish is consumed, nearly or more than half of the essential n-3 LC-PUFA comes as TAG form. Recent studies showed that bioavailability of FA, including LC-PUFA esterified as TAG may be lower than that esterified as PL [58,59], but see [60]. Thus, distribution of LC-PUFA in major lipid classes should be further addressed in studies of nutritional quality of various fish products.

5. Conclusions

The studied fish with similar feeding spectra were identified similarly by a multivariate analysis of FA profiles of total lipids, TAG and PL. Marker FA characteristic of diverse food sources (benthic, terrestrial, etc.), accumulated in nearly similar proportions within TAG and PL, and thereby allow to use both these fractions, as well as total lipids, for identification of food sources of wild fish. The found incorporation of the fatty acid trophic markers in structural polar lipids similarly to that in reserve TAG deserves further studies. Regarding contribution of TAG and PL into content of essential LC-PUFA of the taxonomically closely related fish species of order Salmoniformes, we found that content of EPA+DHA esterified as PL was nearly invariable, presenting presumably a species/taxon-specific optimal level. In contrast, content of EPA+DHA esterified as TAG greatly varied among the studied fish and provided most contribution to total EPA+DHA content in the fatty fish species, charr, whitefish and muksun. We can conclude that EPA+DHA-rich fish species likely accumulate these nutritionally valuable compounds predominately in the TAG form.

Author Contributions: N.N.S.: conceptualization, writing—original draft; O.N.M.: conceptualization, investigation, methodology; A.E.R.: investigation, methodology; L.A.G.: investigation, methodology; S.P.S.: investigation; A.A.K.: investigation; M.I.G.: formal analysis, writing—review & editing. All authors have read and agreed to the published version of the manuscript.

References

1. Dyall, S.C. Long-chain omega-3 fatty acids and the brain: A review of the independent and shared effects of EPA, DPA and DHA. *Front. Aging Neurosci.* **2015**, *7*, 52. [CrossRef]
2. Poudyal, H.; Brown, L. Should the pharmacological actions of dietary fatty acids in cardiometabolic disorders be classified based on biological or chemical function? *Prog. Lipid Res.* **2015**, *59*, 172–200. [CrossRef]
3. Calder, P.C. Very long-chain n-3 fatty acids and human health: Fact, fiction and the future. *Proc. Nutr. Soc.* **2018**, *77*, 52–72. [CrossRef] [PubMed]
4. Hu, Y.; Hu, F.B.; Manson, J.E. Marine omega-3 supplementation and cardiovascular disease: An updated meta-analysis of 13 randomized controlled trials involving 127,477 participants. *J. Am. Heart Assoc.* **2019**, *8*, e013543. [PubMed]

5. Wu, J.H.; Micha, R.; Mozaffarian, D. Dietary fats and cardiometabolic disease: Mechanisms and effects on risk factors and outcomes. *Nat. Rev. Cardiol.* **2019**, *16*, 581–601. [PubMed]

6. Harris, W.S.; Mozaffarian, D.; Lefevre, M.; Toner, C.D.; Colombo, J.; Cunnane, S.C.; Holden, J.M.; Klurfeld, D.M.; Morris, M.C.; Whelan, J. Towards establishing dietary reference intakes for eicosapentaenoic and docosahexaenoic acids. *J. Nutr.* **2009**, *139*, 804S–819S. [CrossRef]

7. Kris-Etherton, P.M.; Grieger, J.A.; Etherton, T.D. Dietary reference intakes for DHA and EPA. *Prostaglandins Leukot. Essent.* **2009**, *81*, 99–104. [CrossRef]

8. Siscovick, D.S.; Barringer, T.A.; Fretts, A.M.; Wu, J.H.Y.; Lichtenstein, A.H.; Costello, R.B.; Kris-Etherton, P.M.; Jacobson, T.A.; Engler, M.B.; Alger, H.M.; et al. Omega-3 polyunsaturated fatty acid (fish oil) supplementation and the prevention of clinical cardiovascular disease. *Circulation* **2017**, *135*, e867. [CrossRef]

9. Tacon, A.G.J.; Metian, M. Fish matters: Importance of aquatic foods in human nutrition and global food supply. *Rev. Fish. Sci.* **2013**, *21*, 22–38. [CrossRef]

10. Gladyshev, M.I.; Sushchik, N.N.; Tolomeev, A.P.; Dgebuadze, Y.Y. Meta-Analysis of factors associated with omega-3 fatty acid contents of wild fish. *Rev. Fish Biol. Fish.* **2018**, *28*, 277–299. [CrossRef]

11. Gladyshev, M.I.; Sushchik, N.N.; Makhutova, O.N. Production of EPA and DHA in aquatic ecosystems and their transfer to the land. *Prostaglandins Lipid Mediat.* **2013**, *107*, 117–126. [CrossRef] [PubMed]

12. Hixson, S.M.; Arts, M.T. Climate warming is predicted to reduce omega-3 long-chain, polyunsaturated fatty acid production in phytoplankton. *Glob. Chang. Biol.* **2016**, *22*, 2744–2755. [CrossRef] [PubMed]

13. Tocher, D.R.; Betancor, M.B.; Sprague, M.; Olsen, R.E.; Napier, J.A. Omega-3 long-chain polyunsaturated fatty acids, EPA and DHA: Bridging the gap between supply and demand. *Nutrients* **2019**, *11*, 89. [CrossRef] [PubMed]

14. Gribble, M.O.; Karimi, R.; Feingold, B.J.; Nyland, J.F.; O'Hara, T.M.; Gladyshev, M.I.; Chen, C.Y. Mercury, selenium and fish oils in marine food webs and implications for human health. *J. Mar. Biol. Assoc. UK* **2016**, *96*, 43–59. [CrossRef]

15. Weil, C.; Lefevre, F.; Bugeon, L. Characteristics and metabolism of different adipose tissues in fish. *Rev. Fish Biol. Fish.* **2013**, *23*, 157–173. [CrossRef]

16. Sargent, J.R. The lipids. In *Fish Nutrition*; Sargent, J.R., Henderson, R.J., Tocher, D.R., Eds.; Academic Press: New York, NY, USA, 1989; pp. 153–218.

17. Vance, D.E.; Vance, J.E. (Eds.) *Biochemistry of Lipids, Lipoproteins and Membranes*; Elsevier Science: Amsterdam, The Netherlands, 1996.

18. Iverson, S. Tracing aquatic food webs using fatty acids: From qualitative indicators to quantitative determination. In *Lipids in Aquatic Ecosystems*; Arts, M.T., Kainz, M., Brett, M.T., Eds.; Springer: New York, NY, USA, 2009; pp. 281–307.

19. Sargent, J.; Bell, G.; McEvoy, L.; Tocher, D.; Estevez, A. Recent developments in the essential fatty acid nutrition of fish. *Aquaculture* **1999**, *177*, 191–199. [CrossRef]

20. Parrish, C.C. Essential fatty acids in aquatic food webs. In *Lipids in Aquatic Ecosystems*; Arts, M.T., Kainz, M., Brett, M.T., Eds.; Springer: New York, NY, USA, 2009; pp. 309–325.

21. Mairesse, G.; Thomas, M.; Gardeur, J.-N.; Brun-Bellut, J. Effects of geographic source, rearing system, and season on the nutritional quality of wild and farmed *Perca fluviatilis*. *Lipids* **2006**, *41*, 221–229. [CrossRef]

22. Gladyshev, M.I.; Sushchik, N.N. Long-Chain omega-3 polyunsaturated fatty acids in natural ecosystems and the human diet: Assumptions and challenges. *Biomolecules* **2019**, *9*, 485. [CrossRef]

23. Gladyshev, M.I.; Sushchik, N.N.; Makhutova, O.N.; Glushchenko, L.A.; Rudchenko, A.E.; Makhrov, A.A.; Borovikova, E.A.; Dgebuadze, Y.Y. Fatty acid composition and contents of seven commercial fish species of genus *Coregonus* from Russian Subarctic water bodies. *Lipids* **2017**, *52*, 1033–1044. [CrossRef]

24. Bogdanov, A.L. Earlier studies, morphology and hydrology of lakes. In *Geographic Characteristic of Lakes of Taimyr Peninsula*; Adamenko, V.N., Egorov, A.N., Eds.; Nauka: Leningrad, Russia, 1985; pp. 184–193. (In Russian)

25. Heintz, R.A.; Wipfli, M.S.; Hudson, J.P. Identification of marine-derived lipids in juvenile Coho Salmon and aquatic insects through fatty acid analysis. *Trans. Am. Fish. Soc.* **2010**, *139*, 840–854. [CrossRef]

26. Young, J.W.; Guest, M.A.; Lansdell, M.; Phleger, C.F.; Nichols, P.D. Discrimination of prey species of juvenile swordfish *Xiphias gladius* (Linnaeus, 1758) using signature fatty acid analyses. *Prog. Oceanogr.* **2010**, *86*, 139–151. [CrossRef]

27. Czesny, S.J.; Rinchard, J.; Hanson, S.D.; Dettmers, J.M.; Dabrowski, K. Fatty acid signatures of Lake Michigan prey fish and invertebrates: Among-Species differences and spatiotemporal variability. *Can. J. Fish. Aquat. Sci.* **2011**, *68*, 1211–1230. [CrossRef]
28. McMeans, B.C.; Arts, M.T.; Lydersen, C.; Kovacs, K.M.; Hop, H.; Falk-Petersen, S.; Fisk, A.T. The role of Greenland sharks (*Somniosus microcephalus*) in an Arctic ecosystem: Assessed via stable isotopes and fatty acids. *Mar. Biol.* **2013**, *160*, 1223–1238. [CrossRef]
29. Hielscher, N.N.; Malzahn, A.M.; Diekmann, R.; Aberle, N. Trophic niche partitioning of littoral fish species from the rocky intertidal of Helgoland, Germany. *Helgol. Mar. Res.* **2015**, *69*, 385–399. [CrossRef]
30. Strandberg, U.; Hiltunen, M.; Jelkänen, E.; Taipale, S.J.; Kainz, M.J.; Brett, M.T.; Kankaala, P. Selective transfer of polyunsaturated fatty acids from phytoplankton to planktivorous fish in large boreal lakes. *Sci. Total Environ.* **2015**, *536*, 858–865. [CrossRef]
31. Chavarie, L.; Howland, K.; Gallagher, C.; Tonn, W. Fatty acid signatures and stomach contents of four sympatric Lake Trout: Assessment of trophic patterns among morphotypes in Great Bear Lake. *Ecol. Freshw. Fish* **2016**, *25*, 109–124. [CrossRef]
32. Wiegand, M.D.; Johnston, T.A.; Porteous, L.R.; Ballevona, A.J.; Casselman, J.M.; Leggett, W.C. Comparison of ovum lipid provisioning among lake whitefish, walleye and northern pike co-habiting in Bay of Quinte (Lake Ontario, Canada). *J. Gt. Lakes Res.* **2014**, *40*, 721–729. [CrossRef]
33. Napolitano, G.E. Fatty acids as trophic and chemical markers in freshwater ecosystems. In *Lipids in Freshwater Ecosystems*; Arts, M.T., Wainman, B.C., Eds.; Springer: New York, NY, USA, 1999; pp. 21–44.
34. Giri, S.S.; Graham, J.; Hamid, N.K.A.; Donald, J.A.; Turchini, G.M. Dietary micronutrients and In Vivo n-3 LC-PUFA biosynthesis in Atlantic salmon. *Aquaculture* **2016**, *452*, 416–425. [CrossRef]
35. Oboh, A.; Kabeya, N.; Carmona-Antoñanzas, G.; Castro, L.F.C.; Dick, J.R.; Tocher, D.R.; Monroig, O. Two alternative pathways for docosahexaenoic acid (DHA, 22:6n-3) biosynthesis are widespread among teleost fish. *Sci. Rep.* **2017**, *7*, 3889. [CrossRef]
36. Tocher, D.R. Fatty acid requirements in ontogeny of marine and freshwater fish. *Aquac. Res.* **2010**, *41*, 717–732. [CrossRef]
37. Olsen, Y. Lipids and essential fatty acids in aquatic food webs: What can freshwater ecologists learn from mariculture. In *Lipids in Freshwater Ecosystems*; Arts, M.T., Wainman, B.C., Eds.; Springer: New York, NY, USA, 1999; pp. 161–202.
38. Litzow, M.A.; Bailey, K.M.; Prahl, F.G.; Heintz, R. Climate regime shifts and reorganization of fish communities: The essential fatty acid limitation hypothesis. *Mar. Ecol. Prog. Ser.* **2006**, *315*, 1–11. [CrossRef]
39. Ahlgren, G.; Vrede, T.; Goedkoop, W. Fatty acid ratios in freshwater fish, zooplankton and zoobenthos—Are their specific optima? In *Lipids in Aquatic Ecosystems*; Arts, M.T., Kainz, M., Brett, M.T., Eds.; Springer: New York, NY, USA, 2009; pp. 147–178.
40. Pérez, M.J.; Rodríguez, C.; Cejas, J.R.; Martín, M.V.; Jerez, S.; Lorenzo, A. Lipid and fatty acid content in wild white seabream (*Diplodus sargus*) broodstock at different stages of the reproductive cycle. *Comp. Biochem. Physiol. B* **2007**, *146*, 187–196. [CrossRef] [PubMed]
41. Snyder, R.J.; Schregel, W.D.; Wei, Y. Effects of thermal acclimation on tissue fatty acid composition of freshwater alewives (*Alosa pseudoharengus*). *Fish Physiol. Biochem.* **2012**, *38*, 363–373. [CrossRef] [PubMed]
42. Kayhan, H.; Bashan, M.; Kaçar, S. Seasonal variations in the fatty acid composition of phospholipids and triacylglycerols of brown trout. *Eur. J. Lipid Sci. Technol.* **2015**, *117*, 738–744. [CrossRef]
43. Suomela, J.-P.; Lundén, S.; Kaimainen, M.; Mattila, S.; Kallio, H.; Airaksinen, S. Effects of origin and season on the lipids and sensory quality of European whitefish (*Coregonus lavaretus*). *Food Chem.* **2016**, *197*, 1031–1037. [CrossRef]
44. Sardenne, F.; Kraffe, E.; Amiel, A.; Fouché, E.; Debrauwer, L.; Ménard, F.; Bodin, N. Biological and environmental influence on tissue fatty acid compositions in wild tropical tunas. *Comp. Biochem. Physiol. A* **2017**, *204*, 17–27. [CrossRef]
45. Suito, T.; Nagao, K.; Hatano, M.; Kohashi, K.; Tanabe, A.; Ozaki, H.; Kawamoto, J.; Kurihara, T.; Mioka, T.; Tanaka, K.; et al. Synthesis of omega-3 long-chain polyunsaturated fatty acid-rich triacylglycerols in an endemic goby, *Gymnogobius isaza*, from Lake Biwa, Japan. *J. Biochem.* **2018**, *164*, 127–140. [CrossRef]
46. Rincon-Cervera, M.A.; Gonzalez-Barriga, V.; Valenzuela, R.; Lopez-Arana, S.; Romero, J.; Valenzuela, A. Profile and distribution of fatty acids in edible parts of commonly consumed marine fishes in Chile. *Food Chem.* **2019**, *274*, 123–129. [CrossRef]

47. Oliveira, A.C.M.; Bechtel, P.J. Lipid analysis of fillets from giant grenadier (*Albatrossia pectoralis*), arrow-tooth flounder (*Atheresthes stomias*), pacific cod (*Gadus macrocephalus*) and walleye pollock (*Theragra chalcogramma*). *J. Muscle Foods* **2006**, *17*, 20–33. [CrossRef]

48. He, C.; Cao, J.; Jiang, X.; Wen, C.; Bai, X.; Li, C. Fatty acid profiles of triacylglycerols and phospholipids of sea-cage cultured *Trachinotus blochii*: A comparative study of head, viscera, skin, bone, and muscle. *J. Food Sci.* **2019**, *84*, 650–658. [CrossRef]

49. Xue, C.; Okabe, M.; Saito, H. Differences in lipid characteristics among populations: Low-Temperature adaptability of ayu, *Plecoglossus altivelis*. *Lipids* **2012**, *47*, 75–92. [CrossRef] [PubMed]

50. Murzina, S.A.; Nefedova, Z.A.; Pekkoeva, S.N.; Veselov, A.E.; Efremov, D.A.; Nemova, N.N. Age-Specific lipid and fatty acid profiles of Atlantic salmon juveniles in the Varzuga River. *Int. J. Mol. Sci.* **2016**, *17*, 1050. [CrossRef] [PubMed]

51. Ahlgren, G.; Blomqvist, P.; Boberg, M.; Gustafsson, I.-B. Fatty acid content of the dorsal muscle—An indicator of fat quality in freshwater fish. *J. Fish Biol.* **1994**, *45*, 131–157.

52. Rojbek, M.C.; Tomkiewicz, J.; Jacobsen, C.; Stottrup, J.G. Forage fish quality: Seasonal lipid dynamics of herring (*Clupea harengus* L.) and sprat (*Sprattus sprattus* L.) in the Baltic Sea. *ICES J. Mar. Sci.* **2014**, *71*, 56–71. [CrossRef]

53. Kainz, M.J.; Hager, H.H.; Rasconi, S.; Kahilainen, K.K.; Amundsen, P.-A.; Hayden, B. Polyunsaturated fatty acids in fishes increase with total lipids irrespective of feeding sources and trophic position. *Ecosphere* **2017**, *8*, e01753. [CrossRef]

54. Leaver, M.J.; Taggart, J.B.; Villeneuve, L.; Bron, J.E.; Guy, D.R.; Bishop, S.C.; Houston, R.D.; Matika, O.; Tocher, D.R. Heritability and mechanisms of n-3 long chain polyunsaturated fatty acid deposition in the flesh of Atlantic salmon. *Comp. Biochem. Physiol. D* **2011**, *6*, 62–69. [CrossRef]

55. Ruffle, B.; Burmaster, D.E.; Anderson, P.D.; Gordon, H.D. Lognormal distributions for fish consumption by the general U.S. population. *Risk Anal.* **1994**, *14*, 395–404. [CrossRef]

56. Williams, M.C.W.; Schrank, C.; Anderson, H.A. Fatty acids in thirteen Wisconsin sport fish species. *J. Gt. Lakes Res.* **2014**, *40*, 771–777. [CrossRef]

57. Gladyshev, M.I.; Glushchenko, L.A.; Makhutova, O.N.; Rudchenko, A.E.; Shulepina, S.P.; Dubovskaya, O.P.; Zuev, I.V.; Kolmakov, V.I.; Sushchik, N.N. Comparative analysis of content of omega-3 polyunsaturated fatty acids in food and muscle tissue of fish from aquaculture and natural habitats. *Contemp. Probl. Ecol.* **2018**, *11*, 297–308. [CrossRef]

58. Ramprasath, V.R.; Eyal, I.; Zchut, S.; Shafat, I.; Jones, P.J. Supplementation of krill oil with high phospholipid content increases sum of EPA and DHA in erythrocytes compared with low phospholipid krill oil. *Lipids Health Dis.* **2015**, *14*, 142. [CrossRef]

59. Cook, C.M.; Hallarake, H.; Sabo, P.C.; Innis, S.M.; Kelley, K.M.; Sanoshy, K.D.; Maki, K.C. Bioavailability of long chain omega-3 polyunsaturated fatty acids from phospholipid-rich herring roe oil in men and women with mildly elevated triacylglycerols. *Prostaglandins Leukot. Essent.* **2016**, *111*, 17–24. [CrossRef] [PubMed]

60. Adkins, Y.; Laugero, K.D.; Mackey, B.; Kelley, D.S. Accretion of dietary docosahexaenoic acid in mouse tissues did not differ between its purified phospholipid and triacylglycerol forms. *Lipids* **2019**, *54*, 25–37. [CrossRef] [PubMed]

9

Fish Oil, but not Olive Oil, Ameliorates Depressive-Like Behavior and Gut Microbiota Dysbiosis in Rats under Chronic Mild Stress

Te-Hsuan Tung [1], Yu-Tang Tung [2], I-Hsuan Lin [3], Chun-Kuang Shih [1], Ngan Thi Kim Nguyen [1], Amalina Shabrina [1] and Shih-Yi Huang [1,2,4,*]

[1] School of Nutrition and Health Sciences, Taipei Medical University, Taipei 11031, Taiwan; derossi83621@gmail.com (T.-H.T.); ckshih@tmu.edu.tw (C.-K.S.); da07107003@tmu.edu.tw (N.T.K.N.); amalina.shabrina@yahoo.com (A.S.)

[2] Graduate Institute of Metabolism and Obesity Sciences, Taipei Medical University, Taipei 11031, Taiwan; f91625059@tmu.edu.tw

[3] Research Center of Cancer Translational Medicine, Taipei Medical University, Taipei 11031, Taiwan; ycl6@tmu.edu.tw

[4] Center for Reproductive Medicine & Sciences, Taipei Medical University Hospital, Taipei 11031, Taiwan

* Correspondence: sihuang@tmu.edu.tw

Abstract: Background: This study investigated the effects of fish oil and olive oil in improving dysbiosis and depressive-like symptoms. Methods and results: Male rats were fed normal, fish oil-rich or olive oil-rich diets for 14 weeks. Chronic mild stress (CMS) was administered from week 2. The sucrose preference test (SPT) and forced swimming test (FST) were used to determine depressive-like behavior. The SPT results revealed that the CMS, CMS with imipramine (CMS+P) treatment, and CMS with olive oil diet (CMS+O) groups exhibited significantly reduced sucrose intake from week 8, whereas the fish oil diet (CMS+F) group exhibited significantly reduced sucrose intake from week 10. The FST results showed that the immobile time of the CMS+F group was significantly less than that of the CMS-only group. Next generation sequencing (NGS) results showed CMS significantly reduced the abundance of *Lactobacillus* and increased that of *Marvinbryantia* and *Ruminiclostridium_6*. However, the CMS+F group showed an increase in the abundance of *Eisenbergiella*, *Ruminococcaceae_UCG_009*, and *Holdemania*, whereas the CMS+O group showed an increase in the abundance of *Akkermansia*. Conclusions: CMS stimuli altered the gut microbiome in depressed rats. Fish oil and olive oil exerted part of a prebiotic-like effect to ameliorate dysbiosis induced by CMS. However, only fish oil ameliorated depressive-like symptoms.

Keywords: chronic mild stress; depression; gut microbiota; fish oil; olive oil

1. Introduction

The World Health Organization has reported that more than 350 million people worldwide have depression. Furthermore, depressive disorder is predicted to be the second leading cause of disability in 2020 [1]. Various therapies have been introduced for treating depression, however, antidepressants have severe side effects that cause low compliance among patients and even patient resistance to regular medical therapy. Consequently, discovering new therapies is extremely urgent.

Some adjunctive therapies have been discovered for the treatment of depression. Probiotic and prebiotic supplements are one potential approach. Kelly et al. [2] reported that depression can be induced in healthy rats by transplanting gut microbiota from major depressive disorder (MDD) patients. This suggests that gut microbiota might modulate brain activity and behavior. Among clinical trials relevant to depression, one randomized controlled study linked treatment with multispecies probiotics to emotional reactions of sad moods, particularly rumination and aggressive thoughts in non-depressed people [3]. Another study proved that prebiotics, which are non-digestible fibers that promote the growth or activity of beneficial microorganisms, have potential antidepressive effects [4]. The role of the microbiome–gut–brain axis in the pathology of depression has been discussed by scholars.

Associations between various types of diet and the pathology of depression have been discovered in the last two decades [5–8]. For instance, evidence has emerged that the Mediterranean-style diet has beneficial effects on neurological disorders, including stroke, depression, and cognitive impairment [9,10]. Fish oil, one of the main lipids in the Mediterranean diet, contains a high percentage of n-3 polyunsaturated fatty acid (PUFA), particularly eicosapentaenoic acid (EPA) and docosahexaenoic acid (DHA). A clinical trial that implemented a Mediterranean-style dietary intervention with fish oil supplementation demonstrated that an increase in n-3 PUFA intake reduced the severity of depression symptoms and improved quality of life [11]. The large amount of olive oil used in the Mediterranean diet has been considered to have health benefits, particularly regarding depression risk [12]. However, the association between the effects of olive oil on gut microorganisms and depressive-like behavior has not yet been elucidated in basic and clinical studies.

Scholars have concluded that the amounts and types of lipids in the diet affect the occurrence of depression [5,10,11]. C57BL/6 mice fed a high-fat diet (60% of energy (kJ) obtained from lipids, with refined palm oil as the main lipid source) for 8 weeks showed significantly decreased sociability and sucrose preference [5]. In another study, a lard-based high-fat and high-sugar diet (36% of energy (kJ) obtained from lipids) significantly reduced the frequency of social behaviors, impaired memory, and altered microbiome composition [13]. However, whether dysbiosis caused by an unhealthy saturated fatty acids-rich diet results in neurobehavioral alteration remains unclear. Therefore, we investigated whether fish oil and olive oil interventions exerted an antidepressive effect in a chronic mild stress (CMS) model and explored the potential effects of these two lipids on the intestinal dysbiosis induced by CMS.

2. Materials and Methods

2.1. Animals and Diets

In this study, male Sprague–Dawley rats (n = 43, 6 weeks old; Bio-LASCO, Taiwan) were used. The rats were housed in a temperature- and humidity-controlled room (22 °C ± 2 °C; humidity: 60%) under a 12 h light–dark cycle (light period: 08:00–20:00) and had free access to food and water. After 2 weeks of acclimation, the rats were divided into five groups (n = 8 or 9 per group): the normal control (N), CMS, CMS treated with a drug (imipramine) (CMS+P), CMS treated with a fish oil diet (CMS+F), and CMS treated with olive oil (CMS+O) groups (Figure 1). The study was conducted in accordance with institutional guidelines and approved by the Institutional Animal Care and Use Committee of Taipei Medical University (LAC-2016-0405).

The animal diets were prepared on the basis of the AIN-93M semi-purified diet composition. Three oil-based diets were used, specifically, the diets of the N, CMS, and CMS+P groups contained 4% (w/w) soybean oil, the diet of the CMS+F group contained 2% fish oil and 2% soybean oil, and the CMS+O group's diet contained 2% olive oil and 2% soybean oil. The fish oil (Chueh Hsin Co., Taipei, Taiwan) contained 20.5% (w/w) EPA and 11.2% DHA, whereas the extra virgin olive oil (EVOO) (Laconia Greece S.A., Sparta, Greece) contained 65% oleic acid.

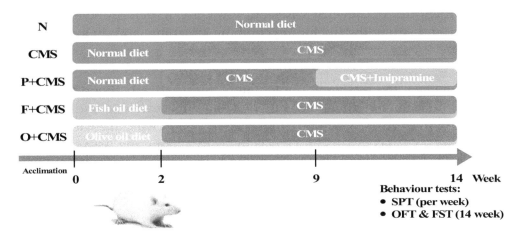

Figure 1. Experimental flow chart. Experimental animals were divided into five groups. CMS—chronic mild stress; SPT—sucrose preference test; OFT:—open field test; FST—forced swimming test.

2.2. Experimental Protocols

The experiment was conducted over 14 weeks (Figure 1). Briefly, the diets containing different dietary oils were administrated during the experimental period. Except for the N group, all groups were subjected to CMS from week 2 to week 14. CMS was exerted every week by randomly applying six out of nine possible stresses. The chronic mild stresses were as follows: (1) Water and food deprivation for 12 h, (2) a 30° cage tilt for 6 h, (3) damp sawdust (250 mL water in sawdust bedding) for 24 h, (4) physical restraint for 1 h, (5) cold swimming for 1 h, (6) blank cages without sawdust for 24 h, (7) reversed rhythm circadian for 2 days, (8) living space limitation for 8 h, and (9) social stress for 12 h. The N group lived normally without being placed under any stress. Food and water were freely available. Imipramine (Sigma-Aldrich Co., Ltd., Taiwan) was administered daily to the CMS+P group through drinking water (20 mg/kg) from week 8 to week 14, as described elsewhere [14].

2.3. Sucrose Preference Test

The sucrose preference test (SPT) is a measure of CMS-induced anhedonia, a key depressive behavior. Briefly, the rats were fasted for 12 h and then given two bottles of water, one containing reverse osmosis water and the other containing 1% sucrose solution. Sucrose preference was calculated as the percentage of the 1% sucrose solution consumed relative to the total liquid intake.

2.4. Open Field Test and Forced Swimming Test

The apparatus for the open field test (OFT) consisted of a square area (50×50 cm^2) with walls 40 cm high constructed from black polyvinyl chloride plastic board. The arena was lit by lights placed 145 cm above the arena and was divided into a central area (25 cm $\times$ 25 cm) and an outer area, which included the peripheral region of the arena and the wall area. During a test session, the total distance traveled and central visit duration were measured. Each test session lasted 5 min and was recorded using a video camera placed 145 cm above the arena. The videos were analyzed using ActualTrackTM software (ActualAnalytics Co., Ltd., UK).

The forced swimming test (FST) is a model of behavioral despair that is considered effective for predicting antidepressant efficacy. The study was conducted using a previously reported method with slight modification [15]. In brief, each rat was placed in a Plexiglas cylinder (37 cm in diameter and 70 cm in height) containing 50 cm of water (24 °C $\pm$ 1 °C). In the pretest session, a rat was placed in the water for 15 min to induce a state of despair and then dried with a towel and warmed in a plastic cage under a heat lamp. After 24 h, the rat was exposed to the same experimental conditions for a 5 min test session, which was recorded by HDR-SR1 (SONY, Tokyo, Japan). The videos were analyzed using software (Forced Swim Scan 2.0, CleverSys, Reston, VA, USA) to determine the time each rat spent immobile, swimming, and struggling (including climbing, escaping, and diving)).

2.5. Corticosterone Assay

After finishing the FST, the rats were immediately anesthetized. Blood was collected directly from the abdominal aorta, stored in prechilled ethylenediaminetetraacetic acid (EDTA)-coated blood collection tubes, and centrifuged (3000 rpm, 10 min, 4 °C). Plasma was taken and immediately stored at −80 °C until analysis. Plasma corticosterone was measured using the AssayMax™ Corticosterone ELISA kit (AssayPro, St. Charles, MO, USA), according to the manufacturer's instructions.

2.6. Lipid Extraction and Fatty Acid Profile Analysis

Selected tissue samples were extracted using a modified Folch method, as previously described [16]. Twenty milligrams of prefrontal cortex (PFC) or hippocampus and 1 mL of phosphate buffered saline (PBS) were completely homogenized, and 1 mL of red blood cells was mixed with 1 mL of water. Methanol (1.5 mL) was added to 200 µL of the sample aliquot and vortexed. Subsequently, 3 mL of chloroform was added, the mixture was left for 1 h at room temperature with gentle shaking, and the liquid was then separated by adding 1.25 mL of water. The extract was incubated for 10 min at room temperature and centrifuged at 3000 rpm for 10 min. The lower (chloroform) phase was collected. Phospholipids were separated using a HybridSPE®-Phospholipid column (Supelco, St. Louis, MO, USA). Fatty acid methylation was performed by heating the samples at 90 °C for 1 h with boron trifluoride–methanol reagent (15%, 0.3 mL) to form fatty acid methyl ether (FAME), and the solvent was then removed using a vacuum pump. The FAME was analyzed using a TRACE™ gas chromatograph (Thermo Fisher Scientific Inc., Milan, Italy) equipped with a 30 m × 0.32 mm inner diameter (I.D.) × 0.20 µm df Rtx-2330 column (Restek, Bellefonte, PA USA) and flame ionization detector. The gas chromatograph oven temperature was initially maintained at 160 °C and then increased at 5 °C/min to 250 °C, where it was maintained for 5 min. The injector and detector were both maintained at 260 °C. Results were obtained according to the retention time of the appropriate standard GLC-455 (Supelco, St. Louis, MO, USA), and the percentage of fatty acid profiles was calculated based on the 12 different fatty acids (Table S1).

2.7. DNA Extraction, Amplification, and Sequencing

Faeces were collected before the rats were sacrificed and immediately stored at −80 °C until analysis. DNA was extracted from 200 mg of faeces by using the PowerSoil® DNA Isolation kit (MO BIO Laboratories, Inc., Carlsbad, CA, USA), according to the manufacturer's instructions.

The Illumina MiSeq system and the MiSeq Reagent Kit v2 500-cycle (San Diego, CA, USA) was used to sequence the V3–V4 regions of the 16s rRNA gene extracted from rat faeces. Universal primers were removed, and low-quality reads were trimmed using cutadapt (v1.15). The paired reads were then processed using the DADA2/phyloseq workflow in the R environment. Briefly, filtering, trimming, dereplication, and denoising of the forward and reversed reads were performed using DADA2 (v1.6.0). The paired reads were then merged, and chimeras were subsequently removed. The inferred amplicon sequence variants were subjected to taxonomy assignment using the SILVA database (v132) as the reference, with a minimum bootstrap confidence of 80. Multiple sequence alignment of the amplicon sequence variants was performed using DECIPHER (v2.6.0), and a phylogenetic tree was constructed from the alignment using RAxML (v8.2.11). The frequency table, taxonomy, and phylogenetic tree information were used to create a phyloseq object, and bacterial community analyses were performed using phyloseq (v1.19.1).

2.8. Statistics

All data were presented as mean ± SD. Statistical analyses were performed using GraphPad Prism software, version 7.0 (San Diego, CA, USA). The data were analyzed using analysis of variance followed by Tukey's post hoc test. The correlation between microbiota abundance and the fatty acid profile was analyzed using Pearson's correlation coefficient; $p < 0.05$ was considered significant. Microbiota enrichment analysis was conducted using the linear discriminant analysis effect size (LEfSe) method and visualized through cladograms obtained using GraPhlAn.

3. Results

3.1. Weight Change under CMS

At the beginning of the experiment, the groups did not differ significantly in body weight [$F(4,38) = 2.003$, $p > 0.05$]. CMS stimuli were administered to the rats from week 2, and the body weight gain of the CMS-treated rats (CMS, CMS+P, CMS+F, and CMS+O groups) slowed during week 4–8 compared with that of the N group (Figure 2). A similar result was observed in week 8–14. In addition, the weight change of the CMS+P group was significantly smaller than that of the other groups during week 8–14.

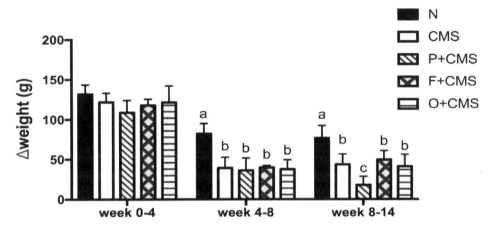

Figure 2. Weight change groups. The entire experimental period was divided into three parts. The group differences are displayed for each part. CMS was administered from week 2, and CMS significantly reduced the increase in weight during week 5–8. Data are expressed as mean ± SD (n = 8 or 9 per group). Values with different superscript letters are significantly different at $p < 0.05$.

3.2. Anxiety-Like Behavior Test

The OFT was used to identify anxiety-like symptoms in this research. We determined that the CMS rats traveled significantly smaller total distances than the N group. Fish oil and olive oil intervention restored this anxiety-like behavior, but imipramine had no effect on anxiety-like behavior in the OFT (Figure 3A). Nonetheless, no significant differences in central visit duration were discovered between all groups (Figure 3B).

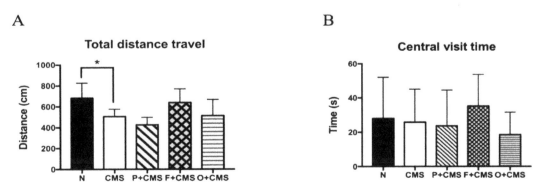

Figure 3. *Cont.*

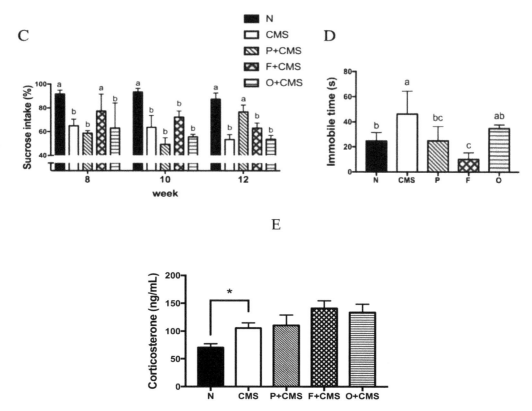

Figure 3. Anxiety-like and depressive-like behavioral tests. (**A**) Total distance traveled and (**B**) central visit duration in the OFT. CMS significantly reduced the total distance traveled. (**C**) Percentage of sucrose water consumed in the SPT. The CMS groups drank significantly less sucrose water. (**D**) Immobile time in the FST and (**E**) corticosterone levels in plasma. CMS significantly elevated the immobile time. Data are expressed as mean ± SD (n = 6 per group). * Significantly different using Student's t test ($p < 0.05$). Values with different superscript letters are significantly different at $p < 0.05$.

3.3. Depressive-Like Behavior Tests

Anhedonia, a depressive-like behavior, was represented by the percentage of sucrose solution intake in the SPT. Compared with the N group, the CMS rats had a significantly lower percentage of sucrose solution intake in week 8, except for the CMS+F group. We also observed that fish oil intervention delayed the onset of depressive-like behavior in the SPT, because the CMS+F group showed significantly reduced sucrose intake at week 10. The antidepressant imipramine was administered to the CMS+P group from week 8 to week 14, and after 4 weeks (at week 12), sucrose intake trend was significantly reversed (Figure 3C). Thus, six weeks of CMS successfully induced depressive-like symptoms in this study.

As illustrated in Figure 3D, the time the rats spent immobile in the FST was analyzed and represented depressive-like symptoms. Compared with the N group, the immobile time of the CMS group significantly increased during the test ($p = 0.001$). Fish oil intervention and imipramine treatment reversed the stress-induced abnormal depressive-like behavior exhibited in the FST ($p < 0.001$).

Overactivity of the hypothalamic-pituitary-adrenal (HPA) axis was discovered in long-term corticosterone levels after the FST (Figure 3). A significant difference was revealed in the corticosterone levels between the N and CMS groups ($p = 0.02$). Fish oil, olive oil, and imipramine did not recover the abnormally high corticosterone under acute stress (Figure 3D).

3.4. Fatty Acid Profiles of Brain and Red Blood Cells

Different dietary lipids, such as fish oil and olive oil, can result in significantly differing fatty acid profiles in both brain regions (the PFC and hippocampus) and red blood cells. The percentage of EPA (C20:5) in the prefrontal cortex (PFC) was significantly higher in the CMS+F group than the other groups, whereas the percentage of DHA (C22:6) was unaffected by the fish oil intervention (Figure 4). We also discovered that the C18:0 in the PFC was significantly higher in the CMS+O group than the N group (Table S1). Moreover, the EPA percentage in the hippocampus of the CMS+F group was significantly higher than that of the other groups, except for the N group. The DHA percentage in the hippocampus of the CMS+F group was significantly higher than that of the N group. We also found that the EPA percentage in the red blood cells of the CMS+F group was significantly higher than that of the other groups (Table S2).

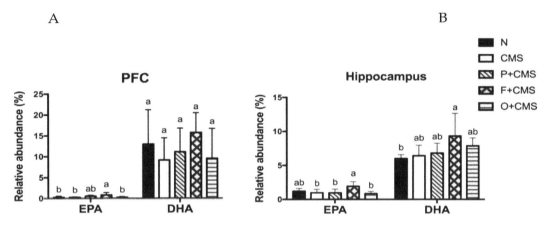

Figure 4. Percentages of eicosapentaenoic acid (EPA) and docosahexaenoic acid (DHA) in the (**A**) prefrontal cortex (PFC) and (**B**) hippocampus. Data are expressed as mean ± SD (n = 8 or 9 per group). Values with different superscript letters are significantly different at $p < 0.05$.

3.5. Microbiota Alteration after Different Dietary Lipid Interventions

After removing bias from variation in the sample read number, sequencing of the microbiota performed using the Miseq led to 2.2 million sequenced reads. In the N and CMS group samples, distinct clustering and separation of the CMS group from the N group were observed in the nonmetric multidimensional scaling plot within the Bray–Curtis distance methods. The results revealed clear separation of the CMS+F and CMS+O groups from the other groups (Figure 5A).

Analysis of relative abundance using the linear discriminant analysis (LDA) effect size (LEfSe) method indicated that the faecal microbiota composition differed significantly between all groups. Bacteroidaceae, Prevotellaceae, and Lactobacillaceae were significantly more abundant in the N group. *Bacteroides, Lactobacillus, Terrisporobacter, Candidatus_Stoquefichus,* and *Proteus* were significantly more abundant in the N group, whereas *Marvinbryantia, Ruminiclostridium_6, Ruminococcaceae_NK4A214,* and *Erysipelotrichaceae_ge* were significantly more abundant in the CMS group. Different dietary lipids resulted in differing microbiota compositions. At the genus level, fish oil (CMS+F group) significantly elevated the relative abundance of *Eisenbergilla, Ruminococcaceae_UCG_009,* and *Holdemania*. Nevertheless, the CMS+O group had significantly higher abundances of *Romboutsia, Akkermansia,* and *Ruminococcaceae_UCG_003* (Figure 5B,C).

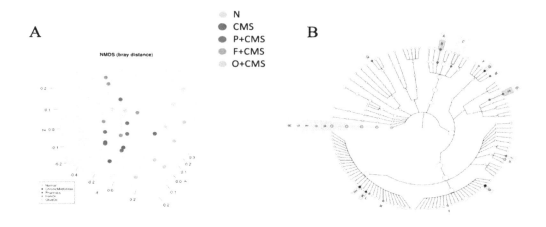

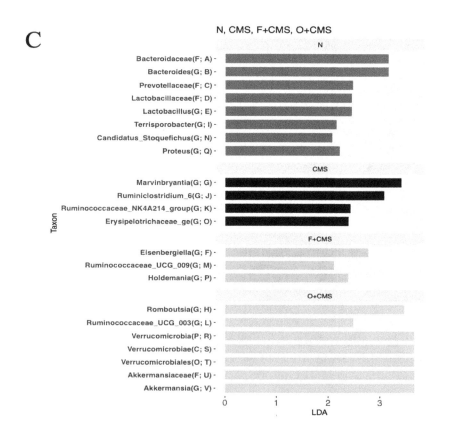

Figure 5. Taxonomic differences in faecal microbiota. Comparison of relative abundance across all groups. (**A**) Non-metric multidimensional scaling Bray distance methods were used to discriminate between the five groups. (**B**) LEfSe was used to identify the most differentially abundant bacteria in all groups, except the CMS+P group. The brightness of each dot is proportional to the effect size. (**C**) Only bacteria meeting a linear discriminant analysis threshold of >2 are shown (n = 5 per group).

3.6. Correlation between Fatty Acid Profiles and Microbiota

We evaluated the correlations between the relative abundance of microbiota and EPA, DHA, and arachidonic acid (AA) percentage by using Pearson coefficient correlation analysis. The results indicated that some bacterial genera were significantly correlated with the percentage of specific fatty acids in the hippocampus. *Lachnospiraceare_UCG006* was positively correlated with EPA and DHA percentages (Figure 6A,B p = 0.0198 and 0.04, respectively). We also determined that *Eryspelatoclostridium* was negatively correlated with EPA percentage (Figure 6C, p = 0.04). Notably, *Ruminiclostridium_5*, which belongs to Ruminococcaceae, negatively correlated with AA percentage (Figure 6D, p = 0.029) and positively correlated with EPA percentage (Figure 6E, p = 0.006).

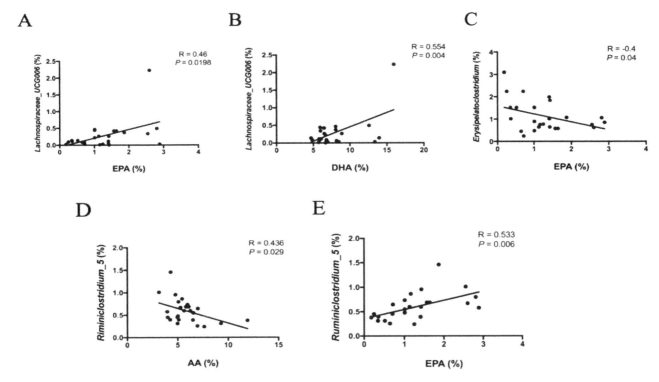

Figure 6. (**A**) and (**B**) Pearson correlation between relative abundance of *Lachnoapiraceae_UCG006* and percentage of EPA and DHA in the hippocampus, respectively. (**C**) Pearson correlation between relative abundance of *Erysipelatoclostridium* and percentage of EPA in the hippocampus. (**D**) and (**E**) Pearson correlation between relative abundance of *Ruminiclostridium_5* and percentage of EPA and arachidonic acid (AA) in the hippocampus, respectively. A value of p < 0.05 indicates a significant correlation.

4. Discussion

The CMS model is recommended as an effective animal model that induces depressive-like symptoms such as anhedonia, hopelessness, and despair [17–19]. This depression animal model simulates the progression of human depression, which is induced by continuous psychological stress. The present data demonstrated that CMS induced a depressive-like state characterized by a decrease in sucrose solution intake in the SPT and an increase in immobile time in the FST.

The CMS model also resulted in lower body weight gain in the experimental rats than the control rats. Cavigelli et al. demonstrated that four weeks of CMS reduced the body weight of rats [20]. Another study reported that six weeks of CMS significantly reduced body weight, and fluoxetine antidepressant treatment improved depressive-like behaviors without causing body weight gain [21]. Our study demonstrated that persistent psychological stress for more than 12 weeks strongly influenced body weight gain. Given that the food consumption was significantly lower after two weeks of CMS, persistent psychological stress may attenuate the appetite of rats. Notably, food consumption slowly recovered from week 10 in all CMS groups except the CMS+P group; however, the body weight of the CMS groups remained lower than that of the N group. Our data revealed that imipramine may have adverse effects such as decreased appetite [22]. In addition, the CMS intervention may have impaired the rats' physiological metabolism [23]. Therefore, the food consumption and body weight of the CMS+P group were significantly lower than those of the other groups (Figure S1).

Various neurobehaviors, such as despair, anhedonia, loss of willingness to explore, and general locomotor activity, are considered to represent depressive-like and anxiety-like behaviors caused by various types of stress [1,21,24,25]. In this study, CMS strongly induced anxiety-like and depressive-like behaviors, including decreased total distance traveled in the OFT, percentage of sucrose solution intake in the SPT, and increased immobile time in the FST. Dietary fish oil intake increased the total distance traveled and percentage intake of sucrose solution and reduced the immobile time; however, the other main lipid source in the Mediterranean diet, olive oil, did not improve depressive-like behaviors. Notably, the total distance traveled and central visit duration of the CMS+P group did not improve at all, indicating that imipramine had a weak anxiolytic effect [26]. However, several studies reported that imipramine can improve anxiety-like behaviors in the elevated-plus-maze test [27,28]. Therefore, the anxiolytic effect of imipramine remains controversial and requires additional research.

Fish oil, which contains large amounts of n-3 PUFA, has been demonstrated to exert antidepressant effects in both preclinical and clinical studies [25,29–31]. Recently, a meta-analysis of 13 randomized clinical trials indicated that fish oil exhibited beneficial effects in patients with major depressive disorder [32]. In the present study, the fish oil intervention slowed the progression of depression characterized by significantly decreased sucrose intake until week 10 for the CMS+F group compared with the N group, which was two weeks later than the other CMS groups. Thus, the fish oil intervention exerted an antidepressant effect.

The most well-known pathogenesis of depression is HPA axis dysregulation [33,34]. Because of a potent endocrine mechanism, the HPA axis influences numerous other parameters related to the pathophysiology of depression, such as monoaminergic neurotransmission, synaptic plasticity, and neuropeptide activity [35]. Furthermore, HPA axis dysregulation in patients with depression was found to be associated with immune function and age-related disease [36]. In the present study, corticosterone, which is a reliable biochemical marker of HPA axis dysregulation, was detected after the FST. The results indicated that after 12 weeks of CMS, the ability of the rats to adapt to acute stress decreased significantly because the corticosterone levels of the CMS-related groups were significantly higher than those of the N group. We also discovered that the antidepressant effects of n-3 PUFA was not associated with corticosterone level, because different dietary lipids could not relieve overactivity of the HPA axis. However, we only measured the corticosterone level of the rats. Whether the various dietary lipids modulate other parts of the HPA axis, such as the functions of glucocorticoid receptors or signal transduction of the central nervous system, remains unclear.

To determine whether the intake of different lipids altered the phospholipid composition of rat cell membranes, we analyzed the fatty acid profile of phospholipids in the PFC and hippocampus. The results revealed that fish oil intake not only ameliorated depressive-like behaviors in the FST and SPT, but also changed the composition of phospholipids in the rats. This suggests that EPA may play a more crucial role than DHA in the antidepressive effect of fish oil. A meta-analysis reported that EPA exerted a stronger effect on patients with depressive disorder than those without [37]. Studies have demonstrated that more than 50% of DHA supplements do not significantly reduce the severity of depression symptoms, whereas supplements containing pure EPA or more than 50% EPA significantly improve these symptoms [30,38]. Notably, although DHA is a critical and fundamental component of the brain, only EPA was elevated significantly in the hippocampus and PFC. Conversely, DHA has various essential functions in the brain and exerted an antidepressant effect on women with postpartum depression [30]. In recent years, some scholars have mentioned that intake of free DHA and triglyceride (TG)-form DHA did not increase DHA levels in the brain or improve depression symptoms because these forms of DHA could not completely cross the blood–brain barrier and simply elevated DHA levels in adipose tissue and the heart, whereas lysophosphatidylcholine-form DHA significantly increased DHA levels in the brain [39,40].

Furthermore, our data suggest that behavior alterations may be regulated by gut microbiota. Recently, microbiota have been proven to contribute to depression through several mechanisms. For example, scholars proved that the gut microbiota directly affects the immune system through activation of the vagus nerve [41,42]. In our study, linear discriminant analysis with effect size measurement was used to determine the major bacteria in the different treatment groups. Intestinal dysbiosis in the CMS group was characterized by identifying significant taxonomical differences compared with the N group. Well-known probiotics, namely *Lactobacillus* and Lactobacillaceae, were significantly decreased in the CMS-exposed rats. The abundances of Bacteroidaceae, *Bacteroides*, and Prevotellaceae were significantly higher in the N group than in those rats with dysbiosis. *Ruminiclostridium_6* was significantly higher in the CMS group in the LEfSe analysis and was previously reported to be increased in both early-diabetic and diabetic mice [43]. The relative abundance of *Ruminiclostridium_6* was positively correlated with the levels of pro-inflammatory factors IL-17A, TNF-α, and lipopolysaccharides (LPS), but negatively correlated with the level of anti-inflammatory cytokine IL-10 in the plasma [43]. Dietary prebiotics significantly improved the inflammatory response and reduced the abundance of *Riminiclostridium_6* [43,44].

In the present study, dietary fish oil prevented depressive-like status by improving gut dysbiosis. Prebiotic chrysanthemum polysaccharide intervention was previously reported to boost the abundance of *Ruminococcaceae UCG_009* [44]. In this research, *Ruminococcaceae UCG_009* increased with the fish oil intervention in the CMS+F group. In clinical trials, high dietary fiber intake increased the abundance of *Holdemania*, which is involved in butyrate production [45]. *Holdemania* abundance also increased in the CMS+F group. Moreover, dietary olive oil changed the composition of the microbiota considerably by significantly increasing the relative abundance of *Akkermansia*, which was demonstrated to have a positive effect on obesity, insulin resistance, and diabetes [46,47]. In the present study, however, olive oil did not exhibit a positive effect on psychological abnormalities.

Pearson's correlation analysis was performed to elucidate the associations between types of fatty acids in the hippocampus and the relative abundance of microbiota to understand the potential effects of dietary lipids on the gut microbiota. Our results indicated that *Lachnospiraceae_UCG006* abundance was positively associated with both EPA and DHA percentages in the hippocampus. We also discovered that *Ruminiclostridium_5* was significantly positively associated with EPA percentage and significantly negatively associated with AA percentage. EPA is an n-3 fatty acid and exerts an anti-inflammatory effect, whereas AA is an n-6 fatty acid and has a pro-inflammatory effect.

Although our study revealed that different dietary lipids resulted in differing gut microbiota composition, which may involve improvement of CMS-induced depression, there were still some limitations in our experiment. First, the olive oil we used only contained 65% oleic acid. Considering that oleic acid and polyphenols are two main beneficial factors of EVOO, we could have used EVOO, which contains higher levels of oleic acid. Also, more studies should be conducted to elucidate the effects of dietary lipids on the composition of microbiota, especially clinical trials. Whether dietary lipids affect the composition of the intestinal microbiota in humans still needs to be demonstrated.

5. Conclusions

In summary, fish oil improved psychiatric status by ameliorating gut dysbiosis in the CMS rat model. Additionally, the two main lipids in the Mediterranean diet, namely, fish oil and olive oil, were discovered to exert part of the effect exerted by prebiotics, preventing CMS-induced dysbiosis. Fish oil exerted a mild preventive effect on depression and improved the severity of depressive-like symptoms, but olive oil exhibited weaker effects than fish oil. Additional studies involving the alteration of lipid metabolites may elucidate the temporal and causal relationships between gut microbiota, depression, and dietary lipids.

Author Contributions: Statement of authorship: The author's responsibilities were as follows: T.-H.T., Y.-T.T., and S.-Y.H. conducted the study design and performed data analyses. T.-H.T., C.-K.S., N.T.K.N., and A.S. performed the brain and RBC fatty acid analyses. T.-H.T. and I-H.L. performed the microbiota analysis. T.-H.T. performed the depressive behavioral analysis. Y.-T.T., A.S., N.T.K.N., and S.-Y.H. assisted with the editing of the manuscript. T.-H.T., Y.-T.T., and S.-Y.H. prepared the initial draft and finalized the manuscript. All authors participated in the analytical discussion of the results and approved the final version of the manuscript.

References

1. Zhang, W.Y.; Guo, Y.J.; Han, W.X.; Yang, M.Q.; Wen, L.P.; Wang, K.Y.; Jiang, P. Curcumin relieves depressive-like behaviors via inhibition of the NLRP3 inflammasome and kynurenine pathway in rats suffering from chronic unpredictable mild stress. *Int. Immunopharmacol.* **2019**, *67*, 138–144. [CrossRef] [PubMed]
2. Kelly, J.R.; Borre, Y.; O' Brien, C.; Patterson, E.; El, A.S.; Deane, J.; Kennedy, P.J.; Beers, S.; Scott, K.; Moloney, G.; et al. Transferring the blues: Depression-associated gut microbiota induces neurobehavioural changes in the rat. *J. Psychiatr. Res.* **2016**, *82*, 109–118. [CrossRef] [PubMed]
3. Steenbergen, L.; Sellaro, R.; van Hemert, S.; Bosch, J.A.; Colzato, L.S. A randomized controlled trial to test the effect of multispecies probiotics on cognitive reactivity to sad mood. *Brain Behav. Immun.* **2015**, *48*, 258–264. [CrossRef] [PubMed]
4. Rieder, R.; Wisniewski, P.J.; Alderman, B.L.; Campbell, S.C. Microbes and mental health: A review. *Brain Behav. Immun.* **2017**, *66*, 9–17. [CrossRef] [PubMed]
5. Hassan, A.M.; Mancano, G.; Kashofer, K.; Frohlich, E.E.; Matak, A.; Mayerhofer, R.; Reichmann, F.; Olivares, M.; Neyrinck, A.M.; Delzenne, N.M.; et al. High-fat diet induces depression-like behaviour in mice associated with changes in microbiome, neuropeptide Y, and brain metabolome. *Nutr. Neurosci.* **2018**, 1–17. [CrossRef] [PubMed]
6. Winpenny, E.M.; van Harmelen, A.L.; White, M.; van Sluijs, E.M.; Goodyer, I.M. Diet quality and depressive symptoms in adolescence: No cross-sectional or prospective associations following adjustment for covariates. *Public Health Nutr.* **2018**, *21*, 2376–2384. [CrossRef] [PubMed]
7. Khayyatzadeh, S.S.; Mehramiz, M.; Mirmousavi, S.J.; Mazidi, M.; Ziaee, A.; Kazemi-Bajestani, S.M.R.; Ferns, G.A.; Moharreri, F.; Ghayour-Mobarhan, M. Adherence to a Dash-style diet in relation to depression and aggression in adolescent girls. *Psychiatry Res.* **2018**, *259*, 104–109. [CrossRef]
8. Matta, J.; Hoertel, N.; Kesse-Guyot, E.; Plesz, M.; Wiernik, E.; Carette, C.; Czernichow, S.; Limosin, F.; Goldberg, M.; Zins, M.; et al. Diet and physical activity in the association between depression and metabolic syndrome: Constances study. *J. Affect. Disord.* **2019**, *244*, 25–32. [CrossRef]
9. Psaltopoulou, T.; Sergentanis, T.N.; Panagiotakos, D.B.; Sergentanis, I.N.; Kosti, R.; Scarmeas, N. Mediterranean diet, stroke, cognitive impairment, and depression: A meta-analysis. *Ann. Neurol.* **2013**, *74*, 580–591. [CrossRef]
10. Masana, M.F.; Haro, J.M.; Mariolis, A.; Piscopo, S.; Valacchi, G.; Bountziouka, V.; Anastasiou, F.; Zeimbekis, A.; Tyrovola, D.; Gotsis, E.; et al. Mediterranean diet and depression among older individuals: The multinational MEDIS study. *Exp. Gerontol.* **2018**, *110*, 67–72. [CrossRef]
11. Parletta, N.; Zarnowiecki, D.; Cho, J.; Wilson, A.; Bogomolova, S.; Villani, A.; Itsiopoulos, C.; Niyonsenga, T.; Blunden, S.; Meyer, B.; et al. A Mediterranean-style dietary intervention supplemented with fish oil improves diet quality and mental health in people with depression: A randomized controlled trial (HELFIMED). *Nutr. Neurosci.* **2019**, *22*, 474–487. [CrossRef] [PubMed]
12. Li, Y.; Lv, M.R.; Wei, Y.J.; Sun, L.; Zhang, J.X.; Zhang, H.G.; Li, B. Dietary patterns and depression risk: A meta-analysis. *Psychiatry Res.* **2017**, *253*, 373–382. [CrossRef] [PubMed]
13. Reichelt, A.C.; Loughman, A.; Bernard, A.; Raipuria, M.; Abbott, K.N.; Dachtler, J.; Van, T.T.H.; Moore, R.J. An intermittent hypercaloric diet alters gut microbiota, prefrontal cortical gene expression and social behaviours in rats. *Nutr. Neurosci.* **2018**, 1–15. [CrossRef] [PubMed]
14. Iwata, M.; Ishida, H.; Kaneko, K.; Shirayama, Y. Learned helplessness activates hippocampal microglia in rats: A potential target for the antidepressant imipramine. *Pharmacol. Biochem. Behav.* **2016**, 138–146. [CrossRef] [PubMed]
15. Guo, Y.R.; Lee, H.C.; Lo, Y.C.; Yu, S.C.; Huang, S.Y. N-3 polyunsaturated fatty acids prevent d-galactose-induced cognitive deficits in prediabetic rats. *Food Funct.* **2018**, *9*, 2228–2239. [CrossRef] [PubMed]
16. Matyash, V.; Liebisch, G.; Kurzchalia, T.V.; Shevchenko, A.; Schwudke, D. Lipid extraction by methyl-tert-butyl ether for high-throughput lipidomics. *J. Lipid Res.* **2008**, *49*, 1137–1146. [CrossRef] [PubMed]

17. Katz, R.J.; Roth, K.A.; Carroll, B.J. Acute and chronic stress effects on open field activity in the rat: Implications for a model of depression. *Neurosci. Biobehav. Rev.* **1981**, *5*, 247–251. [CrossRef]

18. Willner, P. The chronic mild stress (CMS) model of depression: History, evaluation and usage. *Neurobiol. Stress* **2017**, *6*, 78–93. [CrossRef]

19. Calabrese, F.; Brivio, P.; Gruca, P.; Lason-Tyburkiewicz, M.; Papp, M.; Riva, M.A. Chronic, Mild Stress-Induced Alterations of Local Protein Synthesis: A Role for Cognitive Impairment. *ACS Chem. Neurosci.* **2017**, *8*, 817–825. [CrossRef]

20. Cavigelli, S.A.; Bao, A.D.; Bourne, R.A.; Caruso, M.J.; Caulfield, J.I.; Chen, M.; Smyth, J.M. Timing matters: The interval between acute stressors within chronic mild stress modifies behavioral and physiologic stress responses in male rats. *Stress* **2018**, *21*, 453–463. [CrossRef]

21. Pan, Y.; Chen, X.Y.; Zhang, Q.Y.; Kong, L.D. Microglial, NLRP3 inflammasome activation mediates IL-1beta-related inflammation in prefrontal cortex of depressive rats. *Brain, Behav. Immun.* **2014**, *41*, 90–100. [CrossRef] [PubMed]

22. Dunbar, G.C.; Cohn, J.B.; Fabre, L.F.; Feighner, J.P.; Fieve, R.R.; Mendels, J.; Shrivastava, R.K. A comparison of paroxetine, imipramine and placebo in depressed out-patients. *Br. J. Psychiatry* **1991**, *159*, 394–398. [CrossRef] [PubMed]

23. Garcia-Diaz, D.F.; Campion, J.; Milagro, F.I.; Lomba, A.; Marzo, F.; Martinez, J.A. Chronic mild stress induces variations in locomotive behavior and metabolic rates in high fat fed rats. *J. Physiol. Biochem.* **2007**, *63*, 337–346. [CrossRef] [PubMed]

24. Chanana, P.; Kumar, A. GABA-BZD Receptor Modulating Mechanism of Panax quinquefolius against 72-h Sleep Deprivation Induced Anxiety like Behavior: Possible Roles of Oxidative Stress, Mitochondrial Dysfunction and Neuroinflammation. *Front. Neurosci.* **2016**, *10*, 84. [CrossRef] [PubMed]

25. Reus, G.Z.; Maciel, A.L.; Abelaira, H.M.; de Moura, A.B.; de Souza, T.G.; Dos Santos, T.R.; Darabas, A.C.; Parzianello, M.; Matos, D.; Abatti, M.; et al. Omega-3 and folic acid act against depressive-like behavior and oxidative damage in the brain of rats subjected to early- or late-life stress. *Nutrition* **2018**, *53*, 120–133. [CrossRef] [PubMed]

26. Durand, M.; Aguerre, S.; Fernandez, F.; Edno, L.; Combourieu, I.; Mormede, P.; Chaouloff, F. Strain-dependent neurochemical and neuroendocrine effects of desipramine, but not fluoxetine or imipramine, in spontaneously hypertensive and Wistar-Kyoto rats. *Neuropharmacology* **2000**, *39*, 2464–2477. [CrossRef]

27. Simoes, L.R.; Netto, S.; Generoso, J.S.; Ceretta, R.A.; Valim, R.F.; Dominguini, D.; Michels, M.; Reus, G.Z.; Valvassori, S.S.; Dal-Pizzol, F.; et al. Imipramine treatment reverses depressive- and anxiety-like behaviors, normalize adrenocorticotropic hormone, and reduces interleukin-1beta in the brain of rats subjected to experimental periapical lesion. *Pharmacol. Rep.* **2019**, *71*, 24–31. [CrossRef] [PubMed]

28. Ramirez, K.; Sheridan, J.F. Antidepressant imipramine diminishes stress-induced inflammation in the periphery and central nervous system and related anxiety- and depressive- like behaviors. *Brain Behav. Immun.* **2016**, *57*, 293–303. [CrossRef]

29. Nemets, B.; Stahl, Z.; Belmaker, R.H. Addition of omega-3 fatty acid to maintenance medication treatment for recurrent unipolar depressive disorder. *Am. J. Psychiatry* **2002**, *159*, 477–479. [CrossRef]

30. Su, K.P.; Huang, S.Y.; Chiu, T.H.; Huang, K.C.; Huang, C.L.; Chang, H.C.; Pariante, C.M. Omega-3 fatty acids for major depressive disorder during pregnancy: Results from a randomized, double-blind, placebo-controlled trial. *J. Clin. Psychiatry* **2008**, *69*, 644–651. [CrossRef]

31. Robertson, R.C.; Seira, O.C.; Murphy, K.; Moloney, G.M.; Cryan, J.F.; Dinan, T.G.; Paul, R.; Stanton, C. Omega-3 polyunsaturated fatty acids critically regulate behaviour and gut microbiota development in adolescence and adulthood. *Brain Behav. Immun.* **2017**, *59*, 21–37. [CrossRef] [PubMed]

32. Bastiaansen, J.A.; Munafo, M.R.; Appleton, K.M.; Oldehinkel, A.J. The efficacy of fish oil supplements in the treatment of depression: Food for thought. *Transl. Psychiatry* **2016**, *6*, 975. [CrossRef] [PubMed]

33. Touma, C. Stress and affective disorders: Animal models elucidating the molecular basis of neuroendocrine-behavior interactions. *Pharmacopsychiatry* **2011**, *44*, 15–26. [CrossRef] [PubMed]

34. de Kloet, E.R.; Joels, M.; Holsboer, F. Stress and the brain: From adaptation to disease. *Nat. Rev. Neurosci.* **2005**, *6*, 463–475. [CrossRef]

35. Dean, J.; Keshavan, M. The neurobiology of depression: An integrated view. *Asian J. Psychiatr.* **2017**, *27*, 101–111. [CrossRef] [PubMed]

36. Doolin, K.; Farrell, C.; Tozzi, L.; Harkin, A.; Frodl, T.; O'Keane, V. Diurnal Hypothalamic-Pituitary-Adrenal Axis Measures and Inflammatory Marker Correlates in Major Depressive Disorder. *Int J. Mol. Sci.* **2017**, *18*, 2206. [CrossRef] [PubMed]

37. Martins, J.G. EPA but Not, DHA Appears To Be Responsible for the Efficacy of Omega-3 Long Chain Polyunsaturated Fatty Acid Supplementation in Depression: Evidence from a Meta-Analysis of Randomized Controlled Trials. *J. Am. Coll. Nutr.* **2009**, *28*, 525–542. [CrossRef]

38. Jazayeri, S.; Tehrani-Doost, M.; Keshavarz, S.A.; Hosseini, M.; Djazayery, A.; Amini, H.; Jalali, M.; Peet, M. Comparison of therapeutic effects of omega-3 fatty acid eicosapentaenoic acid and fluoxetine, separately and in combination, in major depressive disorder. *Aust. N. Z. J. Psychiatry* **2008**, *42*, 192–198. [CrossRef] [PubMed]

39. Sugasini, D.; Thomas, R.; Yalagala, P.C.R.; Tai, L.M.; Subbaiah, P.V. Dietary docosahexaenoic acid (DHA) as lysophosphatidylcholine, but not as free acid, enriches brain DHA and improves memory in adult mice. *Sci. Rep.* **2017**, *7*, 11263. [CrossRef]

40. Yalagala, P.C.; Sugasini, D.; Dasarathi, S.; Pahan, K.; Subbaiah, P.V. Dietary lysophosphatidylcholine-EPA enriches both EPA and DHA in the brain: Potential treatment for depression. *J. Lipid Res.* **2019**, *60*, 566–578. [CrossRef]

41. Bravo, J.A.; Forsythe, P.; Chew, M.V.; Escaravage, E.; Savignac, H.M.; Dinan, T.G.; Bienenstock, J.; Cryan, J.F. Ingestion of Lactobacillus strain regulates emotional behavior and central GABA receptor expression in a mouse via the vagus nerve. *Proc. Natl. Acad. Sci. USA* **2011**, *108*, 16050–16055. [CrossRef] [PubMed]

42. Perez-Burgos, A.; Wang, B.; Mao, Y.K.; Mistry, B.; McVey Neufeld, K.A.; Bienenstock, J.; Kunze, W. Psychoactive bacteria Lactobacillus rhamnosus (JB-1) elicits rapid frequency facilitation in vagal afferents. *Am. J. Physiol. Gastrointest. Liver Physiol.* **2013**, *304*, G211–G220. [CrossRef] [PubMed]

43. Li, K.; Zhang, L.; Xue, J.; Yang, X.; Dong, X.; Sha, L.; Lei, H.; Zhang, X.; Zhu, L.; Wang, Z.; et al. Dietary inulin alleviates diverse stages of type 2 diabetes mellitus via anti-inflammation and modulating gut microbiota in db/db mice. *Food Funct.* **2019**, *10*, 1915–1927. [CrossRef] [PubMed]

44. Tao, J.H.; Duan, J.A.; Jiang, S.; Feng, N.N.; Qiu, W.Q.; Ling, Y. Polysaccharides from Chrysanthemum morifolium Ramat ameliorate colitis rats by modulating the intestinal microbiota community. *Oncotarget* **2017**, *8*, 80790–80803. [CrossRef] [PubMed]

45. Gomez-Arango, L.F.; Barrett, H.L.; Wilkinson, S.A.; Callaway, L.K.; McIntyre, H.D.; Morrison, M.; Dekker Nitert, M. Low dietary fiber intake increases Collinsella abundance in the gut microbiota of overweight and obese pregnant women. *Gut Microbes* **2018**, *9*, 189–201. [CrossRef] [PubMed]

46. Everard, A.; Belzer, C.; Geurts, L.; Ouwerkerk, J.P.; Druart, C.; Bindels, L.B.; Guiot, Y.; Derrien, M.; Muccioli, G.G.; Delzenne, N.M.; et al. Cross-talk between Akkermansia muciniphila and intestinal epithelium controls diet-induced obesity. *Proc. Natl. Acad. Sci.* **2013**, *110*, 9066–9071. [CrossRef]

47. Anhe, F.F.; Roy, D.; Pilon, G.; Dudonne, S.; Matamoros, S.; Varin, T.V.; Garofalo, C.; Moine, Q.; Desjardins, Y.; Levy, E.; et al. A polyphenol-rich cranberry extract protects from diet-induced obesity, insulin resistance and intestinal inflammation in association with increased Akkermansia spp. population in the gut microbiota of mice. *Gut* **2015**, *64*, 872–883. [CrossRef]

10

Impact of Water Pollution on Trophic Transfer of Fatty Acids in Fish, Microalgae and Zoobenthos in the Food Web of a Freshwater Ecosystem

Shahid Mahboob [1,2,*], **Khalid Abdullah Al-Ghanim** [1], **Fahad Al-Misned** [1], **Tehniat Shahid** [3], **Salma Sultana** [2], **Tayyaba Sultan** [2], **Bilal Hussain** [2] and **Zubair Ahmed** [1]

[1] Department of Zoology, College of Science, King Saud University, Riyadh 11451, Saudi Arabia; kghanim@ksu.edu.sa (K.A.A.-G.); almisned@ksu.edu.com (F.A.-M.); zahmed@ksu.edu.sa (Z.A.)

[2] Department of Zoology, Government College University, Faisalabad-38000, Pakistan; sal545@live.com (S.S.); arif123@yahoo.com (T.S.); prof_bilal@yahoo.com (B.H.)

[3] House No. 423, Block M-1, Street No. 14, Lake city, Lahore 55150, Pakistan; tehniatshahid123@gmail.com

[*] Correspondence: shahidmahboob60@hotmail.com

Abstract: This research work was carried out to determine the effects of water contamination on the fatty acid (FA) profile of periphyton, zoobenthos, two Chinese carps and a common carp (*Hypophthalmichthys molitrix, Ctenopharygodon idella* and *Cyprinus carpio*), captured from highly polluted (HP), less polluted (LP), and non-polluted (NP) sites of the Indus river. We found that the concentration of heavy metals in the river water from the polluted locations exceeded the permissible limits suggested by the World Health Organization (WHO) and the US Environmental Protection Agency (EPA). Fatty acid profiles in periphyton, zoobenthos, *H. molitrix, C. idella*, and *C. carpio* in the food web of river ecosystems with different pollution levels were assessed. Lauric acid and arachidic acids were not detected in the biomass of periphyton and zoobenthos from HP and LP sites compared to NP sites. Alpha-linolenic acid (ALA), eicosadienoic acid and docosapentaenoic acid were not recorded in the biomass samples of periphyton and zoobenthos in both HP and LP sites. Caprylic acid, lauric acid, and arachidic acid were not found in *H. molitrix, C. idella*, and *C. carpio* captured from HP. In this study, 6 and 9 omega series FAs were identified in the muscle samples of *H. molitrix, C. idella* and *C. carpio* captured from HP and LP sites compared to NP sites, respectively. Less polyunsaturated fatty acids were observed in the muscle samples of *H. molitrix, C. idella*, and *C. carpio* collected from HP than from LP. The heavy metals showed significant negative correlations with the total FAs in periphyton, zoobenthos, and fish samples.

Keywords: Fatty acid; fish; food web; periphyton; trophic transfer; water pollution; zoobenthos

1. Introduction

The aquatic fauna and flora of river ecosystems comprise a complex assemblage of different communities and are biologically important because of the interlinking between different trophic levels. These aquatic food chains are very feeble and sensitive to contaminants, especially to the toxicity of exogenous chemicals and heavy metals that are discharged into freshwater reservoirs due to various human activities. Heterotrophic aquatic organisms in food chains consume a variety of metalloids and xenobiotic compounds, which usually cause immutable degradation of the planktonic life at higher concentrations [1,2]. The toxic response in freshwater fish species to contaminated environments has been reported on a global scale [3,4]. The uptake of heavy metals into the aquatic food chain can occur either by dietary or non-dietary routes [5]. Therefore, the concentration of heavy metals in fish normally indicates levels present in sediment and water that is specifically in freshwater reservoirs where the fish is captured from [6], as well as the time of exposure [7]. The concentration of essential metals, if increased above the normal metabolic needs of fish, may become toxic for the fish and for the ultimate consumer, humans [8]. Heavy metals may accumulate in primary producers such as microalgae, where diatoms ultimately transfer them to other trophic levels [9]. Heavy metals are ingested by fish and bio accumulate in the liver, kidneys, and other vital organs through adsorption and absorption [7].

Lipids are considered to be one of the most essential nutrients, which affect the growth and development of fish and other organisms [10], and alleviate immune competence [11]. Essential lipids are nutritionally important for the consumers in the food chain because they promote the growth and development and overall health of aquatic fauna and flora of aquatic communities in freshwater ecosystems [12–14]. Kainz et al. [15] proposed that the trophic movement of fatty acid (FA) in the food chain may be used as a physiological biomarker for monitoring the status of contamination in freshwater ecosystems. Kainz et al. [15] further mentioned that this trophic movement of FA in the food chain may be used as a physiological biomarker for observation of the status of contamination in freshwater ecosystems. Thus, it is necessary to ensure the abundance of microalgae and zoobenthos for trophic transfer into higher levels in the food web to ensure the transfer of FA and polyunsaturated fatty acid (PUFA) to the fish [16]. Currently, there are no comprehensive reports in the literature describing the interlinking trophic movements of PUFA and the impact of water pollution in the river ecosystems. The latter is still poorly understood with reference to FA profiles of periphyton, zoobenthos, and fish, and effect of contaminants. periphyton and zoobenthos can be used as valuable indicators to determine the effect of contamination and the synthesis of FA in freshwater ecosystems [17].

Aquaculture plays an important role in providing good quality animal protein and provides sustainable livelihood opportunities and food security for the ever increasing world population [18,19]. Fish are recognized as an important part of the human diet, owing to its balanced ratio proteins/PUFAs, including omega series FAs [20] which may reduce the risk of heart diseases. Because of the nutritional and pharmaceutical importance of PUFAs, researchers in the discipline of fishery sciences have been paying them increasing attention [21,22].

The bioaccumulation of metals in fish is triggered by the accumulation of these elements in phyto- and zooplankton; however, this event ceases to be the most relevant as long as biomagnification takes place. Biomagnification can have serious impacts on the food chain [23]. Fish has the potential to accumulate more metals from food and water [24]. Kainz and Fisk [25] mentioned that most of the FAs and pollutants move trophically through the food chain, ultimately having effects on the final consumer. This situation warrants an understanding of the fate of FAs and the impact of heavy metals contamination on the variability of FA on the food chain in river ecosystems. Variation in FAs dynamics in the food web is linked to increases in environmental stress and habitat destruction due to water pollution within a freshwater ecosystem [26]. Moreover, the disparity in FAs and pollutant trophic movement in the food chain may give insights into ecological functions and their impact on habitat and environmental stress. Keeping this in mind, it is necessary to investigate the interlink and biotransformation of FAs in the food web and the relationship with water pollution. This requires assessing and contrasting the trophic movement of lipids and pollutants in the aquatic food chain. The main aim of research was (i) to assess the fatty acid profiles in periphyton, zoobenthos, *Hyphpthalmichthys molitrix*, *Ctenopharyngodon idella*, and *Cyprinus carpio* in the food web of river ecosystems with different pollution levels; (ii) to assess their flow in aquatic ecosystems; (iii) to explore their potential for evaluating and monitoring the health of aquatic habitats; (iv) and to apply FA profiles as a possible physical biomarker of environmental stress from heavy metal pollution.

2. Materials and Methods

2.1. Study Area

The Indus river is the longest river in Pakistan. The Indus River originates on the Tibetan Plateau, enters into towards Gilgit-Baltistan from Ladakh, and then flows from Punjab Province and joins into the Arabian Sea. It is the largest river in Pakistan with a total catchment basin of about 1,165,000 km^2 (450,000 m^2) https://en.wikipedia.org/wiki/Indus_River).

The Mianwali District is situated in the province of the Punjab and is about 200 m above sea level (Figure 1; 25). The Mianwali is one of the districts in the province of the Punjab and is about 200 m above sea level [27]. This district is rich in minerals, clay, coal, gypsum, limestone, etc., which are excavated for commercial purposes. In this district there is also a nuclear power plant and the Chashma Hydel power plant, which are adding electricity into the national electricity grid. The temperature ranged between –2 °C and 51 °C with 255 mm of rainfall [28]. The experimental sites were selected in Area 1 (Kalabagh; high pollution (HP) site), Area 2 (Chashma; low pollution (LP) site), and non-polluted (NP) site (Area 3; Attock) along the River, and these sites were 35 km apart from each other.

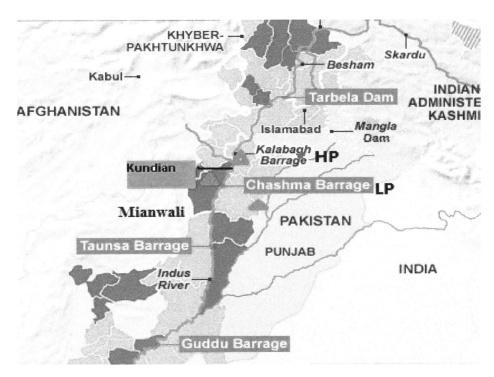

Figure 1. Map of the locations of sampling sites on the Indus River (Kundian Barrage, Kalabagh Barrage, and Chashma Barrage) [25]. Source: OCHA (United Nation Office for the Coordination Humanitarian Affairs).

2.2. Collection and Preparation of Fish Samples

Hypophthalmichthys molitrix is planktivorous and consumes the organisms within lower multiple lower trophic levels across a range of habitats. Grass carp (*Ctenopharyngodon idella*) is a large cyprinid and is a voracious feeder. Small grass carp consume planktonic crustaceans, rotifers, and insect larvae, while the adults are completely vegetarian. *Cyprinus carpio* is a popular benthivorous fish that has larger bottom–up effects than other benthivorous fish. The bottom–up effects of *C. carpio* mainly depend on the incorporation of benthos-derived nutrients and the release of nutrients from bottom sediment during grazing on benthos. Twenty-one specimens of *H. molitrix*, *C. idella*, and *C. carpio* each were captured from HP, LP and NP sites for an evaluation of the fatty acid profiles. A total of 63 fish specimens were procured with the help of fishermen. The average weight ranged from 900 to 1200 g. Fish specimens were transferred live in polyethylene bags to the laboratory. Muscle samples were processed as per the method mentioned by reference [29]. This study was approved by the Ethics and Animal Welfare Committee of the Department of Zoology, GC University, Faisalabad (Ethical code number: GCUF/Zool/EAWC/34).

2.3. Analysis of Water Samples

Water samples were collected in hydrographic bottles of 32 oz capacity at the depth of 30 cm below the surface from the three determined sampling sites for the determination of selected physiochemical parameters and selected heavy metals through an atomic absorption spectrophotometer ("Hitachi polarized Zeeman AAS, 2000 series") by following the procedure as mentioned by reference [30]. The water samples were collected in the morning and these were stored in iceboxes before being taken to the laboratory for analysis. Different dilutions of Hg, Sn, Cr, Pb, Zn, Mn, Cu, and Cd were made to check the accurateness of the equipment during the analysis of samples. The quality control and quality assurance protocol was followed as mentioned in our previous published work [9]. Calibration curves were plotted and validated with their corresponding R^2 values for the detection of each metal. The values of R^2 of the curves were 0.99983, 0.99981, 0.99951, 0.99984, 0.99926, 0.99987, and 0.99982 for Hg, Sn, Cr, Pb, Zn, Mn, Cu, and Cd, respectively.

2.4. Periphyton Sampling

Periphyton samples were obtained from the three experimental locations by following the methodology of references [28,29]. "A 10 × 10 cm steel frame was fixed at the bottom at three points of each location and composite them to collect the periphyton, then the pebble was removed. The periphyton samples was cleaned from the pebble surface using brushes, after which it was washed with river water. Aliquots from this volume were centrifuged at 2500 g for 15 min for the further analysis of metals and fatty acids" [31].

2.5. Zoobenthos Sampling

Zoobenthos samples were obtained from the experimental locations at three points and composite them through a Samples Surber-type kick-bottom sampler as mentioned by reference [31].

2.6. Fatty Acid Profiling

The lipid components were obtained from the fish muscle, periphyton, and zoobenthos samples with help of Soxhlet extractor (Electrothermal EME6 England), as described by reference [9]. "The extracted lipids were transformed to fatty acid methyl esters using methanolic sulfuric acid by an esterification procedure", as described by references [32,33]. The fatty acid profiling was carried out by following the methods of reference [34], through gas chromatograph (Perkin Elmer Model 3920) with flame ionization detector (FID) column 2 m in length and 2 mm in diameter. The chromatograms recorded from all samples were used to observe the retention time of each fatty acid (Fatty acid methyl esters (FAMEs)) and these were compared to the chromatogram of a standard (mixture of pure FAMEs) as described by reference [35].

2.7. Statistical Analysis

The data obtained was processed using Minitab software for analysis of variance (ANOVA) to assess the dissimilarity between various parameters of this study between the three sampling sites. Duncan's multiple range test (DMR test) ($p < 0.05$) was used to compare the means. "Shapiro-Wilk's W test and Levene's test" was used for normality and homogeneity of the data when necessary [36]. Correlation coefficients were calculated to determine the relationship between the concentration of heavy metals and the total FAs profile in fish and planktonic life from three sampling sites.

3. Results

3.1. Physico-Chemical Factors and Heavy Metals

The physico-chemical parameters of the water samples from sampling sites (HP, LP and NP) are presented in Table 1. The level of salinity of HP was about 2%, found to be close to the level of salinity of the open ocean (normally about 3%). pH levels were 12.1 ± 0.36, 8.6 ± 0.12, and 8.1 ± 0.08 in HP, LP, and NP sites, respectively. The pH level was very high at the HP site. The highest biochemical oxygen demand (81.2 ± 1.10 mg/L) and chemical oxygen demand (195.8 ± 1.16 mg/L) were recorded at the HP site. The concentration of total dissolved solids (2445.5 ± 8.41 mg/L) and total suspended solids (329.6 ± 6.41 mg/L) were very high at the HP site. The concentration of phenols and sulfates were highest at the HP site, closely followed by the LP site. The level of phenols at the HP site was 15 times higher than at the NP site (Table 1).

Table 1. Mean Physico-Chemical parameters and metal concentrations ($\pm$ SE) at different sampling locations of Indus River.

Water Quality Characteristics	HP Site	LP Site	NP Site	Permissible Limits
pH	12.1 ± 0.36 a	8.5 ± 0.12 b	8.1 ± 0.08 b	D: 6.5–8.5, P: **
BOD (mg/L)	81.2 ± 1.10 a	48.8 ± 0.41 b	36.7 ± 0.77 c	†D: 30 mg/L, P: **
COD (mg/L)	195.8 ± 1.16 a	71.2 ± 0.90 b	65.5 ± 0.58 c	†D: 250 mg/L, P: **
TDS (mg/L)	2444.5 ± 8.41 a	1319.8 ± 10.62 b	340.3 ± 7.24 c	D: 500 mg/L, P: 2000 mg/L
TSS (mg/L)	329.6 ± 6.41 a	218.6 ± 5.15 b	190.6 ± 4.24 c	D: 100mg/L, P: **
Salinity (mg/L)	1951.2 ± 18.31 a	458.5 ± 7.22 b	242.3 ± 4.90 b	P: <100 mg/L
Conductivity µS/cm	4.1 ± 0.22 a	1.55 ± 0.11 b	0.42 ± 0.051 c	D:650 µS/cm, P: 1055 µS/cm
Phenols (mg/L)	2.49 ± 0.18 a	0.84 ± 0.04 b	0.21 ± 0.01 c	D: 0.001 mg/L, P: 0.002 mg/L
Sulfates (mg/L)	452.3 ± 7.62 a	341.21 ± 0.08 b	97.4 ± 3.90 c	D: 0.001 mg/L, P: 0.002 mg/L
Heavy Metal Contamination				
Sn (mg/L)	0.54 ± 0.02 a	0.03 ± 0.0 b	0.01 ± 0.0 b	D: 0.01 mg/L, P: **
Cr (mg/L)	0.72 ± 0.03 a	0.36 ± 0.02 b	0.05 ± 0.00 c	D: 0.05 mg/L, P: **
Pb (mg/L)	3.02 ± 0.07 a	0.21 ± 0.02 b	0.14 ± 0.01 c	D: 0.05 mg/L, P: **
Zn (mg/L)	0.56 ± 0.02 a	0.251 ± 0.03 b	0.05 ± 0.00 a	D: 5 mg/L, P: 15 mg/L
Mn (mg/L)	2.81 ± 0.12 a	2.05 ± 0.06 a	0.41 ± 0.01 c	D: 0.1 mg/L, P: 0.3 mg/L
Cu (mg/L)	2.05 ± 0.05 a	0.99 ± 0.11 b	0.08 ± 0.00 c	D: 0.05 mg/L, P: 1.5 mg/L
Cd (mg/L)	0.29 ± 0.02 a	0.03 ± 0.00 b	0.00 ± 0.00 b	D: 0.01 mg/L, P: **
Hg (mg/L0	1.51 ± 0.04 b	0.05 ± 0.01 c	< 0.001	D: 0.001 mg/L, P: **

BOD: Biological oxygen demand, COD: Chemical oxygen demand. TDS: Total dissolved solids, TSS: Total suspended solids. Different letters (a, b, c) in the same row represent significant ($p < 0.05$) differences. D; Desirable limits. P; Permissible limits. †; Effluent inland surface water quality standards. ** No relaxation.

The concentration of studied heavy metals are presented in Table 1. These concentrations exhibited significant variations between the three sites. The level of Sn, Cr, Pb, Mn, Cu, Cd, and Hg at the HP in fish muscle, periphyton, and zoobenthos biomass were highest at HP compared to LP and NP site and was above the upper limits stated by reference [34] (Tables 1 and 2). The highest level of Cu in muscle samples was detected in *C. carpio* and plankton from HP, followed by LP sites. The maximum level of metals was recorded in the muscle samples of *C. carpio* captured from HP, followed by LP and NP sites. The lowest concentration of these metals was recorded in the muscles of *C. idella* (Table 2).

3.2. Fatty Acids Profile

The saturated fatty acids (SFAs) were low in the biomass of periphyton and zoobenthos obtained from HP and LP sites, compared to NP sites (Table 3). Lauric acid and arachidic acids were not detected in the biomass of periphyton and zoobenthos from HP and LP. The Environmental Protection Agency (EPA) value was significantly higher in the biomass sampled from NP, compared to HP and LP. The number of monounsaturated fatty acids (MUFAs) was higher in samples of periphyton and zoobenthos from NP (Table 3). Palmitoleic acid, vaccenic acid, oleic acid, eicosenic acid and erucic acid were not detected in periphyton biomass samples from HP and LP. PUFAs level was greater in periphyton and zoobenthos biomass from NP, compared to HP and LP sites. Alpha-linolenic acid (ALA), eicosadienoic acid, docosapentaenoic acid and docosapentaenoic acid were not detected in the biomass of periphyton and zoobenthos sampled from HP and LP. The percentage of EPA and DHA were higher in the periphyton biomass from HP, compared to NP (Table 3).

Table 2. Heavy metal concentrations (mg/kg) in the biomass of periphyton, zoobenthos and in the muscle of fish species from different sampling locations of the Indus River.

Parameter	HP Site	LP Site	NP Site
Periphyton			
Sn	17.61 ± 0.90 a	11.34 ± 0.77 b	4.88 ± 0.61 c
Cr	6.10 ± 0.70 a	1.98 ± 0.55 b	0.15 ± 0.23 c
P b	1.68 ± 0.67 a	0.41 ± 0.10 b	0.15 ± 0.03 c
Zn	14.46 ± 1.35 b	23.11 ± 1.66 a	8.19 ± 0.1.05 c
Mn	20.66 ± 1.41 a	11.44 ± 1.33 a	5.90 ± 0.88 c
Cu	18.43 ± 1.44 a	6.97 ± 0.92 b	4.02 ± 0.55 c
Cd	2.41 ± 0.31 a	0.62 ± 0.12 b	0.16 ± 0.02 c
Hg	3.78 ± 0.40 b	1.70 ± 0.18 c	0.17 ± 0.00 c
Zoobenthos			
Sn	3.43 ± 0.31 a	2.90 ± 0.70 b	1.11 ± 0.40 c
Cr	2.80 ± 0.42 a	1.22 ± 0.18 b	0.26 ± 0.03 c
P b	1.21 ± 0.05 a	0.38 ± 0.01 b	0.15 ± 0.02 c
Zn	6.94 ± 0.77 b	8.06 ± 1.0 a	3.80 ±0.22 c
Mn	3.91 ± 0.48 a	1.30 ± 0.26 b	0.92 ± 0.01 c
Cu	6.02 ± 0.72 a	1.40 ± 0.21 b	0.33 ± 0.03 c
Cd	1.99 ± 0.18 a	0.89 ± 0.02 b	0.36 ± 0.07 c
Hg	1.92 ± 0.31 b	0.86 ± 0.18 c	0.09 ± 0.00 c
Hypophthalmichthys molitrix			
Sn	1.95 ± 0.41 a	1.32 ± 0.36 b	0.67± 0.16 c
Cr	3.01 ± 0.60 a	1.71 ± 0.22 b	0.82 ± 0.15 c
P b	0.81 ± 0.16 a	0.49 ± 0.08 b	0.21 ± 0.01 c
Zn	5.67 ± 0.67 a	3.62 ± 0.41 b	1.62 ± 0.33 c
Mn	2.70 ± 0.41 a	1.92 ± 0.38 b	0.79 ± 0.08 c
Cu	4.93 ± 0.62 a	1.08 ± 0.15 b	0.49 ± 0.05 c
Cd	1.41 ± 0.28 a	0.76 ± 0.12 b	0.35 ± 0.05 c
Hg	1.37 ± 0.22 a	0.87 ± 0.09 b	0.06 ± 0.00 c
Ctenopharyngodon idella			
Sn	1.81 ± 0.3 a	1.12 ± 0.18 b	0.61 ± 0.06 c
Cr	2.28 ± 0.27 a	1.57 ± 0.20 b	0.71 ± 0.15 c
P b	0.74 ± 0.10 a	0.49 ± 0.06 b	0.23 ± 0.01 c
Zn	5.62 ± 0.72 a	3.01 ± 0.47 b	1.52 ± 0.31 c
Mn	2.50 ± 0.38 a	1.62 ± 0.21 b	0.69 ± 0.17 c
Cu	4.70 ± 0.52 a	1.02 ± 0.20 b	0.42 ± 0.05 c
Cd	1.34 ± 0.31 a	0.70 ± 0.21 b	0.31 ± 0.06 c
Hg	1.41 ± 0.22 a	0.87 ± 0.16 b	0.09 ± 0.00 c
Cyprinus carpio			
Sn	2.41 ± 0.4 a	1.61 ± 0.17 b	0.72 ± 0.06 c
Cr	3.44 ± 0.40 a	1.90 ± 0.23 b	0.95 ± 0.16 c
P b	0.98 ± 0.12 a	0.69± 0.07 b	0.40 ± 0.03 c
Zn	6.75 ± 0.88 a	3.96 ± 0.60 b	1.88 ± 0.44 c
Mn	2.90 ± 0.40 a	1.80 ± 0.24 b	0.77 ± 0.18 c
Cu	5.48 ± 0.80 a	1.57 ± 0.28 b	0.58 ± 0.02 c
Cd	1.62 ± 0.40 a	0.88 ± 0.16 b	0.45 ± 0.05 c
Hg	1.69 ± 0.21 a	0.92 ± 0.18 b	0.22 ± 0.01 c

Values (Mean ± SE) are averages of five samples analyzed in triplicate. Different letters (a, b, c) in the same row represent significant ($p < 0.05$) differences.

Table 3. Fatty acids (% ±SE) in periphyton and zoobenthos from three sampling sites at different pollution levels in the Indus River.

Pytoperiphyton			
Fatty Acids	**HP Site**	**LP Site**	**NP Site**
SFAs			
C8:0	——	——	——
C10:0	3.33 ± 0.11 c	6.78 ± 0.88 b	8.12 ± 0.98 a
C12:0	——	——	
C14:0	1.67 ± 0.08 c	4.41 ± 0.16 b	5.69 ± 0.77 a
C16:0	16.69 ± 2.77 c	19.63 ± 2.55 b	21.89 ± 2.44 a
C18:0	12.47 ± 1.63 c	13.12 ± 1.20 b	15.76 ± 2.89 a
C20:0	——	——	0.67 ± 0.11
MUFAs			
C16:1(n-7)	——	——	——
C16:1(n-9)	——	——	
C18:1(n-7)	——	——	1.99± 0.16 a
C18:1(n-9)	——	——	0.97 ± 0.11
C20:1(n-9)	——	——	——
C22:1(n-9)	3.11± 0.13 c	5.12 ± 0.34 b	7.89 ± 0.25 a
PUFAs			
C18:2(n-6)	0.43 ± 0.04 c	0.66 ± 0.19 b	1.15 ± 0.22 a
C18:3(n-3)	8.31 ± 0.70 b	9.11 ± 0.90 a	9.44 ± 0.88 a
C18:4(n-3)	——	——	——
C20:2(n-6)	——	——	2.24 ± 0.05 a
C20:4(n-6)	6.11 ± 0.55 c	8.02 ± 0.71 b	9.98 ± 0.41 a
C20:5(n-6)	——	4.11 ± 0.22 b	5.23 ± 0.28 a
C20:5(n-3)	6.78 ± 0.90 a	5.97 ± 0.60 a	4.66 ± 0.70 c
C22:4(n-6)	——	——	
C22:5(n-6)	——	——	2.79 ± 0.24 a
C22:5(n-3)	6.57 ± 0.41 c	7.33 ± 0.70 b	8.95± 0.66 a
C22:6(n-3)	5.44 ± 0.25 a	4.76 ± 0.41 a	3.77 ± 0.33 b

Zoobenthos			
Fatty acids	**Highly polluted water**	**Less polluted water**	**Non-polluted site**
SFAs			
C8:0	——	——	——
C10:0	1.44 ± 0.23 c	4.11 ± 0.71 b	6.32 ± 0.71 a
C12:0	——	——	——
C14:0	0.99 ± 0.08 c	2.98 ± 0.27 b	4.22 ± 0.55 a
C16:0	13.22 ± 1.66 c	15.38 ± 2.80 b	18.45 ± 2.88 a
C18:0	7.23 ± 1.66 b	10.45 ± 1.76 b	14.77 ± 2.18 a
C20:0	——	——	0.1 ± 0.11
MUFAs			
C16:1(n-7)	——	——	——
C16:1(n-9)	——	——	
C18:1(n-7)	——	——	1.66 ± 0.07 a
C18:1(n-9)	——	——	0.77 ± 0.08
C20:1(n-9)	——	——	——
C22:1(n-9)	1.72 ± 0.14 c	3.77 ± 0.53 b	4.99 ± 0.66 a
PUFAs			
C18:2(n-6)	0.20 ± 0.02 c	0.41 ± 0.28 b	0.81 ± 0.30 a
C18:3(n-3)	4.10 ± 0.7 c	6.22 ± 0.41 b	7.69 ± 0.44 a
C18:4(n-3)	——	——	——
C20:2(n-6)	——	——	2.89 ± 0.14
C20:4(n-6)	3.00 ± 0.49 c	4.77 ± 0.79 b	6.67 ± 0.51 a

Table 3. *Cont.*

Pytoperiphyton			
Fatty Acids	**HP Site**	**LP Site**	**NP Site**
C20:5(n-6)	———	1.90 ± 0.27 b	3.69 ± 0.47 a
C20:5(n-3)	4.89 ± 0.71 a	3.81 ± 0.20 a	2.44 ± 0.22 b
C22:4(n-6)	———	———	———
C22:5(n-6)	0.60 ± 0.09 b	1.00 ± 0.11 b	1.89 ± 0.47 a
C22:5(n-3)	4.12 ± 0.70 c	6.22 ± 0.60 b	7.52 ± 0.70 a
C22:6(n-3)	4.78 ± 0.31 a	3.83 ± 0.54 a	2.77 ± 0.40 c

SFAs: Saturated fatty acids; MUFAs; Monounsaturated fatty acids; PUFAs: Polyunsaturated fatty acids; Values (Mean ± SE) are averages of five samples for each fish species analyzed in triplicate. Different letters (a, b, c) in the same row represent significant ($p < 0.05$) differences.

The fish captured from HP exhibited lower FAs and SFAs compared to the fish captured from LP (Table 4). The percentage of PUFAs in *H. molitrix, C. idella*, and *C. carpio* captured from HP was 32.32 ± 0.65, 7.19 ± 0.35, and 26.13 ± 0.82%, respectively. The percentage of PUFAs in *H. molitrix, C. idella*, and *C. carpio* captured from NP was 48.65 ± 1.11, 41.55 ± 0.97, and 44.15 ± 1.90%, respectively. The percentage of MUFAs and SFAs in *H. molitrix, C. idella*, and *C. carpio* captured from HP were 6.74 ± 0.29, 5.14 ± 0.17, and 5.46 ± 0.54 and 43.38 ± 2.45, 62.94 ± 3.05, and 74.07 ± 4.14%, respectively. The total MUFAs and SFA profiles in *H. molitrix, C. idella*, and *C. carpio*, captured from LP showed a similar trend of fluctuations to fish from HP (Table 4).

Table 4. Fatty acid profile % (±SE) of fish muscle from three sites at different pollution levels.

Less Polluted Site (LP)			
Fatty Acids	***H. molitrix***	***C. idella***	***C. carpio***
SFAs			
C8:0	———	———	———
C10:0	1.02 ± 0.07 a	0.41 ± 0.01a b	1.44 ± 0.03 a
C12:0	0.01 ± 0.00 a b	0.01 ± 0.03 a	0.74 ± 0.06 a
C14:0	1.54 ± 0.04 a	1.37 ± 0.03 a	0.53 ± 0.02 b
C16:0	41.22 ± 2.70 a	37.66 ± 2.61 a	43.98 ± 2.14 a b
C18:0	0.47 ± 0.05 b	24.66 ± 3.40 a b	29.51 ± 2.67 b
C20:0	———	———	1.45 ± 0.03
MUFAs			
C16:1(n-7)	0.44 ± 0.04 a	———	———
C16:1(n-9)	0.71 ± 0.10 a b	0.79 ± 0.02 b	0.29 ± 0.01 a b
C18:1(n-7)	0.14 ± 0.08 b	0.08 ± 0.01a b	0.40 ± 0.04 b
C18:1(n-9)	3.95 ± 0.33 b	4.79 ± 0.04 a	0.11 ± 0.02 a b
C20:1(n-9)	0.77 ± 0.03	———	———
C22:1(n-9)	0.90 ± 0.06 b	0.41 ± 0.03 a	0.32 ± 0.04 a b
PUFAs			
C18:2(n-6)	0.70 ± 0.06 a b	1.29 ± 0.3 a	0.41 ± 0.07 a b
C18:3(n-3)	12.66 ± 0.79 a	3.82 ± 0.04 a	0.003 ± 0.00 a b
C18:4(n-3)	3.24 ± 0.19 a	0.44 ± 0.05 a	2.60 ± 0.02 a b
C20:2(n-6)	———	———	———
C20:4(n-6)	0.87 ± 0.09 a	0.39 ± 0.02 a	0.82 ± 0.01 a
C20:5(n-6)	12.06 ± 0.54 a	8.02 ± 0.24 a	9.16 ± 0.42 a
C20:5(n-3)	2.42± 0.30 b	0.37 ± 0.18 c	3.61 ± 0.22 a
C22:4(n-6)	———	———	———
C22:5(n-6)	16.77 ± 0.66 a	4.70 ± 0.12 b	5.44 ± 0.22 b
C22:5(n-3)	1.17± 0.10 b	0.02 ± 0.00 c	1.88 ± 0.06 b
C22:6(n-3)	4.02 ± 0.33 a b	2.94 ± 0.22 a	3.71 ± 0.14 a b

Table 4. *Cont.*

Highly polluted site (HP)

Fatty acids	*H. moiltrix*	*C. idella*	*C. carpio*
SFAs			
C8:0	———	———	———
C10:0	0.18 ± 0.03 c	1.97 ± 0.11 b	6.93 ± 0.77 a
C12:0	———	———	———
C14:0	0.44 ± 0.05 c	1.83 ± 0.21 b	2.65 ± 0.22 a
C16:0	34.25 ± 4.66 c	44.25 ±5.26 b	50.12 ± 4.77 a
C18:0	27.18 ± 3.16 b	35.69 ± 3.75 a	12.24 ± 1.77 c
C20:0	———	0.25 ± 0.01	———
MUFAs			
C16:1(n-7)	———	———	———
C16:1(n-9)	———	———	———
C18:1(n-7)	5.22 ± 0.52	———	———
C18:1(n-9)	0.62 ± 0.06 a	———	0.49 ± 0.01 a
C20:1(n-9)	———	———	———
C22:1(n-9)	0.32 ± 0.02 c	5.61 ± 0.17 a	4.11 ± 0.22 b
PUFAs			
C18:2(n-6)	———	———	———
C18:3(n-3)	16.55 ± 0.54 a	3.66 ± 0.32 b	0.24 ± 0.01 c
C18:4(n-3)	———	———	———
C20:2(n-6)	———	0.87 ± 0.11 a	0.59 ± 0.02 b
C20:4(n-6)	4.72 ± 0.33 b	7.42 ± 0.22 a	3.13 ± 0.07 c
C20:5(n-6)	———	———	0.92 ± 0.11
C20:5(n-3)	0.68± 0.07	———	———
C22:4(n-6)	———	———	———
C22:5(n-6)	0.52 ± 0.11 b	0.17 ± 0.01 c	0.88 ± 0.11 a
C22:5(n-3)	6.12 ± 0.44 a	4.12 ± 0.22 b	———
C22:6(n-3)	3.62 ± 0.11 a	0.21 ± 0.01 c	1.76 ± 0.22 b

Non-polluted site (NP)

Fatty acids	*H. moiltrix*	*C. idella*	*C. carpio*
SFAs			
C8:0	0.80 ± 0.05 a	0.01 ± 0.01 a	———
C10:0	0.31 ± 0.00 b	3.15 ± 0.03 a	3.67 ± 0.08 a
C12:0	1.21± 0.15 c	3.10 ± 0.40 a	2.16 ± 0.05 b
C14:0	0.45 ± 0.01 c	4.68 ± 0.55 b	7.79± 0.44 a
C16:0	17.10 ± 0.61 a	14.78 ± 1.12 b	12.42 ± 0.55 b
C18:0	6.01 ± 0.41 a	3.80 ± 0.25 c	8.90 ± 0.75 a
C20:0	1.42 ± 0.17 a	0.32 ± 0.01 b	0.38 ± 0.01 b
MUFAs			
C16:1(n-7)	———	———	0.72 ± 0.04
C16:1(n-9)	4.75 ± 0.16 a	2.41 ± 0.04 b	2.77 ± 0.05 b
C18:1(n-7)	4.81 ± 0.45 a	3.60 ± 0.02 b	1.81 ± 0.11 b
C18:1(n-9)	13.12 ± 0.82 a	12.87 ± 0.66 a	10.94 ± 0.42 a
C20:1(n-9)	1.44 ± 0.07 b	4.14 ± 0.51 a	3.88 ± 0.22 a
C22:1(n-9)	0.01 ± 0.00 c	0.61 ± 0.01 a	0.70 ± 0.02 a
PUFAs			
C18:2(n-6)	4.07 ± 0.06 a	3.42 ± 0.11 a	0.01 ± 0.00 b
C18:3(n-3)	3.44 ± 0.23 b	3.22 ± 0.22 b	5.94 ± 0.41 a
C18:4(n-3)	2.41 ± 0.11 a	2.05 ± 0.07 a	0.71 ± 0.02 b
C20:2(n-6)	0.91 ± 0.02 a	0.88 ± 0.01 a	0.57 ± 0.02 b
C20:4(n-6)	14.26 ± 0.87 a	12.14 ± 0.28 b	9.87 ± 0.71 c

Table 4. *Cont.*

Highly polluted site (HP)			
Fatty acids	*H. moiltrix*	*C. idella*	*C. carpio*
C20:5(n-6)	6.44 ± 0.28 a	4.12 ± 0.20 b	0.21 ± 0.01 c
C20:5(n-3)	0.40 ± 0.01 c	5.06 ± 0.31 b	8.23 ± 0.60 a
C22:4(n-6)	1.21 ± 0.02 a	0.60 ± 0.00 b	0.50 ± 0.00 b
C22:5(n-6)	3.44 ± 0.32 b	4.02 ± 0.09 b	6.68 ± 0.25 a
C22:5(n-3)	5.05 ± 0.33 a	3.60 ± 0.22 b	2.39 ± 0.04 c

SFAs: Saturated fatty acids; MUFAs; Monounsaturated fatty acids; PFAs: Polyunsaturated fatty acids; Values (Mean ± SE) are averages of five samples for each fish species analyzed in triplicate. Different letters (a, b, c) in the same row represent significant ($p < 0.05$) differences.

The maximum percentage of SFAs in *C. carpio* was observed in HP. A decrease in the abundance of *C. carpio* was noticed during the study period (Table 4). Caprylic acid (C8:0), lauric acid (C12:0) and C20:0 arachidic acid were not found in *H. molitrix, C. idella,* and *C. carpio* from HP. A very small amount of lauric acid (C12:0) and C20:0 arachidic acid was recorded in *H. molitrix, C. idella,* and *C. carpio* from LP. Eicosapentaenoic acid was not detected in any of the fish species collected from HP sites. In this study, 6 and 9 omega series FAs were found in muscle samples of *H. molitrix, C. idella,* and *C. carpio* from HP and LP, respectively. Linoleic acid (C18:4(n-3), eicosadienoic acid (C20:2 (n-6), and docosapentaenoic acid (C22:4 (n-6) were not recorded in fish from HP. Eicosapentaenoic acid (C20:5 (n-3) was detected only in the muscle samples of *H. molitrix* from HP. Total 11 omega series FA were recorded in muscle of *H. molitrix, C. idella,* and *C. carpio* from NP sites. Caprylic acid was not detected in *H. molitrix, C. idella,* and *C. carpio* from LP sites (Table 4). Myristic acid (C14:0) was recorded as 0.44 ± 0.05, 1.826 ± 0.21, and 2.651 ± 0.22% in *H. molitrix, C. idella,* and *C. carpio,* respectively, from HP. Myristic acid was determined as 1.54 ± 0.04, 1.37 ± 0.03, and 0.53 ± 0.02 and 0.14 ± 0.01, 4.68 ± 0.55, and 7.79 ± 0.44% in *H. molitrix, C. idella,* and *C. carpio* from LP and NP sites, respectively. Arachidic acid was not found in *H. molitrix* and *C. carpio* from HP sites. Oleic acid (C18:1 (n-9) was not detected in *C. idella* collected from HP sites. C16:1 (n-7) (palmitoleic acid), C16:1 (n-9) (Cis-7 hexadecenoic acid), and C20:1 (N-9) (Eicosenoic acid) were not found in these fish species captured from HP. C18:1 (n-7) (cis-vaccenic acid) was detected only in the muscle sample of *H. molitrix* from HP. C16:1 (n-7) was detected only in the muscle samples of *H. molitrix* from LP sites. The concentration of C16:1 (n-7) was only determined as 0.72 ± 0.04 in *C. carpio* from NP sites (Table 4).

Correlation indices that were calculated among the concentrations of total FAs in periphyton, zoobenthos, fish muscle, and heavy metals in water samples are presented in Table 5. It has been observed that Sn, Cr, Pb, Zn, Mn, Cu, Cd, and Hg indicated significantly negative correlations with total FA profile in periphyton, zoobenthos, and fish samples from HP and LP (Table 5). Highly significant negative correlations were observed among Cr, Zn, Mn, and Cu and the total fatty acid profile samples of periphyton, zoobenthos, and fish collected from the HP site. The variation in FA found positively correlated with the level of contamination of these heavy metals in the food web. The health of the aquatic system was found to be significantly affected by the water quality of the HP and LP sites of the river compared to the NP site, which possibly causes decreases in the abundance of periphyton and fish populations in the aquatic system.

Table 5. Correlation matrix for metal concentrations with total fatty acids in periphyton, zoobenthos, and fish in three sites at different pollution levels.

Metals	HP-PP	LP-PP	NP-PP	HP-ZB	LP-ZB	NP-ZB	HP-HM	LP-HM	NP-HM	HP-GC	LP-GC	NP-GC	HP-CP	LP-CP	NP-CP
Sn	−0.24 *	−0.13	0.07	−0.37 *	−0.28 *	0.05	−0.46 **	−0.24 *	0.06	−0.30 *	−0.31 *	0.08	−0.60 **	−0.27 *	0.05
Cr	−0.56 **	−0.33 *	0.10	−0.58 **	−0.33 *	0.02	−0.33 *	−0.24 *	0.04	−0.58 **	−0.31 *	−0.07	−0.54 **	−0.28 *	−0.11
Pb	−0.37 *	−0.26 *	−0.06	−0.58 **	−0.28 *	−0.05	−0.44 **	−0.25 *	0.03	−0.40 **	−0.22 *	0.01	−0.560 **	−0.27 *	−0.01
Zn	−0.51 **	−0.23 *	0.03	−0.53 **	−0.25 *	−0.07	−0.34 *	−0.23 *	0.01	−0.28 *	−0.27 *	−0.03	−0.57 **	−0.24 *	0.09
Mn	−0.60 **	−0.28 *	0.02	0.56 **	−0.27 *	−0.09	−0.48 **	−0.24 *	−0.02	−0.64 **	−0.26 *	0.04	−0.57 **	0.28 *	−0.05
Cu	−0.60 **	−0.27 *	0.05	−0.57 **	−0.44 **	−0.08	−0.53 **	−0.25 *	0.03	−0.37 *	−0.28 *	−0.06	−0.57 **	−0.29 *	0.04
Cd	−0.27 *	−0.24 *	0.10	−0.33 *	−0.25 *	−0.07	−0.41 *	−0.27 *	−0.12	−0.42 **	−0.28 *	0.02	−0.51 **	−0.25 *	−0.05
Hg	−0.34 *	−0.22 *	0.01	−0.27 *	0.12	0.01	−0.28 *	−0.22 *	0.004	−0.35 *	−0.27 *	−0.01	−0.48 **	−0.26 *	0.003

*significant at 0.05 level; ** significant at 0.01 level; HP: highly polluted site; LP: low polluted site; NP: Non-polluted site; PP: periphyton; ZB: zoobenthos; HM: *H. molitrix*; CI: *C. idella*; CP: *C. carpio*.

4. Discussion

The trophic transfer of FAs from periphyton to the organisms at higher trophic levels is important for their health and growth [14,37]. This movement of important nutrients in the food chain may be affected by different contaminants in the freshwater ecosystem [15]. The metals and metalloids, phenols, and organic contaminants in freshwater ecosystems enter the food of aquatic animals from various sources, including anthropogenic activities, and accumulate in planktonic life and fish. The heavy metals which accumulate can cause physiological stress on FA at different trophic levels in the food chain, and ultimately in humans [16,37].

Fish are used as a bioindicator for different organic and inorganic pollutants in freshwater ecosystems due to their presence in different trophic levels, because of their age, size, and mode of nutrition [15]. Various factors have effects on the distribution of aquatic fauna and flora in freshwater reservoirs [38,39]. Abiotic parameters are considered to mostly affect the pattern of distribution and richness of planktonic life [39,40]. The metals assessed in this study accumulated in fish directly from the water and planktonic life in the Indus River in the study area. In this study, higher concentrations of salinity, sulfates, phenol and heavy metals were the driving force which decreased the abundance of phytoplankton and zooplankton, and their FA profile. The phenol, sulfate, total dissolved solids (TDS), and TS values clearly indicated difference in their concentration at HP and LP sites. The level of salinity at HP was very close to that of brackish waters. The higher concentration of total TDS and TS at HP may be due to high turbidity. The presence of different metals in freshwater ecosystems varied with the physico-chemical factors of the corresponding ecosystem, particularly the pH and redox state. Reference [41] reported that the decrease in pH at high river discharges may release metals from complexes in the river and streams, which may be toxic to the aquatic fauna and flora in the ecosystem.

The levels of heavy metals in the water samples collected from HP passed the upper limits recommended by reference [37]. The heavy metals level in the water samples and in the muscles of *H. molitrix, C. idella*, and *C. carpio*, and planktonic biomass collected from HP. The bioaccumulation of heavy metals is known to influence the FA profile of fish. Reference [42,43] mentioned that metals stimulate cellular synthesis and metabolism of FA through β-oxidation, while pharmaceutical products act as peroxisomal proliferators [43–45]. Very limited information is available about the influence of heavy metals on Proliferation Activate Receptors (PPARs) expression and the transcription factors of FA metabolism in fish [46]. Elements in these fish species captured from HP and LP sites were accumulated by bio-concentration, and through food and water [16,37]. The increased concentration of heavy metals along with salinity and phenols at HP and LP sites compared to N P site probably are major factors which caused physiological variation in the food web and disturb the biosynthesis of FAs in *H. molitrix, C. idella*, and *C. carpio* [16,38]. The concentration of many heavy metals decreased in higher trophic levels in the food web [45]. In this study, similar results were obtained for most of the heavy metals, except for Hg [47,48]. This was particularly so in the higher trophic levels, and ultimately affected terrestrial ecosystems through fish [16,38].

PUFAs enter at the first trophic level of the food chain via FA synthesis in periphyton. Reference [49] has mentioned that light causes multiple effects on periphyton lipid metabolism and FA profiles. In general, higher light intensity normally causes oxidative damage to PUFA. In addition to the contamination, low light intensity and poor water quality at HP and LP sites influenced the abundance of periphyton producing high quality FA, thereby affecting PUFAs. The movement of FAs from periphyton to the fish level was found to be increasing with the pH at HP and LP sites. Thus, alkaline pH stress promoted an accumulation of TAG (Triacylglycerols) and a proportionally decrease in membrane lipids [50] In this research work, the changes in physio-chemical factors influenced the production of lipids in the planktonic life at HP and LP sites. The current findings seem to agree with the results of reference [51]. They had mentioned that phytoplanktonic abundance and their diversity were affected by eutrophication, which influence the FAs production due to interspecific variations in periphyton FA levels. However, there remains very little information on the molecular mechanisms involved in these abiotic environmental stressors.

The concentration of EPA, DHA and PUFAs was greater in the microalgae at LP compared to HP, which may due to the higher biomass of microalgae. The increase in microalgae growth is promoted by the higher concentration of nutrients, which might have promoted the synthesis of EPA [31,52]. Reference [53] reported that fluctuations in nutrient availability in the food chain affect on FAs profiles of periphyton. The fluctuations in FA profiles in the trophic levels of the aquatic food chain are probably due variations in the periphyton community composition. Our results of increases in the percentage of EPA were not in line with the results of reference [54]. Total PUFA and PUFA:SAFA ratios were reduced in periphyton and zoobenthos with increase in pollution at the HP site compared to the NP site. The level of PUFA in zoobenthos relies on various biotic and abiotic factors [50] such as food types and levels of contamination [31]. Reference [55] reported that increases in the concentration of Cd decreased the production FA profiles in *Chlorella vulgaris*. However, more Cd accumulated under N stress, which reduced the production of triglycerols in algae. DHA is necessary for the good growth of these freshwater fish species in aquatic ecosystems. The low level of DHA in planktonic food may affect the growth and development of different organs in freshwater fish species [31,56]. Here the reduction in the level of DHA was detrimental to the fatty acid profiles of *H. molitrix*, *C. idella*, and *C. carpio* from HP sites. The accumulation of PUFA in zoobenthos depends on various biotic and abiotic factors [52,57], food types [58] and pollution levels [37].

Fish are considered to be the best source of animal protein, globally. However, deterioration in their quality and losses in FAs cannot be recouped. Differences were non-significant for the FA profile in *C. carpio* sampled from HP and LP sites compared to NP sites, which exhibit an identical response to the chemical pollutants. The maximum percentage of SFAs in *C. carpio* was recorded in the fish procured from HP. Fish with a high concentration of SFAs need more energy for their movement and to search for food [59]. The SFAs C8:0, C12:0 and C20:0 were not recorded for *H. molitrix*, *C. idella*, and *C. carpio* captured from HP sites. The higher levels of heavy metals at HP and LP sites in the river adversely affected the synthesis of FAs in the three fish species. The higher SFA levels are probably due to de novo synthesis within these fish species. The heavy metals accumulate towards the bottom of the river, and *C. carpio* feeds on a variety of benthic organisms and macrophytes, thereby exposing it to high proportions of heavy metals [60]. The total MUFA concentrations recorded were supported by the findings of reference [61].

A significant lower percentage of PUFAs was noticed in *H. molitrix*, *C. idella*, and *C. carpio* from HP and LP. However, H. molitrix exhibited higher levels of ω-3 FAs and a large loss of ω-6 fatty acids. Eicosapentaenoic acid (C20:5n3) was not recorded in *C. idella* and *C. carpio* from HP. The concentrations of EP and C20: 5n3 were lower than those of menhaden oil. Identical results were reported by reference [62]. Linoleic acid, eicosadienoic acid and docosapentaenoic acid were not detected in fish procured from HP. Eicosapentaenoic acid was detected only in the muscle samples of H. molitrix from HP. The reduction in the production of PUFAs in *H. molitrix*, *C. idella*, and *C. carpio* from HP and LP may be due to increased levels of metals in the water at these locations in the river [63]. The zooplankton is a source of EPA and DHA for fish in the aquatic ecosystem [26]. The alterations in the food web, linked with an increase in environmental stress in freshwater ecosystems, invasive species, and habitat deterioration, may cause a significant variation in pollutant and lipid trophic transfer [26,64]. *C. idella* captured from HP and LP sites surprisingly responded to the general environment for FA profiles, although they feed on aquatic vegetation. We did not work on the FA profile of aquatic vegetation, and suspect that the alteration in the FA profile was due to an increased water pollution. The higher metal concentration might have affected the FA profile of the aquatic vegetation. This aspect may be verified in future studies. The alterations in the food web, linked with an increase in environmental stress in freshwater ecosystems, invasive species, and habitat deterioration, may cause a significant variation in pollutant and lipid trophic transfer [16]. The variations in FA and heavy metals trophic transfers in the food chain can provide insights into ecological functioning and the fallout of environmental stressors on the FA profile of different organisms in freshwater food webs.

5. Conclusions

Lipids play a significant role in the bioaccumulation of lipophilic pollutants in freshwater fish. The increase of heavy metals in the waters of the Indus River has produced trophic transfers to periphyton, zoobenthos, and fish in highly polluted (HP) and less polluted (LP) sites. Polyunsaturated Fatty Acids (PUFAs) level was greater in periphyton and zoobenthos biomass from non-polluted (NP) sites, compared to HP and LP sites. Fatty acids in the fish muscles were affected by the level of contamination due to the alterations in the food web, linked with an increase in environmental stress, invasive species, and habitat deterioration. It has been inferred that abiotic factors and chemical pollutants induced the trophic transfer in the food, and ultimately the loss of essential fatty acids (FAs) in fish meat. The variations in FA and heavy metals trophic transfers in the food chain can provide insights into ecological functioning and the fallout of environmental stressors on the FA profile of different organisms in freshwater food webs.

It is proposed that FAs may be used to evaluate trophic relationships among water, planktonic life forms, and fish in the food web in order to provide information to consumers about the safety of fish meat. Thus, the variation in FA profiles may be used as a biomarker to assess the status of the health of the ecosystem, and possibly to identify the causes of decreases in the abundance of fish populations.

Author Contributions: Conceptualization, S.M. S.S., T.S. (Tayyaba Sultan) and B.H.; methodology; T.S. (Tehniat Shahid) and B.H.; software, B.H.; validation, K.A.A.-G., T.S. (Tehniat Shahid) and F.A.-M.; formal analysis, S.M.; investigation, B.H.; resources, K.A.A.-G.; data curation, B.H.; writing—original draft preparation, S.M.; writing—review and editing, S.S., T.S. (Tayyaba Sultan); and F.A.-M.; visualization, Z.A.; supervision, S.M.; project administration, F.A.-M.; funding acquisition, K.A.A.-G.

References

1.   Beasley, G.; Kneale, P. Investigating the influence of heavy metals on macroinvertebrate assemblages using Partial Canonical Correspondence Analysis (pCCA). *Hydrol. Earth Syst. Sci.* **2003**, *7*, 221–233. [CrossRef]

2.   Dahl, J.; Johnson, R.K.; Sandin, L. Detection of organic pollution of streams in southern Sweden using benthic macroinvertebrates. *Hydrobiologia* **2004**, *516*, 161–172. [CrossRef]

3.   Korkmaz, G.F.; Keser, R.; Akcay, N.; Dizman, S. Radioactivity and heavy metal concentrations of some commercial fish species consumed in the Black Sea Region of Turkey. *Chemosphere* **2012**, *87*, 356–361. [CrossRef] [PubMed]

4.   Petrovic, Z.; Teodrorovic, V.; Dimitrijevic, M.; Borozan, S.; Beukovic, M.; Milicevic, D. Environmental Cd and Zn concentration in liver and kidney of European hare from different Serbian region: Age and tissue difference. *Bull. Environ. Contamin. Toxicol.* **2013**, *90*, 203–207. [CrossRef] [PubMed]

5.   Riberio, O.C.A.; Vollaire, Y.; Sanchez-Chardi, A.; Roche, H. Bioaccumulation and the effects of organochlorine pesticides PAH and heavy metals in the eel (Anguilla anguilla) at the Camargue Nature Reserve, France. *Aquat. Toxicol.* **2005**, *74*, 53–69. [CrossRef] [PubMed]

6.   Nhiwatiwa, T.; Barson, M.; Harrison, A.P.; Utete, B.; Cooper, R.G. Metal concentrations in water, sediment and sharp tooth catfish Clarias gariepinus from three periurban rivers in the upper Manyame catchment, Zimbabwe. *Afr. J. Aquat. Sci.* **2011**, *36*, 243–252. [CrossRef]

7.   Annabi, A.; Said, K.; Messaoudi, I. Cadmium: Bioaccumulation, histopathology and detoxifying mechanisms in fish. *Am. J. Res. Commun.* **2013**, *1*, 60–79.

8.   Canli, M. Natural occurrence of metallothionein like proteins in the hepatopancreas of the Norway lobster, *Nephros norvegicus* and effects of Cd, Cu and Zn exposures on levels of the metal bound on metallothionein. *Turk. J. Zool.* **1995**, *19*, 313–321.

9.   Hussain, B.; Sultana, T.; Sultana, S.; Iqbal, Z.; Nadeem, S.; Mahboob, S. Habitat induced mutational effects and fatty acid profile changes in bottom dweller Cirrhinus mrigala inhabitant of river Chenab. *Grasas y Aceites* **2015**, *66*, e075. [CrossRef]

10. Müller-Navarra, D.C.; Brett, M.T.; Liston, A.M.; Goldman, C.R. A highly unsaturated fatty acid predicts carbon transfer between primary producers and consumers. *Nature* **2000**, *403*, 74–77.

11. Kiron, V.; Fukuda, H.; Takeuchi, T.; Watanabe, T. Essential fatty acid nutrition and defense mechanisms in rainbow trout *Oncorhynchus mykiss*. *Comp. Biochem. Physiol. A Physiol.* **1995**, *111*, 361–367. [CrossRef]

12. Bec, A.; Martin-Creuzburg, D.; von Elert, E. Trophic upgrading of autotrophic picoplankton by the heterotrophic nano-flagellate *Paraphysomonas* sp. *Limnol. Oceanogr.* **2006**, *51*, 1699–1707. [CrossRef]

13. Klein Breteler, W.C.M.; Schogt, N.; Baas, M.; Schouten, S.; Kraay, G.W. Trophic upgrading of food quality by protozoans enhancing copepod growth: Role of essential lipids. *Mar. Biol.* **1999**, *135*, 191–198. [CrossRef]

14. Tocher, D.R. Metabolism and functions of lipids and fatty acids in teleost fish. *Rev. Fish. Sci.* **2003**, *11*, 107–184. [CrossRef]

15. Borgå, K.; Fisk, A.T.; Hoekstra, P.F.; Muir, D.C.G. Biological and chemical factors of importance in the bioaccumulation and trophic transfer of persistent organochlorine contaminants in arctic marine food webs. *Environ. Toxicol. Chem.* **2004**, *23*, 367–2385. [CrossRef]

16. Kainz, M.; Telmer, K.; Mazumder, A. Bioaccumulation patterns of methyl mercury and essential fatty acids in the planktonic food web and fish. *Sci. Total Environ.* **2006**, *368*, 271–282. [CrossRef] [PubMed]

17. Lowe, R.L.; Pan, Y. Benthic algal communities as biological indicators. In *Algal Ecology: Freshwater Benthic Ecosystems*; Stevenson, R.J., Bothwell, M.L., Lowe, R.L., Eds.; Academic Press: San Diego, CA, USA, 1996; pp. 705–739.

18. Jabeen, F.; Chaudhry, A.S. Nutritional composition of seven fish species and the use of cluster analysis as a tool for their classification. *J. Anim. Plant Sci.* **2016**, *26*, 282–290.

19. Dawood, M.A.O.; Koshio, S. Recent advances in the role of probiotics and prebiotics in carp aquaculture: A review. *Aquaculture* **2016**, *454*, 243–251. [CrossRef]

20. Dawood, M.A.O.; Koshio, S.; Abdel-Daim, M.M.; Doan, H.V. Probiotic application for sustainable aquaculture. *Rev. Aquac.* **2018**. [CrossRef]

21. ElShehawy, S.M.; Gab-Alla, A.A.; Mutwally, H.M. Amino acids pattern and fatty acids composition of the most important fish species of Saudi Arabia. *Int. J. Food Sci. Nutr. Eng.* **2016**, *6*, 32–41.

22. Razak, Z.K.A.; Basri, M.; Dzulkefly, K.; Razak, C.N.A.; Salleh, A.B. Extraction and characterization of fish oil from Monopterus albus. *Malay J. Anal. Sci.* **2001**, *7*, 217–220.

23. Zhang, W.; Wang, W.X. Large-scale spatial and interspecies differences in trace elements and stable isotopes in marine wild fish from Chinese waters. *J. Hazard. Mater.* **2012**, *215*, 65–74. [CrossRef] [PubMed]

24. Pont, D.; Hugueny, B.; Beier, U.; Goffaux, D.; Melcher, A.; Noble, R.; Rogers, C.; Roset, N.; Schmutz, S. Assessing river biotic condition at a continental scale: A European approach using functional metrics and fish assemblages. *J. Appl. Ecol.* **2006**, *43*, 70–80. [CrossRef]

25. Kainz, M.J.; Fisk, A.T. Integrating lipids and contaminants in aquatic ecology and ecotoxicology. In *Lipids in Aquatic Ecosystems*; Kainz, M., Brett, M., Arts, M., Eds.; Springer: New York, NY, USA, 2009.

26. Kelly, E.N.; Schindler, D.W.; St. Louis, V.L.; Donald, D.B.; Vlaclicka, K.E. Forest fire increases mercury accumulation by fishes via food web restructuring and increased mercury inputs. *Proc. Natl. Acad. Sci.* **2006**, *103*, 19380–19385. [CrossRef] [PubMed]

27. Al-Ghanim, K.A.; Mahboob, S.; Seema, S.; Sultana, S.; Sultana, T.; Al-Misned, F.; Ahmed, Z. Monitoring of trace metals in tissues of *Wallago attu* (lanchi) from the Indus River as an indicator of environmental pollution. *Saudi J. Biol. Sci.* **2015**. [CrossRef] [PubMed]

28. Jabeen, F.; Chaudhry, A.S. Monitoring trace metals in different tissues of *Cyprinus carpio* from the Indus River in Pakistan. *Environ. Monit. Assess.* **2010**, *170*, 645–656. [CrossRef] [PubMed]

29. Mahboob, S.; Al-Balwai, H.F.A.; Al-Misned, F.; Ahmad, Z. Investigation on the genotoxicity of mercuric chloride to freshwater *Clarias gariepinus*. *Pak. Vet. J.* **2014**, *34*, 100–103.

30. Boyd, E.C. *Water Quality in Warm Water Fishponds*, 2nd ed.; Craftmaster Printers Inc.: Auburn, AL, USA, 1981; pp. 213–247.

31. Gladyshev, M.I.; Anishchenko, O.V.; Sushchnik, N.N.; Kalacheva, G.S.; Gribovskaya, I.V.; Ageev, A.V. Influence of anthropogenic pollution on content of essential polyunsaturated fatty acids in links of food chain of river ecosystem. *Contemp. Probl. Ecol.* **2012**, *5*, 376–385. [CrossRef]

32. Folch, J.; Lees, M.; Stanely, S.G.H. A simple method for the isolation and purification of total lipids from animal tissues. *J. Biol. Chem.* **1957**, *226*, 497–509.

33.   Bell, J.G.; McVicar, A.H.; Park, M.T.; Sargent, R.J. Effect of high dietary linoleic acid on fatty acid compositions of individual phospholipids from tissues of Atlantic salmon (*Salmo salar*): Association with a novel cardiac lesion. *J. Nutr.* **1991**, *121*, 1163–1172. [CrossRef]

34.   Kiessling, A.; Pickova, J.; Johansson, L.; Asgard, T.; Storebakken, T.; Kiessling, K.H. Changes in fatty acid composition in muscle and adipose tissue of farmed rainbow trout (*Oncorhynchus mykiss*) in relation to ration and age. *Food Chem.* **2001**, *73*, 271–284. [CrossRef]

35.   Hussain, B.; Sultana, T.; Sultana, S.; Al-Ghanim, K.A.; Mahboob, S. Effect of pollution on DNA damage and essential fatty acid profile in Cirrhinus mrigala from River Chenab. *Chin. J. Oceanol. Limnol.* **2016**. [CrossRef]

36.   SAS. *Statistical Analysis Systems*; S.A.S. Institute Inc.: Cary, NC, USA, 1995.

37.   United States Environmental Protection Agency. Drinking Water and Health. 2002. Available online: http://water.epa.gov/drink/index.cfm (accessed on 20 October 2013).

38.   Hussain, B.; Sultana, T.; Sultana, S.; Al-Ghanim, K.A.; Al-Misned, F.; Mahboob, S. Influence of habitat degradation on the fatty acid profiles of fish, microalgae, and zoobenthos in a river ecosystem. *Proc. Saf. Environ. Prot.* **2019**, *122*, 24–32. [CrossRef]

39.   Daly, H.V. General classification and key to the orders of aquatic and semi aquatic insects. In *An Introduction to the Aquatic Insects of North America*; Merrit, R.W., Cummins, K.W., Eds.; Kendall/Hunt: Dubuque, IA, USA, 1984; pp. 76–81.

40.   Carol, J.; Benejam, L.; Alcaraj, C.; Vila-Gispert, A.; Zamora, L.; Navarro, E.; Armengol, J.; Garcia-Berthou, E. The effects of limnological features on fish assemblages of 14 Spanish reservoirs. *Ecol. Freshw. Fish.* **2006**, *15*, 66–77. [CrossRef]

41.   Gundersen, P.; Olsvik, P.A.; Steinnes, E. Variations in heavy metal concentrations and speciation in two mining-polluted streams in central Norway. *Environ. Toxicol. Chem.* **2001**, *20*, 978–984. [CrossRef] [PubMed]

42.   Burger, J.; Gochfeld, M. Heavy metals in commercial fish in New Jersey. *Environ. Res.* **2005**, *99*, 403–412. [CrossRef] [PubMed]

43.   Staels, B.; Dallongeville, J.; Auwerx, J.; Schoonjans, K.; Leitersdorf, E.; Fruchart, J.C. Mechanism of action of fibrates on lipid and lipoprotein metabolism. *Circulation* **1998**, *19*, 2088–2093. [CrossRef]

44.   Gervois, P.; Torra, I.P.; Furchart, J.C.; Staels, B. Regulation of lipid and lipoprotein metabolism by PPAR activators. *Clin. Chem. Lab. Med.* **2000**, *38*, 3–11. [CrossRef]

45.   Suga, T. Hepatocarcinogenesis by peroxisome proliferators. *J. Toxicol. Sci.* **2004**, *29*, 1e12. [CrossRef]

46.   Olivares-Rubio, H.F.; Vega-Lopez, A. Fatty acid metabolism in fish species as a biomarker for environmental Monitoring: Review. *Environ. Pollut.* **2016**, *218*, 297e312. [CrossRef]

47.   Campbell, L.M.; Norstrom, R.J.; Hobson, K.A.; Muir, D.C.G.; Backus, S.; Fisk, A.T. Mercury and other trace elements in a pelagic Arctic marine food web (North water Polynya, Baffin Bay). *Sci. Total Environ.* **2005**, *351*, 247–263. [CrossRef] [PubMed]

48.   Campbell, L.M.; Fisk, A.T.; Wang, X.; Köck, G.; Muir, D.C.G. Evidence of biomagnification of rubidium in aquatic and marine food webs. *Can. J. Fish. Aquat. Sci.* **2005**, *62*, 1161–1167. [CrossRef]

49.   Harwood, J.L. Membrane lipids in algae. In *Lipids in Photosynthesis: Structure, Function and Genetics*; Siegenthaler, P.-A., Murata, N., Eds.; Kluwer: Dordrecht, The Netherlands, 1998; pp. 53–64.

50.   Guckert, J.B.; Cooksey, K.E. Triglyceride accumulation and fatty acid profile changes in Chlorella (Chlorophyta) during high pH-induced cell cycle inhibition. *J. Phycol.* **1990**, *26*, 72–79. [CrossRef]

51.   Hartwich, M.; Straile, D.; Gaedke, U.; Wacker, A. Use of ciliate and phytoplankton taxonomic composition for the estimation of eicosapentaenoic acid concentration in lakes. *Freshw. Biol.* **2012**, *57*, 1385–1398. [CrossRef]

52.   Gladyshev, M.I.; Sushchik, N.N.; Makhutova, O.N.; Dubovskaya, O.P.; Kravchuk, E.S.; Kalachova, G.S. Correlations between fatty acid composition of seston and zooplankton and effects of environmental parameters in a eutrophic Siberian reservoir. *Limnologica* **2010**, *40*, 343–357. [CrossRef]

53.   Mayzaud, P.; Claustre, H.; Augier, P. Effect of variable nutrient supply on fatty acid composition of phytoplankton grown in an enclosed experimental ecosystem. *Mar. Ecol. Prog. Ser.* **1990**, *60*, 123–140. [CrossRef]

54.   Torres-Ruiz, M.; Wehr, J.D.; Perrone, A.A. Trophic relations in a stream foodweb: Importance of fatty acids for macroinvertebrate consumers. *J. N. Am. Benthol. Soc.* **2007**, *26*, 509–522. [CrossRef]

55.   Mathias, A.C.; Lombardi, A.T.; Grac, M.D.; Mela, G.; Parrish, C.C. Effects of cadmium and nitrogen on lipid composition of Chlorella vulgaris (Trebouxiophyceae, Chlorophyta). *Eur. J. Phycol.* **2013**, *48*, 1–11.

56. Descroix, A.; Bec, A.; Bourdier, G.; Sargos, D.; Sauvanet, J.; Misson, B.; Desvilettes, C. Fatty acids as biomarkers to indicate main carbon sources of four major invertebrate families in a large River (the Allier, France). *Fundam. Appl. Limnol.* **2010**, *177*, 39–55. [CrossRef]

57. Masclaux, H.; Bec, A.; Kainz, M.J.; Desvilettes, C.; Jouve, L.; Bourdier, G. Combined effects of food quality and temperature on somatic growth and reproduction of two freshwater cladocerans. *Limnol. Oceanog.* **2009**, *54*, 1323–1332. [CrossRef]

58. Taipale, S.; Kankaala, P.; Hamalainen, H.; Jones, R.I. Seasonal shifts in the diet of lake zooplankton revealed by phospholipid fatty acid analysis. *Freshw. Biol.* **2009**, *54*, 90–104. [CrossRef]

59. Kandemir, Ş.; Polat, N. Seasonal variation of total lipid and total fatty acid in muscle and liver of rainbow trout (*Oncorhynchus mykiss*) reared in Derbent Dam Lake. *Turk. J. Fish. Aquat. Sci.* **2007**, *7*, 27–31.

60. Koca, S.; Koca, Y.B.; Yildiz, S.; Gürcü, B. Genotoxic and histopathological effects of water pollution on two fish species, Barbus capito pectoralis and *Chondrostoma nasus* in the Büyük Menderes River. *Turk. Biol. Trace Elem. Res.* **2008**, *122*, 276–291. [CrossRef] [PubMed]

61. Kwetegyeka, J.; Mpango, O.G.; Niesen, G. Variation in fatty acid composition in muscle and heart tissues among species and populations of tropical fish in Lakes Victoria and Kyoga. *Lipids* **2008**, *43*, 1017–1029. [CrossRef] [PubMed]

62. Masa, P.; Ogwok, J.H.; Muyonga, J.; Kwetegyeka, V.; Makokha, D. Fatty acid composition of muscle, liver and adipose tissue of freshwater fish from Lake Victoria, Uganda. *J. Aquat. Food Prod. Technol.* **2011**, *20*, 64–72. [CrossRef]

63. Abedi, E.; Sahari, M.A. Long-chain polyunsaturated fatty acid sources and evaluation of their nutritional and functional properties. *Food Sci. Nutr.* **2014**, *2*, 443–463. [CrossRef] [PubMed]

64. Li, D.; Hu, X. Fish and its multiple human health effects in times of threat to sustainability and affordability: Are there alternatives? *APJCN* **2009**, *218*, 553–563.

11

Variation in ω-3 and ω-6 Polyunsaturated Fatty Acids Produced by Different Phytoplankton Taxa at Early and Late Growth Phase

Sami Taipale [1],*, Elina Peltomaa [2,3] and Pauliina Salmi [4]

[1] Department of Biological and Environmental Science, Nanoscience center, University of Jyväskylä, P.O. Box 35 (YA), 40014 Jyväskylä, Finland

[2] Institute of Atmospheric and Earth System Research (INAR)/Forest Sciences, University of Helsinki, P.O. Box 64, 00014 Helsinki, Finland; elina.peltomaa@helsinki.fi

[3] Helsinki Institute of Sustainability Science (HELSUS), University of Helsinki, P.O. Box 4 (Yliopistonkatu 3), 00014 Helsinki, Finland

[4] Faculty of Information Technology, University of Jyväskylä, P.O. Box 35, FI-40014 Jyväskylän, Finland; pauliina.u.m.salmi@jyu.fi

* Correspondence: sami.taipale@jyu.fi

Abstract: Phytoplankton synthesizes essential ω-3 and ω-6 polyunsaturated fatty acids (PUFA) for consumers in the aquatic food webs. Only certain phytoplankton taxa can synthesize eicosapentaenoic (EPA; 20:5ω3) and docosahexaenoic acid (DHA; 22:6ω3), whereas all phytoplankton taxa can synthesize shorter-chain ω-3 and ω-6 PUFA. Here, we experimentally studied how the proportion, concentration (per DW and cell-specific), and production ($\mu g\ FA\ L^{-1}\ day^{-1}$) of ω-3 and ω-6 PUFA varied among six different phytoplankton main groups (16 freshwater strains) and between exponential and stationary growth phase. EPA and DHA concentrations, as dry weight, were similar among cryptophytes and diatoms. However, *Cryptomonas erosa* had two–27 times higher EPA and DHA content per cell than the other tested cryptophytes, diatoms, or golden algae. The growth was fastest with diatoms, green algae, and cyanobacteria, resulting in high production of medium chain ω-3 and ω-6 PUFA. Even though the dinoflagellate *Peridinium cinctum* grew slowly, the content of EPA and DHA per cell was high, resulting in a three- and 40-times higher production rate of EPA and DHA than in cryptophytes or diatoms. However, the production of EPA and DHA was 40 and three times higher in cryptophytes and diatoms than in golden algae (chrysophytes and synyrophytes), respectively. Our results show that phytoplankton taxon explains 56–84% and growth phase explains ~1% of variation in the cell-specific concentration and production of ω-3 and ω-6 PUFA, supporting understanding that certain phytoplankton taxa play major roles in the synthesis of essential fatty acids. Based on the average proportion of PUFA of dry weight during growth, we extrapolated the seasonal availability of PUFA during phytoplankton succession in a clear water lake. This extrapolation demonstrated notable seasonal and interannual variation, the availability of EPA and DHA being prominent in early and late summer, when dinoflagellates or diatoms increased.

Keywords: polyunsaturated fatty acids; phytoplankton; freshwater; nutritional value

1. Introduction

Phytoplankton, the microscopic primary producers, are central transformers and cyclers of energy and biomolecules in aquatic food webs [1]. The ability of phytoplankton to synthesize different biomolecules influences their nutritional values and reflects their productivity throughout the aquatic food web [2–4]. Among all biomolecules synthesized by phytoplankton, alfa-linolenic acid (ALA; 18:3ω3) and linoleic acid (LIN, 18:2ω6) can be considered as essential polyunsaturated fatty acids

(PUFAs) since consumers cannot synthesize these de novo [5]. These medium-chain ω-3 and ω-6 PUFA are precursors for eicosapentaenoic acid (EPA, 20:5ω-3), docosahexanoic acid (DHA, 22:6ω-3), and arachidonic acid (ARA, 20:4w6), but due to the consumers' limited ability to bioconvert them from ALA or LIN, they can be considered as physiologically essential [5,6]. The physiological importance of long-chain ω-3 and ω-6 PUFA varies by consumers. Usually, DHA appears to be the most retained FA for copepods and many fish, whereas EPA is the most retained FA for *Daphnia* and some benthic invertebrates [6–10]. However, *Daphnia* can grow and reproduce without EPA, whereas total ω-6 availability may negatively affect somatic growth of *Daphnia* [11]. The egg production and hatching success of marine copepods from the genus *Acartia* have been reported to be highly positively correlated with ALA, EPA, and DHA and negatively correlated with SDA and LIN [12,13]. More precisely, ALA had less effect on egg production and hatching success than EPA and DHA, and DHA had higher effect than EPA [13]. Nevertheless, EPA and DHA are not the only important PUFA for zooplankton, and thus, production of medium chain ω-3 and ω-6 PUFA can promote consumers' optimal health.

Although phytoplankton can synthesize many different biomolecules (e.g., amino acids, sterols, carotenoids) [11], species containing high amounts of EPA and DHA are considered high-quality food for zooplankton [2,14]. Among freshwater phytoplankton, cryptophytes, dinoflagellates, golden algae, diatoms, and raphidophytes have been identified as EPA-synthesizing taxa and cryptophytes, dinoflagellates, golden algae, and euglenoids as DHA-synthesizing taxa [15–17]. In addition, some marine green algae and eustigmatophytes can synthesize EPA, and cryptophytes synthesize DHA [12]. Even though cyanobacteria and freshwater green algae cannot synthesize EPA or DHA, some cyanobacteria strains and all green algae can synthesize ALA and stearidonic acid (SDA, 18:4ω3) [16,18] and can contribute much or all their FA. In addition to long-chain and medium-chain PUFA, green algae and diatoms can synthesize 16 PUFA, which does not have physiological importance for aquatic consumers [19].

There is a gap in the knowledge on how efficient different phytoplankton groups are in producing different PUFAs and on how much PUFA content per cell varies among phytoplankton species and within phytoplankton groups. Current knowledge of production efficiency comes from biofuel studies and other applications and majorly focuses on fast growing taxa, e.g., non-EPA- and DHA-producing green algae, or in the optimization of PUFA production of specific species in certain growth conditions, utilizing, for example, industrial side streams [20,21]. These results are therefore not directly applicable when implemented to phytoplankton field data. Studies on laboratory cultures have shed light on the effects of environmental conditions on different phytoplankton taxa to synthesize PUFA [22]. The nutritional value of phytoplankton has shown to be dependent on growth rate regulated by ambient temperature and irradiance [23,24] or on nutrient stress experienced by the phytoplankton. Mitchell et al. [25] reported three–four times higher importance of phytoplankton taxa in relation to environmental conditions on PUFA contributions. However, they were not able to define how much the PUFA content (per biomass or cell) varied within phytoplankton groups or by environmental conditions. Taipale et al. [26] studied the nutritional values of natural phytoplankton communities in 107 boreal lakes sampled once for two summers. They found a negative pattern along nutrient concentration and nutritional value of phytoplankton; however, the variation in the predictability was rather high, suggesting that there are other factors influencing phytoplankton PUFA content.

The main aim of the current research was to study the connections between phytoplankton taxa and the production of ω-3 and ω-6 PUFA along their growth. Furthermore, we wanted to study how the nutritional value of phytoplankton changes when ω-3 and ω-6 PUFA content is calculated per cell instead of per biomass. For this experimental study, we cultured 16 strains from six phytoplankton main groups isolated from boreal and temperate freshwaters. We also studied how the abundance of certain phytoplankton groups influences the production of ω-3 and ω-6 PUFA in eutrophic lake by a calculation of PUFA concentrations based on phytoplankton biomasses. We hypothesized that strains belonging to cryptophytes, dinoflagellates, chrysophytes, and diatoms display higher concentrations—both proportion and cell specific—of ω-3 and ω-6 PUFAs than green algae and cyanobacteria both in early and late growth phases. Additionally, we hypothesized that production rates of the former algae group were higher than that of the latter.

2. Materials and Methods

2.1. Phytoplankton Culturing

To study how phytoplankton taxa and growth influence the contribution, content, and production of ω-3 and ω-6 PUFAs, we cultured 16 freshwater phytoplankton strains belonging to six phytoplankton main groups (Table 1). From now on, we refer to the strains by their main groups or genus for readability. Each phytoplankton strain was pre-cultured using MWC medium [27,28] with AF6 vitamins [29] at a temperature of 18 ± 1 °C, under 14 h:10 h light:dark cycle with a light intensity of 50–70 μmol m^{-2} s^{-1}. For the actual experiment, we used 200 mL tissue tubes with 75 mL inoculum of pre-cultured algae and 125 mL of fresh MWC with AF6 vitamins. Each strain was cultured in three replicates. Cell density of phytoplankton cultures were measured prior and during the experiment by using an electronic cell counter (Casy, Omni Life Science, Bremen, Germany) with 60 μm capillary (measurement range 1.2–40 μm). Samples for fatty acid analyses were harvested by filtering 20–100 mL of phytoplankton culture onto cellulose nitrate membrane filters (pore size 3 μm, Whatman, Maidstone, Kent, UK).

The specific rates of increase (r_n, divisions day^{-1}) for all strains were calculated for the exponential growth phase using Equation (1):

$$r_n = \ln(N_t/N_0)/t \tag{1}$$

where N_0 is a population at the beginning of the experiment, N_t is the population size at the time t that was determined as the exponential growth phase at the time when the first fatty acid samples were harvested.

2.2. Lipid Extraction and Fatty Acid Methylation

Lipids were extracted from the filters using a chloroform:methanol 2:1 mixture and then sonicated for 10 min, after which 0.75 mL of distilled water was added. Samples were mixed by vortexing and centrifuged (2000 rpm) in Kimax glass tubes, after which the lower phase was transferred to a new Kimax tube. The solvent was evaporated to dryness. Fatty acids of total fraction were methylated using acidic conditions. Toluene and sulfuric acid were used for the transesterification of fatty acid methyl esters (FAMEs) at 50 °C for 16 h, which is the optimal method for methylation PUFA [30]. FAMEs were analyzed with a gas chromatograph (Shimadzu Ultra, Kyoto, Japan) equipped with mass detector (GC-MS) and using helium as a carrier gas (linear velocity = 36.3 cm s^{-1}). The temperature of the injector was 270 °C and we used a splitless injection mode (for 1 min). Temperatures of the interface and ion source were 250 °C and 220 °C, respectively. Phenomenex® (Torrance, CA, USA) ZB-FAME column (30 m × 0.25 mm × 0.20 μm) with 5 m Guardian was used with the following temperature program: 50 °C was maintained for 1 min, then the temperature was increased at 10 °C min^{-1} to 130 °C, followed by 7 °C min^{-1} to 180 °C, and 2 °C min^{-1} to 200 °C. This temperature was held for 3 min, and finally, the temperature increased 10 °C min^{-1} to 260 °C. The total program time was 35.14 minutes and solvent cut time was 9 minutes. Fatty acids were identified by the retention times (RT) and using specific ions [18], which were also used for quantification. Fatty acid concentrations were calculated using calibration curves based on known standard solutions (15 ng, 50 ng, 100 ng and 250 ng) of a FAME standard mixture (GLC standard mixture 566c, Nu-Chek Prep, Elysian, MI, USA) and using recovery percentage of internal standards. The Pearson correlation coefficient was >0.99 for each individual fatty acid calibration curve. Additionally, we used 1,2-dinonadecanoyl-sn-glycero-3-phosphatidylcholine (Larodan, Malmö, Sweden) and free fatty acid of $C_{23:0}$ (Larodan, Malmö, Sweden) as internal standards and for the calculation of the recovery percentages.

Table 1. Cultured phytoplankton strains (taxa, order, species, and strain number), their mean size (diameter by electronic cell counter, μm), and growth phase rate (divisions d^{-1}) for the exponential phase (P1, sampling point 1) and stationary phase (P2, sampling point 2)

Taxa	Order	Species	Strain	Nr.	Size (μm)	Growth P1	Growth P2
Chlorophyceae (green algae)	Chlamydomonadales	*Chlamydomonas reinhardtii*	NIVA K-1016	1	6.1	0.14 ± 0.00	−0.02 ± 0.01
	Chlamydomonadales	*Haematococcus pluvialis*	NIVA K-0084	2	17	0.38 ± 0.10	0.07 ± 0.03
	Sphaeropleales	*Acutodesmus* sp.	University of Basel	3	5	0.11 ± 0.00	−0.12 ± 0.00
	Sphaeropleales	*Monoraphidium griffithii*	NIVA-CHL 8	4	4.6	0.11 ± 0.04	0.07 ± 0.03
Cyanophyceae (cyanobacteria)	Chroococcales	*Microcystis* sp.	NIVA-CYA 642	5	4.1	0.21 ± 0.01	−0.06 ± 0.02
	Synechococcales	*Snowella lacustris*	NIVA-CYA 339	6	2	0.05 ± 0.00	−0.08 ± 0.00
Cryptophyceae (cryptophytes)	Cryptomonadales	*Cryptomonas erosa*	CPCC 446	7	6.14	0.09 ± 0.04	0.06 ± 0.00
	Pyrenomonadales	*Rhodomonas lacustris*	NIVA 8/82	8	11.04	0.12 ± 0.03	0.08 ± 0.00
Synyrophyceae (golden algae)	Synurales	*Mallomonas caudata*	CCAP 929/8	9	12.5	0.05 ± 0.00	−0.08 ± 0.00
	Synurales	*Synura petersenii*	CCAP 960/3	10	8.8	0.05 ± 0.00	0.07 ± 0.01
Chrysophyceae (golden algae)	Chromulinales	*Dinobryon bavaricum*	CCAC 2950B	11	5.6	0.12 ± 0.02	0.09 ± 0.00
Bacillariophyceae	Chromulinales	*Uroglena* sp.	CPCC 278	12	8.3	0.14 ± 0.00	0.02 ± 0.01
	Bacillariales	*Nitzchia* sp.		13	6.09	0.56 ± 0.02	0.04 ± 0.01
	Tabellariales	*Diatoma tenuis*	CPCC 62	14	6.14	0.45 ± 0.01	0.10 ± 0.02
	Tabellariales	*Tabellaria fenestrata*	CPCC 619	15	5.94	0.50 ± 0.07	0.07 ± 0.04
Dinophyceae (dinoflagellates)	Peridianales	*Peridinium cinctum*	SCCAP K-1721	16	32.29	0.13 ± 0.00	0.04 ± 0.00

2.3. Quantitation of Fatty Acids

Here, we focused on two medium chain ω-3 (ALA, SDA) and two ω-6 (LIN, GLA) PUFA and two long-chain ω-3 (EPA, DHA) and ω-6 (ARA, DPA) PUFA. However, we calculated the contribution of these PUFA from all quantified fatty acids. In addition to the contribution of PUFA, we calculated their content per phytoplankton dry weight biomass and per cell. The fatty acid content (μg in mg) was calculated based on the following Equation (2):

$$\frac{Q_{FA} \times V_{vial}}{DW_1 \times R_p} \tag{2}$$

where Q_{FA} is the concentration of the fatty acid (μg μL^{-1}) based on calibration curves of GLC-566C (Nu-Chek Prep, Elysian, MN, USA) for each fatty acid, V_{vial} denotes the running volume of the samples (μL), DW_1 is dry weight of the sample, and R_p denotes the recovery percentage based on internal standards.

We calculated ω-3 and ω-6 PUFA content per phytoplankton carbon biomass. The fatty acid content (μg in mg C) was calculated based on Equation (3):

$$\frac{Q_{FA} \times V_{vial}}{V_{filtered} \times TCBM \times R_p} \tag{3}$$

where Q_{FA} is the concentration of the fatty acid (μg μL^{-1}), V_{vial} denotes the running volume of the samples (μL), $V_{filtered}$ is the total volume of filtered lake water (L), $TCBM$ denotes the total phytoplankton carbon biomass (μg C L^{-1}) of the corresponding sample, and R_p denotes the recovery percentage based on internal standards.

The cell-specific fatty acid concentration (pg in cell) was calculated based on Equation (4):

$$\frac{Q_{FA} \times V_{vial}}{V_{filtered} \times Cell \times R_p} \tag{4}$$

where Q_{FA} is the concentration of the fatty acid (μg μL^{-1}), V_{vial} denotes the running volume of the samples (μL), $V_{filtered}$ is the total volume of filtered of cultured phytoplankton (L), *Cell* is the number of cells of the culture, and R_p denotes the recovery percentage based on internal standards.

Additionally, daily production of PUFA (μg L^{-1} Day^{-1}) was calculated based on Equation (5):

$$\frac{Q_{FA} * V_{vial}}{DW_1 * R_p} \times \frac{DW_2 / V_{filtered}}{Days} \tag{5}$$

where Q_{FA} is the concentration of the fatty acid (μg μL^{-1}), V_{vial} denotes the running volume of the samples (μL), DW_1 is dry weight of the sample, and R_p denotes the recovery percentage based on internal standards. DW_2 is dry weight of the phytoplankton samples between time 1 (e.g., initial) and 2 (e.g., exponential phase), $V_{filtered}$ is the total volume of filtered of cultured phytoplankton (L), and Days cites to the number of culturing days between time 1 and 2.

2.4. Statistical Methods

Bray Curtis similarity matrix of fatty acid data was created using Primer 7[81] (Plymouth Routines In Multivariate Ecological Research, Primer E) of which a non-metric multidimensional scaling

(NMDS) plot was created. CLUSTER analysis (Hierarchical Cluster analysis) was used to create 70% similarities in the NMDS ordination. PERMANOVA (Permutational multivariate analysis of variance [31]) was used to test if differences in the ω-3 and ω-6 PUFA composition, biomass, and cell content and production were statistically significant between phytoplantkon groups and growth phase. PERMANOVA was run with unrestricted permutation of raw data and type III sums of squares. Similarity percentages (SIMPER) were used to detect how different units influence the similarity within phytoplankton group and to identify the characteristic fatty acids of each phytoplankton group. We used PERMDISP (Distance-based test for homogeneity of multivariate dispersions [32]) to investigate the within-class variation in ω-3 and ω-6 PUFA composition, biomass, and cell content and production.

2.5. Implementing Laboratory Culturing Data on Field Data

To scrutinize the phenology of PUFA availability in a well-studied urban lake, phytoplankton data from the Enonselkä basin of Lake Vesijärvi, Central Finland (WGS84 61°2.2'N, 25°31.7'E), were taken from the Hertta database of the Finnish Environment Institute (requires registration, https://www.syke.fi/avointieto). Phytoplankton countings saved in the database were done using accredited method (EN 16695, 2015) by the Finnish Environment Institute. Lake Vesijärvi is a eutrophic, clear water lake (total phosphorus 27 μg L^{-1} and water color 10 mg Pt L^{-1}, Finnish Environment Institute, Water Framework Directive classification and status assessment) regularly experiencing blooms of cyanobacteria and diatoms.

Phytoplankton biomasses (mg C L^{-1}) from open water seasons 2015–2018 (five–six samplings in May–November), including contrasting cyanobacteria-dominant years and years without cyanobacteria blooms, were used to form comparisons with the experimental design. For this, the counted phytoplankton taxa were divided into main taxa: cryptophytes, cyanobacteria, diatoms, dinoflagellates, golden algae, and green algae that included also conjugatophytes. Other reported algae were classified as "other." Phytoplankton biomasses were converted to PUFA availabilities by using the amount of each compound in the experimental study as an average dry weight per mg in exponential and stationary phase. A coefficient of 0.45 was used to convert dry weight to carbon biomass based on our previous measurements [33]. If the experimentally studied main taxon included several tested strains, such as cryptophytes, included the Cryptomonas and Rhodomonas species, the average of the two strains was used. This was based on the analysis of experimental data, illustrating that the main taxa explained most of the variation in the fatty acid composition as μg FA per mg dry weight.

3. Results

3.1. Growth Rate

Cell abundance was highest (2.5×10^7) with cultured cyanobacteria strains but remained low ($<2.5 \times 10^4$ cell mL^{-1}) throughout 22 days in cultures of Mallomonas. Growth rate (Table 1, Figure 1) between initial and the middle of exponential growth phase was highest with all three strains of diatoms (Nitzchia, Tabellaria and Diatoma) and second-highest with Haematococcus (green algae; 0.38 divisions d^{-1}) and Microcystis (cyanobacteria; 0.21 divisions d^{-1}), even though Haematococcus culture did not reach high density. Growth rates were slowest with strains of golden algae of Synura, Mallomonas, and Uroglena, and then with dinoflagellate Peridinium. Diatoms reached stationary phase already in eight–13 days, whereas it took 51 days for Uroglena to reach the stationary phase.

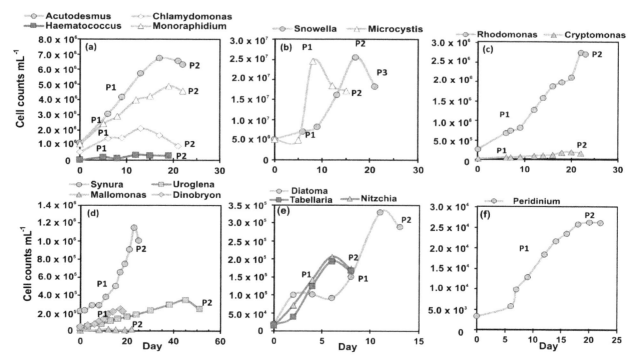

Figure 1. Growth curves for 16 cultures of phytoplankton strains classified by phytoplankton groups: (**a**) green algae, (**b**) cyanobacteria, (**c**) cryptophytes, (**d**) golden algae including chrysophytes and synyrophytes, (**e**) diatoms, and (**f**) dinoflagellate. P1 cites to sampling point during exponential growth phase and P2 cites to the sampling point in stationary phase.

3.2. Phytoplankton Taxa and Growth Phase Impact on the Contribution of ω-3 and ω-6 PUFA

The contribution of ω-3 and ω-6 PUFA of 16 phytoplankton strains varied by the phytoplankton group (Figure 2), but also by growth phase (Figure 3). All strains of green algae and cyanobacteria contained ALA, SDA, and LIN, excluding *Snowella* that did not contain any SDA. The contribution of GLA was highest in *Microcystis*, whereas trace amounts were found among golden algae, diatoms, and green algae. In addition to medium-chain ω-3 and ω-6 PUFA, diatoms, golden algae, and the dinoflagellate contained also EPA and DHA. The absolute contribution of ALA was highest in green algae and *Snowella* (~30% of all FA), whereas cryptophytes and *Dinobryon* had the highest (~26% of all FA) contribution of SDA among all phytoplankton strains. Octadecapentaenoic acid (OPA, 18:5ω3) was found only from the dinoflagellate *Peridinium cinctum* (~4% of all FA). The contribution of LIN was highest (~10% of all FA) in *Haematococcus*, *Uroglena*, *Mallomonas*, and *Synura*, whereas diatoms and the dinoflagellate had only a minor contribution of LIN (<1% of all FA). All strains of cryptophytes, diatoms, and the dinoflagellate had equal contribution of EPA (~13% of all FA), whereas the contribution of DHA was highest (18.4 ± 0.2 % of all FA) in *Peridinium*. Additionally, cryptophytes and golden algae contained also docosapentaenoic acid (ω-6 DPA).

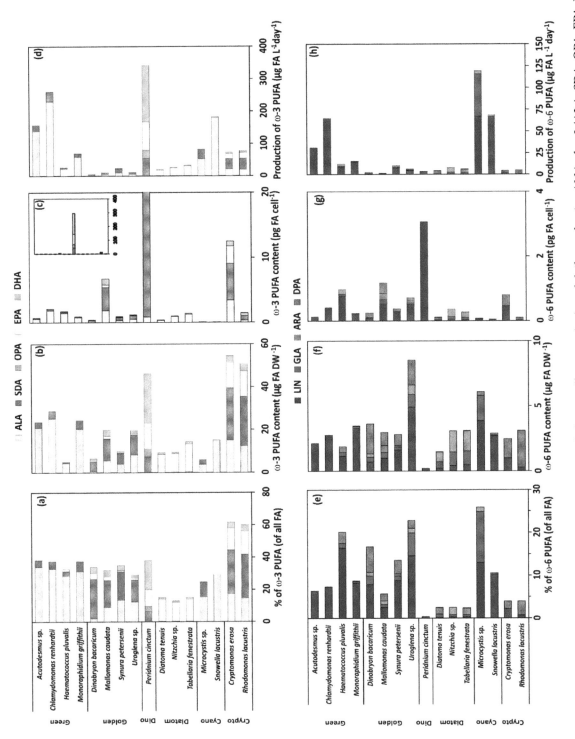

Figure 2. The proportion of all fatty acids (**a,e**), per biomass content (**b,f**), per cell content (**c,g**), and daily production (**d,h**) of ω-3 (ALA, SDA, OPA, EPA, DHA) and ω-6 (LIN, GLA, ARA, DPA) PUFA in cultured 16 phytoplankton strains.

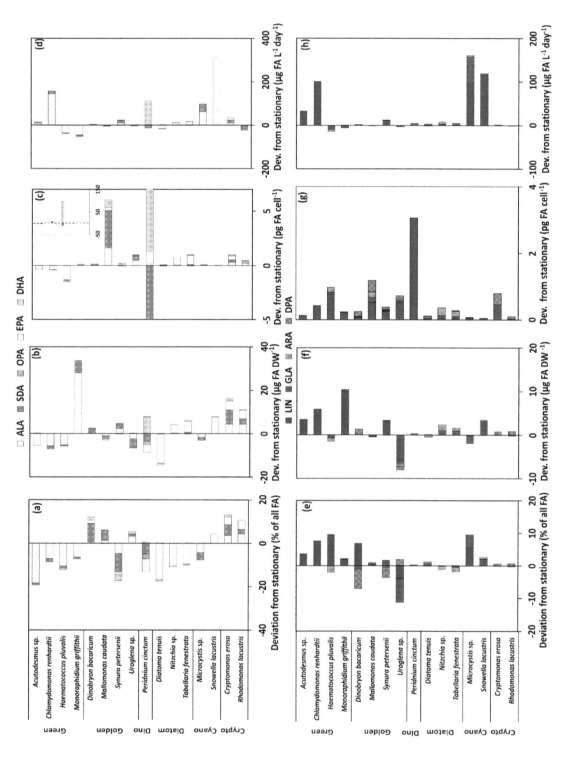

Figure 3. The deviation of ω-3 (ALA, SDA, OPA, EPA, DHA) and ω-6 (LIN, GLA, ARA, DPA) PUFA between exponential and stationary phases: proportion of all fatty acids (**a,e**), per biomass content (**b,f**), per cell content (**c,g**), and daily production (**d,h**).

According to the PERMANOVA (Table 2) the contribution of ω-3 and ω-6 PUFA differed between strains by the taxa, but also by the growth phase. Taxa explained 84% of all variation, but growth phase explained only 1% of the variation. Pairwise PERMANOVA (t = 2.58–27.8, P(MC) < 0.003) showed that the contribution of ω-3 and ω-6 PUFA differed among phytoplankton main groups. However, non-metric multidimensional scaling analysis (Figure 4) clustered (CLUSTER analysis) *Snowella* with green algae and *Microcystis* with exponential phase of *Uroglena* together by 70% similarity excluding. Furthermore, NMDS output of percentages of ω-3 and ω-6 PUFA separated strains by growth phase. Pairwise PERMANOVA (t = 3.7–7.1, P(MC) = 0.001) showed statistical difference between exponential and stationary phase for green algae, diatoms, dinoflagellates, and cryptophytes, but not for cyanobacteria or golden algae (t = 0.75–1.01, P(MC) = 0.35–0.55). The contribution of ω-3 PUFA was higher in exponential phase in green algae, dinoflagellates, and diatoms, whereas cryptophytes and chrysophytes (excluding *Synura*) had higher contribution of different ω-3 PUFA in stationary phase (Figure 3). The contribution of LIN in green algae and cyanobacteria was higher in stationary phase than in exponential phase. Otherwise, similar clear trends were not seen in the contribution of ω-6 PUFA with other taxa. Permutational analysis of multivariate dispersions (PERMDISP) showed lowest dispersion among cryptophytes and green algae, whereas dispersion was highest within cyanobacteria (Figure 5) reflecting high variation among these phytoplankton classes (Figure 3a).

Table 2. Pseudo-F and Monte Carlo p-values (P(MC) for PERMANOVA analysis of ω-3 and ω-6 PUFA of phytoplankton strains by the phytoplankton group and phase and mix of them as factors.

Unit	Factors	Df	PERMANOVA		
			Pseudo-F	exp %	P(MC)
Contribution	Group	5	141.46	84	**0.001 ***
	Phase	1	7.2967	1	**0.001**
	GroupxPhase	5	7.8303	5	**0.001**
Biomass content	Group	5	39.307	69	**0.001**
	Phase	1	1.6199	1	0.154
	GroupxPhase	5	1.075	2	0.345
Cell content	Group	5	33.402	65	**0.001**
	Phase	1	1.2592	0	0.233
	GroupxPhase	5	1.3961	2	0.12
Production	Group	5	40.176	66	**0.001**
	Phase	1	2.9217	1	0.019
	GroupxPhase	5	3.7874	6	**0.001**

* bold value means statistically significant different.

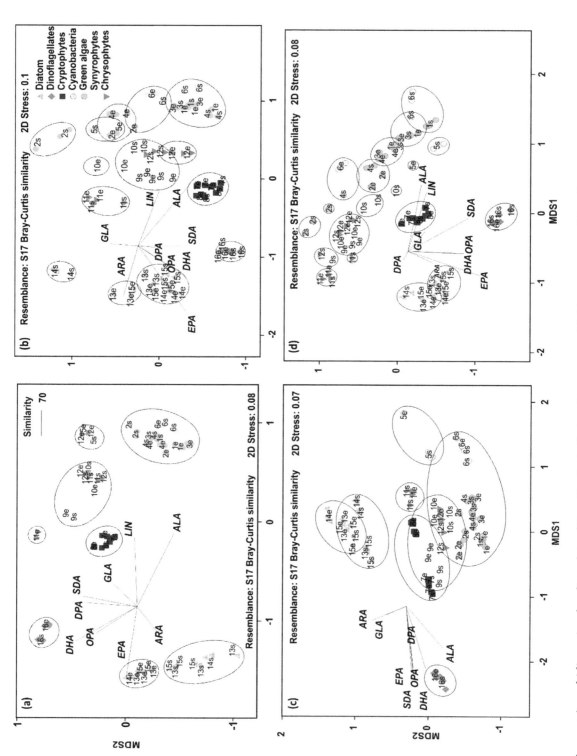

Figure 4. Non-metric multidimensional scaling plots of Bray Curtis similarity of percentages (**a**), biomass content (**b**), cell content (**c**), and daily production (**d**) of ω-3 (ALA, SDA, EPA, DHA) and ω-6 (LIN, GLA, ARA, DPA) PUFA in cultured 16 phytoplankton strains (see Table 1). Golden algae are divided here into Synyrophytes and Chrysophytes to demonstrate the difference in these classes. Abbreviations after strain number: e = exponential phase and s = stationary phase.

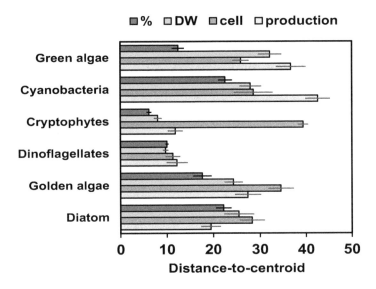

Figure 5. Permutational Analysis of Multivariate dispersion (PERMDISP) of ω-3 and ω-6 PUFA across each phytoplankton class (contribution (%), biomass content (DW), cell content (cell), and production.

3.3. Phytoplankton Taxa and Growth Phase Impact on the Content of ω-3 and ω-6 PUFA

The biomass (DW) and cell content of individual ω-3 and ω-6 PUFA varied greatly among 16 phytoplankton strains (Figure 2). According to the PERMANOVA (Table 2) the content (per biomass and cell) of ω-3 and ω-6 PUFA differed by the phytoplankton group, but not by the growth phase. Phytoplankton taxa explained 69% and 65% of all variation for biomass and cell contents, respectively. Pairwise PERMANOVA (t = 2.3–9.6, P(MC) = 0.001–0.008) comparison showed that all phytoplankton groups differed from each other when PUFA content was calculated per cell but not between cyanobacteria and green algae when PUFA content was calculated per biomass (t = 1.615, P(MC) = 0.071). Total biomass content of ω-3 PUFA was highest in cryptophytes (Figure 2), but when ω-3 PUFA content was calculated per cell the dinoflagellate *Peridinium* had 24-fold content of ω-3 PUFA in relation to any phytoplankton strain (Figure 2). More specifically, green algae excluding *Haematococcus* had highest ALA content per biomass, cryptophytes had the highest SDA content and cryptophytes, diatoms, and dinoflagellates had the highest EPA content. *Peridinium* had seven times higher DHA content than in any other phytoplankton strains. Total ω-6 PUFA biomass content was highest among *Uroglena* and *Microcystis*, which had both especially high LIN and GLA. Additionally, all cryptophytes and golden algae had relatively high ω-6 DPA content.

Dispersion (PERMDISP) of ω-3 and ω-6 PUFA per DW was low (Figure 5) and group similarity was high (SIMPER; Table 3) only among cryptophytes and dinoflagellates (including only one species of exponential and stationary). When using per cell PUFA concentrations in PERMDISP analysis, dispersion was high and similarity low among all phytoplankton. This trend was especially seen with golden algae and cryptophytes: cell ω-3 and ω-6 PUFA content was relatively higher in *Mallomonas* and *Cryptomonas* than in other species of golden algae or cryptophytes, respectively. The output of non-metric multidimensional scaling of ω-3 and ω-6 PUFA content (Figure 4b,c) also showed that dissimilarity within phytoplankton group is higher when PUFA content is calculated per cell than per biomass. This was especially seen between golden algae and cryptophytes that clustered separately in NMDS when using per biomass content but did not differ in NMDS when per cell content was used. We found logarithmic regression (y = 2.9093ln(x) + 8.0141; r^2 = 0.645) between cell size and ω-3 PUFA content per cell. The per biomass content of ω-3 and ω-6 PUFA of phytoplankton strains in exponential and stationary phase varied greatly within phytoplankton groups, and cryptophytes were the only group in which both strains had higher PUFA content in stationary than in exponential phase. When the ω-3 PUFA content was calculated per cell, all cultured strains excluding *Acutodesmus*,

Chlamydomonas, and *Haematococcus* had equal or higher ω-3 PUFA content per cell in stationary than in exponential phase (Figure 3).

Table 3. Similarity percentages of SIMPER analysis used to assess similarity within phytoplankton class/group by the different units of the ω-3 and ω-6 PUFA abundance and main PUFAs, explaining most of the similarity. *n* = strain number within taxa + number of growth phases.

Taxa	Unit	SIMPER	
		Average Sim. (%)	Main PUFA
Diatom	Contribution	70.7	EPA
(*n* = 3 + 2)	Biomass content	64.8	EPA
	Cell content	60.7	EPA
	Production	72.7	EPA
Golden algae	Contribution	74.6	SDA, ALA, LIN
(*n* = 4 + 2)	Biomass content	65.8	SDA, ALA, LIN
	Cell content	52.3	SDA, ALA, LIN
	Production	60.7	SDA, ALA, LIN
Dinoflagellate	Contribution	86.5	DHA, EPA
(*n* = 1 + 2)	Biomass content	85.8	DHA, EPA
	Cell content	83.1	DHA, EPA
	Production	81.3	DHA, EPA, SDA
Cryptophytes	Contribution	91.3	SDA, ALA, EPA
(*n* = 2 + 2)	Biomass content	88.5	SDA, ALA, EPA
	Cell content	51.8	SDA, ALA, EPA
	Production	83.3	SDA, ALA, EPA
Cyanobacteria	Contribution	70.7	ALA, LIN
(*n* = 2 + 2)	Biomass content	62.3	ALA, LIN
	Cell content	51.8	ALA, LIN
	Production	41.3	ALA, LIN
Green algae	Contribution	82.3	ALA
(*n* = 4 + 2)	Biomass content	57.1	ALA
	Cell content	64.6	ALA
	Production	49.2	ALA

3.4. Phytoplankton Taxa and Growth Phase Impact on the Production of ω-3 and ω-6 PUFA

The production of medium-chain and long-chain ω-3 and ω-6 PUFA differed (PERMANOVA, Table 2) according to phytoplankton class (Figure 2d,h), within the phytoplankton main group (PERMDISP and SIMPER; Figure 5, Table 3), and by the growth phase (Table 2). However, growth phase explained only 1% of the variation, whereas phytoplankton taxa explained 66% of all PUFA variation. Pairwise PERMANOVA (t = 4.80–10.37; P(MC) = 0.001) showed that all phytoplankton groups, excluding cyanobacteria and green algae, differed from each other (t = 1.39, P(MC) = 0.124). Production of ω-3 and ω-6 PUFA differed by growth phase among diatoms and cyanobacteria (Pairwise PERMANOVA: t = 1.93–3.38; P(MC) = 0.001–0.041). The production of ALA was highest with green algae (*Chlamydomonas, Acutodesmus*) and cyanobacteria (*Snowella*), whereas dinoflagellate (*Peridinium*) and cryptophytes had the highest production of SDA per day. The dinoflagellate *Peridinium* produced three and 33 times more EPA and DHA per day (μg PUFA L^{-1} day^{-1}), respectively, than any other phytoplankton strain. Diatoms had highest production values for EPA and cryptophytes for DHA after *Peridinium*. Furthermore, diatoms and cryptophytes had 87 and 34 times higher production of EPA than chrysophytes, respectively. Production of LIN was highest in cyanobacteria and *Chlamydomonas* and *Acutodesmus*, whereas *Microcystis* alone had highest production of GLA. Cryptophytes and golden algae produced highest amount of ω-6 DPA in a day, even though it was relatively low in comparison with the production of LIN produced by green algae and cyanobacteria. Similarity analysis (SIMPER) showed that similarity in the production ω-3 and ω-6 PUFA was highest among cryptophytes and diatoms, whereas the similarity (SIMPER) was lowest with green algae and cyanobacteria. Green algae and cyanobacteria also clustered together in the NMDS plot. Production of ω-3 and ω-6 PUFA did not differ statistically between the exponential and stationary growth phase at the main group level, but some strains, e.g., *Chlamydomonas*, *Microcystis*, and *Snowella*, had a relatively higher production of ALA and LIN at the stationary phase (Figure 4d,h).

3.5. Extrapolation to Field Data

The community composition in Lake Vesijärvi had no clear pattern during the study years (Figure 6, Figures S1 and S2). However, the proportion of dinoflagellates was generally highest in spring. In June 2015, cryptophytes and golden algae increased and were followed by diatoms and cyanobacteria in autumn. On the contrary, years 2016 and 2018 were dominated by cyanobacteria from June until autumn, whereas in 2017, cryptophytes and diatoms increased in mid-summer and cyanobacteria in autumn.

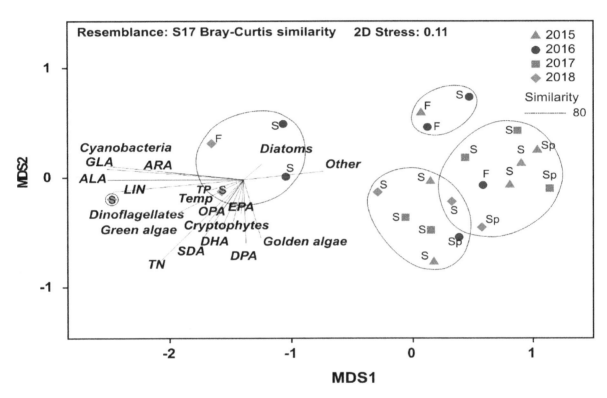

Figure 6. Non-metric multidimensional scaling plots of Bray Curtis similarity of ω-3 (ALA, SDA, EPA, DHA) and ω-6 (LIN, GLA, ARA, DPA) PUFA concentration of phytoplankton (μg PUFA L^{-1}), main phytoplankton groups and physico-chemical parameters in Lake Vesijärvi in years 2015–2018. TN—Total Nitrogen, TP—Total Phosphorus, Temp—temperature in the epilimnion. S = summer, F = fall, Sp = Spring.

Converted to fatty-acid availabilities, the concentration of ω-3 and ω-6 PUFA did not differ between years (PERMANOVA: Pseudo-F = 1.49, p = 0.195), but field data demonstrated notable seasonal and interannual variation (PERMANOVA: Pseudo-F = 4.36, p = 0.007). According to the two factor PERMANOVA, the season explained 24% of all variation in the PUFA concentrations. Generally, non-metric multidimensional scaling clustered phytoplankton and corresponding PUFA concentrations in four groups with 80% similarity (Figure 6). NMDS1 correlated strongly negatively (r = −0.98) with cyanobacteria. One point was close with cyanobacteria, and four points related closely with diatoms and all other sampling points were in the right side of the NMDS output. Different PUFA showed a strong relationship with certain phytoplankton groups. Cyanobacteria-dominance was reflected as the high concentration and proportion of ALA, LIN, and GLA throughout the growing season (Pearson correlation: r = 0.93–0.97, p < 0.001), which peaked after midsummer. The relative proportion of DHA was highest in early summer, when biomass of dinoflagellates was relatively high

(Figure 6, Figure S1). The concentration of DHA showed a strong correlation with the biomass of dinoflagellates (Pearson correlation: $r = 0.957, p < 0.001$), whereas the concentration of EPA was most closely related with diatoms (Pearson correlation: $r = 0.648, p < 0.001$) and cryptophytes (Pearson correlation: $r = 0.671, p < 0.001$). However, NMDS output separated diatoms as their own group, and EPA was more closely related with golden algae and cryptophytes than with diatoms. The abundance of green algae showed strong correlation with biomass of cryptophytes and dinoflagellates (Pearson correlation: $r = 0.47–0.59, p < 0.004–0.022$), as can be seen in the NMDS output (Figure 6), resulting in a strong inter-correlation with the concentration of EPA and DHA (Pearson correlation: $r = 0.57–0.72, p < 0.0001–0.005$). Whereas total phosphorus (TP) was positively related with cyanobacteria in NMDS output, temperature was positively related with dinoflagellates, cryptophytes, golden algae, and green algae. However, a negative relationship between cyanobacteria and TP was not statistically significant ($r = 0.406, p = 0.055$).

4. Discussion

The experimental setup of this study consisted of six main groups of phytoplankton (cryptophytes, dinoflagellates, golden algae, diatoms, green algae, and cyanobacteria), which were sampled at early and late growth phase to understand how phytoplankton nutritional value and production of ω-3 and ω-6 PUFA may vary along phytoplankton growth. Inclusion of one–four different strains in each main group facilitated scrutinization of variation inside taxonomic main groups. Briefly, even though the ability to synthesize different ω-3 and ω-6 PUFA follows strictly phylogenetical groups [15,16,22,34], the PUFA content per cell and the production of PUFA can vary greatly within phytoplankton groups.

Typically, the studies on phytoplankton fatty acids report the contribution of different PUFA together with the total concentrations of PUFA (e.g., per dry weight or carbon) [15,16]. Deviating from the previous studies, we determined the cell-specific fatty acid content and production rates for the main freshwater phytoplankton groups. Proportions, dry weights, and cell-specific concentrations were calculated for both exponential and stationary growth phase. Our results revealed that cell-specific PUFA content differed greatly from biomass-specific PUFA content and the variation in cell PUFA content within phytoplankton group was high likely due to the variable size of phytoplankton. Comparison of the different metrics demonstrated risk of being misled if scrutinizing only one type of concentration and making ecological extrapolation. Proportion and concentration as dry weight can give only restricted amount of information on PUFA and might be of more interest in biofuel production [19]. However, information of the cell-specificity is important, because in plankton communities, secondary consumers feed on a diverse phytoplankton community, and the size of the animal is proportional to the size of the phytoplankton that it can ingest [1,15].

In this study, Peridium had a large cell diameter and relatively slow specific growth rate, both characteristics typical of K-strategists displaying resource-efficiency in traditional r/K classification [1]. In Lake Vesijärvi, dinoflagellates occurred at the time typical for cells displaying these functional traits. However, DHA content per DW was seven times higher in Peridium than in any other phytoplankton strain, whereas DHA content per cell in Peridium was ~200 fold in relation to any other phytoplankton strain. This makes a many-fold difference for filter-feeding zooplankton grazers, and explains why dinoflagellates are the preferable diet for copepods [35]. Daphnia do not grow well with Peridinium, maybe due to the armoring and low amounts of sterols [11]. However, according to the fatty acid modeling, Daphnia's diet consisted of ~20% dinoflagellates in Lake Vesijärvi in year 2016 [26]. Therefore, it seems that dinoflagellates can fuel EPA and DHA demand of both zooplankton groups and the whole food web as seen earlier in a strong correlation between the biomass of dinoflagellates and DHA content of perch [17]. Even though Peridinium grew slowly, we found that the production of DHA was 40 times higher with Peridinium than any other phytoplankton strain, which emphasizes the role of this non-toxic freshwater dinoflagellate in the synthesis of DHA. Therefore, even a small increase in the biomass of Peridinium can significantly increase the production of DHA in boreal lakes. However, some dinoflagellates species, e.g., *Ceratium*, are too large for zooplankton to ingest, and thus, high DHA content in them is not available for zooplankton.

Herbivorous cladoceran can have a high proportion of EPA, whereas DHA is nearly absent in them [10,36,37]. Therefore, the production of EPA is important for herbivorous cladocerans (e.g., *Daphnia*). Diatoms and cryptophytes are crucial producers of EPA in freshwaters [38]. Meanwhile, the percentage and biomass content of EPA is similar with cryptophytes and diatoms. Our results showed that cell content of EPA varies greatly between these two phytoplankton groups. Meanwhile, *Cryptomonas* had a higher EPA content per cell than any of the studied diatoms. We found the lowest cell EPA content in *Rhodomonas*. Moreover, since diatoms grow faster than cryptophytes, we found 2.3 times higher production of EPA with diatoms than with cryptophytes. These two phytoplankton groups equally influenced the concentration of EPA in Lake Vesijärvi, showing the importance of diatoms, especially in spring and autumn, while cryptophytes' importance was largely shown in summer. Furthermore, our previous fatty acid-based modeling on the composition of *Daphnia* diets also showed that cryptophytes and diatoms are the two main dietary sources of this key herbivorous zooplankton in Lake Vesijärvi [26]. However, the size and form of diatoms vary greatly, and they have silica frustules that might be difficult for *Daphnia* to ingest. Therefore, digestibility of diatoms varies greatly. Furthermore, diatoms can form large colonies and blooms, which are not ingestible for zooplankton, resulting in poor utilization of the diatom-produced EPA. Moreover, previous studies have shown that the EPA content of different species and by habitat is highly variable [13,39,40].

Even though ω-3 and ω-6 PUFA profiles of cyanobacteria and green algae differ at some level, the biomass and cell content of these PUFA did not differ markedly but were clustered together in NMDS output. This results from the fact that both phytoplankton groups grow fast and have a high ALA and LIN content. Our results also showed that these two groups were superior in producing ALA and LIN, which is one reason why they have been used for biofuel production. However, in terms of efficient transfer of these medium-chain PUFA in aquatic food webs, phytoplankton need to be digestible for zooplankton, and zooplankton need to have the ability to bioconvert EPA or DHA from ALA or ARA from LIN. Generally, it has been assumed that zooplankton does not feed on especially large-sized cyanobacteria, whereas other studies suggest that zooplankton can feed on cyanobacteria [34,35]. In Lake Vesijärvi, cyanobacteria (e.g., *Planktothrix*, *Snowella*, *Aphanizomenon*, *Microcystis*) can form blooms that can last throughout summer, as were seen in 2016. According to the fatty acid-based modeling [26], cyanobacteria formed less than 10% of the diet of *Daphnia*, and when the model uncertainties were considered, it could be noted that cyanobacteria were an insignificant diet source for *Daphnia*. Therefore, it seems that cyanobacteria may contain much of ALA and LIN but remain an inaccessible resource for zooplankton. Secondly, it should be noted that *Daphnia* has a poor ability to bioconvert EPA from ALA [41,42].

Here, we focused on phytoplankton phylogeny and growth phase and were unable to extrapolate the environmental conditions' impact on production of ω-3 and ω-6 PUFA, since we converted phytoplankton biomass to fatty-acid availabilities in our field data. However, our field data showed a positive relationship between total phosphorus and ALA, LIN, and GLA production by cyanobacteria, whereas increased temperature and total nitrogen was related with the production of SDA, EPA, DHA, and ω-6 DPA by cryptophytes, golden algae, and dinoflagellates. In addition to changes in phytoplankton composition, environmental conditions can affect phytoplankton PUFA content [43–45], and thus, potentially, also their production. Our recent study [46] with 107 boreal lakes showed that intensified eutrophication decreases the nutritional value of phytoplankton. The high difference in temperature between freshwater and brackish and marine phytoplankton strains resulted in a 10-fold difference in the production of EPA [39]. Another study [43] with green algae, cryptophytes, and diatoms showed that the light and temperature increase (from 10 to 25 °C) have a relatively minor impact on PUFA content in green algae. Surprisingly, in our study, slow growing cryptophytes and golden algae had higher EPA contribution in stationary phase, whereas fast growing diatoms and slower growing dinoflagellates and synurophytes had higher EPA contribution in the exponential than in the stationary phase. The same trend was also seen in the biomass and cell PUFA content of cryptophytes, dinoflagellate, and diatoms, excluding *Diatoma*, which had minimal PUFA content in the stationary phase. However, the effect of the growth phase on the EPA production of cryptophytes, golden algae, dinoflagellate, and diatoms was ambiguous, showing that the production of EPA can vary within phytoplankton groups. The growth phase had a small impact on the *Peridinium* biomass and cell EPA content, but *Peridinium* had two times higher DHA content per cell and production of DHA in stationary than in exponential phase. The contribution of ALA and SDA of green algae was higher in stationary than in exponential phase; however, the biomass and cell content and the production of ALA and SDA varied greatly by green algae strains. The growth phase affected the contribution, content, and production of ALA and SDA differently. Altogether, it seemed that the growth phase together with the environmental parameters could affect PUFA content and production of freshwater phytoplankton.

Based on their capability to overcome and adapt to environmental constrains, phytoplankton can be categorized into functional groups [44,45,47]. Functional classification may include growth and morphometric traits that determine how easily a phytoplankter is eaten by a consumer [47]. This could be an important approach, because it includes both environmental conditions and phytoplankton physiological traits, and modern food web models typically use functional rather than phylogenetic phytoplankton inputs [48]. Here, we focused on growth rate and cell size; however, future studies might benefit from using phytoplankton strains from different functional groups.

5. Conclusions

In conclusion, for understanding the synthesis and transfer of ω-3 and ω-6 PUFA, calculations of PUFA content per phytoplankton cell are beneficial in addition to biomass content. Our results showed that phytoplankton PUFA per biomass content varies from the cell PUFA content due to the positive impact of cell size on PUFA content. Therefore, larger cells have a higher PUFA content than smaller cells, but too large cells are not digestible for herbivorous zooplankton, and subsequently, are not utilized or transferred in the freshwater food web. Our laboratory culturing emphasized that different ω-3 and ω-6 PUFA are synthesized by certain phytoplankton taxa. Extrapolation on field phytoplankton data demonstrated how the availability of PUFA differed inter- and intra-annually. Dinoflagellates were superior producers of DHA, whereas diatoms and cryptophytes were crucial producers of EPA in boreal lakes. Our results also demonstrated that phytoplankton PUFA content and production varied by growth phase; however, this change is difficult to predict due to the high variation between strains within the same phytoplankton groups.

Supplementary Materials:
Figure S1: Development of phytoplankton biomass (as mg C L^{-1}) in L. Vesijärvi (data from the database of Finnish Environment Institute) and derived PUFA availability per liter. Figure S2: Development of phytoplankton biomass (as mg C L^{-1}) in L. Vesijärvi (data from the database of Finnish Environment Institute) and the derived PUFA availability per phytoplankton carbon content. Table S1: FA profiles of cultured phytoplankton strains.

Author Contributions: Conceptualization, methodology, and writing—original draft preparation, S.T., writing—review and editing, E.P. and P.S. All authors have read and agreed to the published version of the manuscript.

Acknowledgments: The authors would like to thank Gabriella Chebli for editing the English of this manuscript.

References

1.   Reynolds, C.S. *The Ecology of Phytoplankton*; University Press: Cambridge, UK, 2006.
2.   Brett, M.; Müller-Navarra, D. The role of highly unsaturated fatty acids in aquatic foodweb processes. *Freshwat. Biol.* **1997**, *38*, 483–499. [CrossRef]

3.	Ravet, J.L.; Brett, M.T.; Arhonditsis, G.B. The effects of seston lipids on zooplankton fatty acid composition in Lake Washington, Washington, USA. *Ecology* **2010**, *91*, 180–190. [CrossRef]

4.	Strandberg, U.; Taipale, S.J.; Hiltunen, M.; Galloway, A.; Brett, M.T.; Kankaala, P. Inferring phytoplankton community composition with a fatty acid mixing model. *Ecosphere* **2015**, *6*, 1–18. [CrossRef]

5.	Das, U.N. Essential fatty acids-a review. *Curr. Pharm. Biotechnol.* **2006**, *7*, 467–482. [CrossRef] [PubMed]

6.	Goedkoop, W.; Sonesten, L.; Ahlgren, G.; Boberg, M. Fatty acids in profundal benthic invertebrates and their major food resources in lake erken, sweden: Seasonal variation and trophic indications. *Can. J. Fish Aquat. Sci.* **2000**, *57*, 2267–2279. [CrossRef]

7.	Lau, D.C.; Vrede, T.; Pickova, J.; Goedkoop, W. Fatty acid composition of consumers in boreal lakes–variation across species, space and time. *Freshwat. Biol.* **2012**, *57*, 24–38. [CrossRef]

8.	Makhutova, O.N.; Gladyshev, M.I.; Sushchik, N.N.; Dubovskaya, O.P.; Buseva, Z.F.; Fefilova, E.B.; Semenchenko, V.P.; Kalachova, G.S.; Kononova, O.N.; Baturina, M.A. Comparison of fatty acid composition of cladocerans and copepods from lakes of different climatic zones. *Contemp. Probl. Ecol.* **2014**, *7*, 474–483. [CrossRef]

9.	Makhutova, O.N.; Shulepina, S.P.; Sharapova, T.A.; Dubovskaya, O.P.; Sushchik, N.N.; Baturina, M.A.; Pryanichnikova, E.G.; Kalachova, G.S.; Gladyshev, M.I. Content of polyunsaturated fatty acids essential for fish nutrition in zoobenthos species. *Freshw. Sci.* **2016**, *35*, 1222–1234. [CrossRef]

10.	Hiltunen, M.; Strandberg, U.; Taipale, S.J.; Kankaala, P. Taxonomic identity and phytoplankton diet affect fatty acid composition of zooplankton in large lakes with differing dissolved organic carbon concentration. *Limnol. Oceanogr.* **2015**, *60*, 303–317. [CrossRef]

11.	Peltomaa, E.T.; Aalto, S.L.; Vuorio, K.M.; Taipale, S.J. The importance of phytoplankton biomolecule availability for secondary production. *Front. Ecol. Evol.* **2017**, *5*, 128. [CrossRef]

12.	Jónasdóttir, S.H. Effects of food quality on the reproductive success of acartia tonsa and acartia hudsonica: Laboratory observations. *Mar. Biol.* **1994**, *121*, 67–81. [CrossRef]

13.	Chen, M.; Liu, H.; Chen, B. Effects of dietary essential fatty acids on reproduction rates of a subtropical calanoid copepod, acartia erythraea. *Mar. Ecol. Prog. Ser.* **2012**, *455*, 95–110. [CrossRef]

14.	Guo, F.; Kainz, M.J.; Sheldon, F.; Bunn, S.E. The importance of high-quality algal food sources in stream food webs–current status and future perspectives. *Freshwat. Biol.* **2016**, *61*, 815–831. [CrossRef]

15.	Taipale, S.; Strandberg, U.; Peltomaa, E.; Galloway, A.W.; Ojala, A.; Brett, M.T. Fatty acid composition as biomarkers of freshwater microalgae: Analysis of 37 strains of microalgae in 22 genera and in seven classes. *Aquat. Microb. Ecol.* **2013**, *71*, 165–178. [CrossRef]

16.	Taipale, S.J.; Hiltunen, M.; Vuorio, K.; Peltomaa, E. Suitability of phytosterols alongside fatty acids as chemotaxonomic biomarkers for phytoplankton. *Front. Plant Sci.* **2016**, *7*, 212. [CrossRef]

17.	Taipale, S.J.; Vuorio, K.; Strandberg, U.; Kahilainen, K.K.; Järvinen, M.; Hiltunen, M.; Peltomaa, E.; Kankaala, P. Lake eutrophication and brownification downgrade availability and transfer of essential fatty acids for human consumption. *Env. Int.* **2016**, *96*, 156–166. [CrossRef]

18.	Los, D.A.; Mironov, K.S. Modes of fatty acid desaturation in cyanobacteria: An update. *Life* **2015**, *5*, 554–567. [CrossRef]

19.	Arts, M.T.; Brett, M.T.; Kainz, M. Lipids in Aquatic Ecosystems. Springer: New York, NY, USA, 2009.

20.	Raheem, A.; Prinsen, P.; Vuppaladadiyam, A.K.; Zhao, M.; Luque, R. A review on sustainable microalgae based biofuel and bioenergy production: Recent developments. *J. Clean. Prod.* **2018**, *181*, 42–59. [CrossRef]

21.	Mathimani, T.; Pugazhendhi, A. Utilization of algae for biofuel, bio-products and bio-remediation. *Biocatal. Agric. Biotechnol.* **2019**, *17*, 326–330. [CrossRef]

22.	Galloway, A.W.; Winder, M. Partitioning the relative importance of phylogeny and environmental conditions on phytoplankton fatty acids. *PLoS ONE* **2015**, *10*, e0130053. [CrossRef]

23.	Grima, E.M.; Pérez, J.S.; Sánchez, J.G.; Camacho, F.G.; Alonso, D.L. EPA from isochrysis galbana: Growth conditions and productivity. *Process Biochem.* **1992**, *27*, 299–305. [CrossRef]

24.	Thompson, P.A.; Guo, M.; Harrison, P.J.; Whyte, J.N. Effects of variation in temperature. ii. on the fatty acid composition of eight species of marine phytoplankton. *J. Phycol.* **1992**, *28*, 488–497. [CrossRef]

25.	Mitchell, S.F.; Trainor, F.R.; Rich, P.H.; Goulden, C.E. Growth of daphnia magna in the laboratory in relation to the nutritional state of its food species, chlamydomonas reinhardtii. *J. Plankton. Res.* **1992**, *14*, 379–391. [CrossRef]

26. Taipale, S.J.; Vuorio, K.; Aalto, S.L.; Peltomaa, E.; Tiirola, M. Eutrophication reduces the nutritional value of phytoplankton in boreal lakes. *Env. Res.* **2019**, *179*, 108836. [CrossRef]

27. Guillard, R.R. Culture of Phytoplankton for Feeding Marine Invertebrates. In *Culture of Marine Invertebrate Animals*; Smith, W.L., Chanley, M.H., Eds.; Springer: Boston, MA, USA, 1975; pp. 29–60.

28. Guillard, R.R.; Lorenzen, C.J. Yellow-green algae with chlorophyllide c 1, 2. *J. Phycol.* **1972**, *8*, 10–14. [CrossRef]

29. Media for Freshwater, Terrestrial, Hot Spring and Salt Water Algae. Available online: https://mcc.nies.go.jp/medium/en/media_web_e.html (accessed on 1 April 2020).

30. Schlechtriem, C.; Henderson, R.J.; Tocher, D.R. A critical assessment of different transmethylation procedures commonly employed in the fatty acid analysis of aquatic organisms. *Limnol. Oceanogr.-Meth.* **2008**, *6*, 523–531. [CrossRef]

31. Anderson, M.; Gorley, R.; Clarke, K.; Anderson, M.J.; Gorley, R.N.; Clarke, K.R.; Anderson, M.; Gorley, R.; Andersom, M.J. *PERMANOVA for PRIMER. Guide to Software and Statistical Methods*; PRIMER-E: Plymouth, UK, 2008.

32. Anderson, M.J. Distance-based tests for homogeneity of multivariate dispersions. *Biometrics* **2006**, *62*, 245–253. [CrossRef]

33. Taipale, S.J.; Peltomaa, E.; Hiltunen, M.; Jones, R.I.; Hahn, M.W.; Biasi, C.; Brett, M.T. Inferring phytoplankton, terrestrial plant and bacteria bulk δ^{13}C values from compound specific analyses of lipids and fatty acids. *PLoS ONE* **2015**, *10*, e0133974. [CrossRef]

34. Jónasdóttir, S.H. Fatty acid profiles and production in marine phytoplankton. *Mar. Drugs* **2019**, *17*, 151. [CrossRef]

35. Santer, B. Nutritional suitability of the dinoflagellate ceratium furcoides for four copepod species. *J. Plankton Res.* **1996**, *18*, 323–333. [CrossRef]

36. Brett, M.T.; Müller-Navarra, D.C.; Persson, J. Crustacean Zooplankton Fatty Acid Composition. In *Lipids in Aquatic Ecosystems*; Kainz, M., Brett, M.T., Arts, M.T., Eds.; Springer: New York, NY, USA, 2009; pp. 115–146.

37. Persson, J.; Vrede, T. Polyunsaturated fatty acids in zooplankton: Variation due to taxonomy and trophic position. *Freshwat. Biol.* **2006**, *51*, 887–900. [CrossRef]

38. Sushchik, N.N.; Gladyshev, M.I.; Makhutova, O.N.; Kalachova, G.S.; Kravchuk, E.S.; Ivanova, E.A. Associating particulate essential fatty acids of the ω3 family with phytoplankton species composition in a siberian reservoir. *Fresh Biol.* **2004**, *49*, 1206–1219. [CrossRef]

39. Peltomaa, E.; Hällfors, H.; Taipale, S.J. Comparison of diatoms and dinoflagellates from different habitats as sources of PUFAs. *Mar. Drugs* **2019**, *17*, 233. [CrossRef] [PubMed]

40. Sushchik, N.N.; Gladyshev, M.I.; Kalachova, G.S.; Kravchuk, E.S.; Dubovskaya, O.P.; Ivanova, E.A. Particulate fatty acids in two small siberian reservoirs dominated by different groups of phytoplankton. *Freshwat. Biol.* **2003**, *48*, 394–403. [CrossRef]

41. von Elert, E. Determination of limiting polyunsaturated fatty acids in daphnia galeata using a new method to enrich food algae with single fatty acids. *Limnol. Oceanogr.* **2002**, *47*, 1764–1773. [CrossRef]

42. Guschina, I.A.; Harwood, J.L. Algal Lipids and Effect of the Environment on their Biochemistry. In *Lipids in Aquatic Ecosystems*; Kainz, M., Brett, M.T., Arts, M.T., Eds.; Springer: New York, NY, USA, 2009; pp. 1–24.

43. Piepho, M.; Martin-Creuzburg, D.; Wacker, A. Phytoplankton sterol contents vary with temperature, phosphorus and silicate supply: A study on three freshwater species. *Eur. J. Phycol.* **2012**, *47*, 138–145. [CrossRef]

44. Reynolds, C.S.; Huszar, V.; Kruk, C.; Naselli-Flores, L.; Melo, S. Towards a functional classification of the freshwater phytoplankton. *J. Plankton. Res.* **2002**, *24*, 417–428. [CrossRef]

45. Padisák, J.; Crossetti, L.O.; Naselli-Flores, L. Use and misuse in the application of the phytoplankton functional classification: A critical review with updates. *Hydrobiologia* **2009**, *621*, 1–19. [CrossRef]

46. Taipale, S.J.; Kahilainen, K.K.; Holtgrieve, G.W.; Peltomaa, E.T. Simulated eutrophication and browning alters zooplankton nutritional quality and determines juvenile fish growth and survival. *Ecol. Evol.* **2018**, *8*, 2671–2687. [CrossRef]

47. Salmaso, N.; Naselli-Flores, L.; Padisak, J. Functional classifications and their application in phytoplankton ecology. *Freshwat. Biol.* **2015**, *60*, 603–619. [CrossRef]

48. Boit, A.; Martinez, N.D.; Williams, R.J.; Gaedke, U. Mechanistic theory and modelling of complex food-web dynamics in lake constance. *Ecol. Lett.* **2012**, *15*, 594–602. [CrossRef] [PubMed]

12

Fatty Acid Content and Composition of the Yakutian Horses and their Main Food Source: Living in Extreme Winter Conditions

Klim A. Petrov [1], **Lyubov V. Dudareva** [2], **Vasiliy V. Nokhsorov** [3], **Kirill N. Stoyanov** [4] and **Olesia N. Makhutova** [4,5,*]

[1] Institute for Biological Problems of Cryolithozone of Siberian Branch of the Russian Academy of Sciences, 41 Lenina av., Yakutsk 677000, Russia; kap_75@bk.ru

[2] Siberian Institute of Plant Physiology and Biochemistry, Siberian Branch of Russian Academy of Sciences, 132 Lermontova str., Irkutsk 664033, Russia; laser@sifibr.irk.ru

[3] North-Eastern Federal University, 48 Kulakovskogo str., Yakutsk 677000, Russia; nohvasyavas@mail.ru

[4] Siberian Federal University, 79 Svobodny pr., Krasnoyarsk 660041, Russia; ikirill97@gmail.com

[5] Institute of Biophysics of Federal Research Center "Krasnoyarsk Science Center" of Siberian Branch of Russian Academy of Sciences, Akademgorodok, Krasnoyarsk 660036, Russia

[*] Correspondence: makhutova@ibp.krasn.ru

Abstract: For the first time, seasonal changes in the content of total lipids (TLs) and phospholipids (PLs) were studied in fodder plants growing in Central Yakutia—a perennial cereal, smooth brome (*Bromopsis inermis* L.), and an annual cereal, common oat (*Avena sativa* L.). Both species have concentrated TLs and PLs in autumn under cold hardening. In addition, a significant increase in the content of fatty acids (FAs) of *B. inermis* was observed during the autumn decrease in temperature. The Yakutian horses, which fed on cereals enriched with nutrients preserved by natural cold (green cryo-fodder), accumulated significant amounts of 18:2n-6 and 18:3n-3, the total content of which in cereals was 75% of the total FA content. We found differences in the distribution of these two FAs in different tissues of the horses. Thus, liver was rich in 18:2n-6, while muscle and adipose tissues accumulated mainly 18:3n-3. Such a distribution may indicate different roles of these FAs in the metabolism of the horses. According to FA content, meat of the Yakutian horses is a valuable dietary product.

Keywords: essential polyunsaturated fatty acids; linoleic acid; alpha-linolenic acid; food quality; muscle tissue; subcutaneous adipose tissue; liver; green cryo-fodder

1. Introduction

The Republic of Sakha (Yakutia), located between $105°32'–162°55'$ E and $55°29'–76°46'$ N, occupies the territory of 3103.2 thousand km^2 and lies completely in the permafrost zone in Russia. During the short growing season, plants are exposed to high activity of solar radiation, moisture deficiency, and short-term frosts on the soil surface in early summer and autumn. Native plant species growing in such extreme conditions adapt to going through all the stages of ontogenesis in a shorter time period [1,2]. At different stages of ontogenesis, the ability of plants to adapt to cold hardening is not the same: the closer the plant is to the reproductive phase, the lower its ability to adapt to cold [3]. More than 2000 species of higher vascular plants grow in the permafrost zone of Yakutia, which is an unusual phenomenon [4]. Some of them play an important role as a food source for herbivores.

A specific feature of the seasonal growth and development of the bulk of vegetation in the permafrost zone is that its intensive growth occurs in the first half of summer. However, at this time, northern meadow plant communities are often covered with floodwaters and are also subjected to grazing and haying. After traumatic regeneration, the plants do not have time to go through the full cycle of growth and development, produce fully developed seeds, and stay in a green frozen state under the snow cover in winter (green cryo-fodder). The basis of cool-season grass is cereals, which preserve up to 80% of herbage under snow, as well as sedge, cotton grass, and some horsetails [5,6]. The wintergreen parts of the above families of fodder plants retain higher contents of proteins, carbohydrates, and fats for the winter compared to warm-season grass [7,8].

Green cryo-fodder is the basis of nutrition for many animals, including the Yakutian horses. This breed is considered a direct successor and descendant of the horses brought from the Baikal region in the 13th–15th centuries AD [9–11]. The origin of the horses was confirmed by molecular genetic methods [12–15].

The Yakutian horse demonstrates unique adaptation to long-term low-temperature stress, which has been achieved in a short evolutionary period [16]. The reasons for such good adaptation have not been fully studied. Feeding on green cryo-fodder may help animals survive in extremely cold winters [16].

The aim of the present work was to study lipid accumulation in a perennial cereal (*Bromopsis inermis* L.) and an annual cereal (*Avena sativa* L.) cultivated at different temperatures. Additionally, we aimed to study the content and composition of fatty acids (FAs) in liver, muscle, and subcutaneous adipose tissues of Yakutian horses, which have *B. inermis* and *A. sativa* as part of their staple diet.

2. Materials and Methods

The annual cereal common oat (*Avena sativa* L., Nyurbinsky type) was sown on 31 May 2014 (control) and on 15 July 2014 (treatment). The perennial cereal smooth brome (*Bromopsis inermis* L., Ammachaan type) was mowed after spring growth to allow the aftergrowth in the middle of summer (15 July 2014)—the treatment, and it was compared with the unmown plants—the control. The experiments were carried out in field plots in the conditions of Central Yakutia (near Yakutsk, 62° N, 130° E). Samples of the control and treatment plants were taken, depending on the phases of development and hardening, 4–5 times during the growing season.

For the analysis of total lipids of the common oat, the control samples were taken 4 times from July 7 to July 25, 2014; and the treatment samples were taken 4 times from July 25 to September 30, 2014. For the analysis of total lipids of the smooth brome, the control samples were collected 4 times from June 6 to July 25, 2014; and the treatment samples were collected 5 times from July 25 to September 30, 2014. To analyze phospholipids of the common oat and the smooth brome, the controls were sampled on July 25 and June 16, respectively; and the treatments of both plants were sampled on October 3, 2014. To analyze FA composition of the smooth brome, samples of the control were collected on July 7, 2013 and those of the treatment on September 25, 2013. The FA composition of the common oat (the control and the treatment) was reported in a study by Petrov et al. (2016) [17].

Sampling took place in the first half of the day in three biological replicates. Samples were fixed with liquid nitrogen immediately after their collection, in situ, and transported in Dewar vessels to the laboratory.

The samples of liver, muscle and subcutaneous adipose tissues were collected in November 2017 and 2018, from female and male Yakutian horses, most of which were less than 1 year old. Four female horses were seven, eight, and eighteen months old and five years old; and two male horses were eight months old, and one male horse was seven months old. The horses were feeding on green cryo-fodder during three months before sampling. Muscle and adipose tissues were carved from the costal part of the animals. The samples were collected from horses inhabiting Oymyakonsky, Verhoyansky, Megino-Kangalassky, Churapchinsky, Olekminsky, and Suntarsky districts of Yakutia.

Large pieces of horse tissues (200–300 g) were immediately frozen and kept at −20 °C at the slaughter site. Then, in approximately 2 weeks, frozen tissues were transported to the laboratory. In the laboratory, samples were taken from the frozen horse tissues, placed into vials with chloroform and methanol (2:1, *v/v*), and kept at −20 °C for further analysis.

2.1. Conditions of Keeping and Feeding the Horses

The absolute annual temperature difference in the breeding area of Yakutian horses exceeds 100 °C (the maximum summer and winter temperatures are +38 °C and −70 °C, respectively). The frost period lasts 7–8 months a year. In such conditions, the herds of Yakutian horses (12–15 individuals) are kept in the open. The horses are mainly fed on cereal grains and sedge frozen by natural cold. Horse breeders feed only weakened, emaciated individuals and mares. The weight of a breeding stallion reaches 430–520 kg and the weight of a mare 415–480 kg. In our study, we mainly used tissue samples from 6–8-month-old horses taken from local horse breeders. At this age, horse's tissues have a high nutritional value. Mass slaughter was conducted in November, when horses reached an average of 120–150 kg of live weight, having accumulated the largest amount of fat. For most of their lives, horses fed exclusively on warm-season grass and green cryo-fodder. The biochemical content and the composition of blood of Yakutian horses are described in detail in the literature [18].

2.2. Analyses of Lipids and FAs of Plants

A weighed portion of plant material (0.5 g) was ground to obtain homogeneous mass [19]. Then, it was supplemented with 10 mL of the chloroform: methanol mixture (2:1, *v/v*), and ionol was added as antioxidant (0.00125 g per 100 mL of the chloroform: methanol mixture). The resulting mixture was thoroughly stirred and left for 30 min until the lipids completely diffused into the solvent. The solution was transferred quantitatively to a separatory funnel through a paper skim filter (9 cm in diameter, Khimreaktivkomplekt); the mortar was washed three times using the same solvent mixture. For better delamination, water was added.

For the analysis of total lipids, the chloroform fraction was separated. Chloroform was removed from the lipid extract using an RVO-64 rotary vacuum evaporator (Czech Republic). Nonadecanoic acid (C19:0) was used to control the extractability of lipids (%), with its known amount added at the stage of homogenization. Methyl ethers of fatty acids (FAMEs) were obtained using the method [20]. For additional purification of FAMEs, TLC method was used in a chromatographic chamber with benzene as the mobile phase (R_f = 0.71–0.73) on glass plates with silica gel. The FAME zone was removed from the plate with a spatula and eluted from the silica gel with (*n*)-hexane. The FAME analysis was performed on the gas chromatograph Agilent-6890N coupled to an Agilent-5973 quadrupole mass spectrometric detector (Agilent Technologies, USA). The ionization method used was electron impact; the ionization energy was 70 eV. The analysis was carried out in the recording mode of the total ion current. An HP-INNOWAX capillary column (30 m × 250 µm × 0.50 µm) with a polyethylene glycol stationary phase was used to separate the FAME mixture. The carrier gas was helium, the rate of gas flow was 1 mL/min.

The temperature of injection port was 250 °C, the temperature of the ion source was 230°C and that of a quadrupole was 150 °C. Scanning was performed in the range of 41–450 atomic mass units. The volume of the injected sample was one µL, the flow divider was 5:1. The separation of the FAME mixture was carried out in isothermal mode at 200 °C. The duration of the chromatographic course was 60 min. For identification of FAs, the NIST 08 and WILEY7 mass spectral libraries were used. The relative content of FAs was determined by the method of internal normalization, i.e., as weight percent (wt.%) of their total content in the sample, taking into account the response factor of FAs. The absolute content of total lipids and FAMEs was determined by weighing them on GR-120 electronic scale (A&N Company Ltd., Japan) after drying the samples to constant weight.

Separation of PL fractions into individual lipids was carried out by thin layer chromatography (TLC) on Sorbfil PTLC-AF-V-UV chromatographic plates (10 × 10 cm, Russia). For the detection and identification of phospholipids in plant material, specific reagents were used: molybdenum blue for phosphorus-containing components [21], Dragendorf reagent prepared according to the method described by Wagner et al. [22] for choline-containing lipids, and a 0.2% solution of ninhydrin in acetone for amino-containing lipids [23].

Quantitative determination of phospholipid content was carried out according to the Vaskovsky method [21]. The polar lipids were separated using a two-dimensional system: the mobile phase in the first direction—chloroform—methanol—benzene—28% NH_4OH, 65:30:10:6, and the mobile phase in the second direction—chloroform—methanol—acetic acid—acetone—benzene—water, 70:30:4:5:10:1. To determine the phosphorus content in phospholipids separated by TLC, the silica gel from the zones containing separated phospholipids was transferred with a micro spatula into the tubes; 0.05 mL of 72% perchloric acid was added to each and heated at 180–200 °C for 15–20 min, placing the tubes in a heated aluminum block so that the top of the tube served as an air cooler for perchloric acid vapors. After cooling, 0.45 mL of working reagent was added to the tubes: a mixture of 5.5 mL of universal molybdate reagent, 26 mL of 1N sulfuric acid, and 68.5 mL of distilled water. The reagent was used for one week. The mixture in the tube was thoroughly mixed using a shaker. The tubes were placed in boiling water bath for 15 min and then cooled; the absorbance value was measured at 815 nm. An aliquot of the solvent containing the lipid extract was taken as a blank sample [21]. The air temperature in the experimental area was recorded using a DS 1922L iButton thermograph (Dallas Semiconductor, USA).

2.3. Analyses of FAs in Animal Tissues

The samples (0.2–1.3 g) of intercostal muscle, subcutaneous adipose tissue, and liver were homogenized, and lipids were extracted with chloroform and methanol (2:1, *v/v*). Dry lipids were then supplemented with 1 mL of sodium methylate solution in methanol (8 g/L). The mixture was heated for 15 min at 90 °C. The tubes were cooled, supplemented with 1.3 mL of methanol: H_2SO_4 (97:3, *v/v*), and methylated for 10 min at 90 °C. The FAMEs were extracted from the mixture with 2 mL hexane and washed three times with 5 mL of saturated NaCl solution. The hexane extract containing FAMEs was dried by passing it through a layer of anhydrous Na_2SO_4, and then the layer of anhydrous Na_2SO_4 was washed with 6 mL of hexane. Hexane was evaporated on a rotary vacuum evaporator. FAMEs were resuspended in 0.1 to 0.3 mL hexane prior to chromatographic analysis.

Analysis of fatty acid methyl esters was conducted using a gas chromatograph with a mass spectrometric detector (Model 7000 QQQ, Agilent Technologies, USA), which was equipped with a 30 m capillary HP-FFAP column with the internal diameter of 0.25 mm. The conditions of the analysis were as follows: the velocity of the helium carrier gas was 1.2 mL/min; the temperature of the injection port was 250 °C; the temperature of the heater was programmed from 120 to 180 °C at a rate of 5 °C/min for 10 min isothermally, then to 220 °C with a rate of 3 °C/min for 5 min isothermally, and then to 230 °C at a rate of 10 °C/min for 20 min isothermally; the temperature of the chromatography/mass interface was 270 °C; the temperature of the ion source was 230 °C and that of the quadrupole was 180 °C; the ionization energy of the detector was 70 eV; and scanning was performed in the range of 45–500 atomic units with a rate of 0.5 sec/scan [24]. The data were analyzed and counted by the MassHunter Software (Agilent Technologies). The peaks of fatty acid methyl esters were identified by the mass spectra obtained. The content of fatty acids in the biomass was quantified based on the peak value of the internal standard, nonadecanoic acid (Sigma-Aldrich, USA), a certain amount of which was supplemented to the samples before the extraction of lipids.

2.4. Desaturase and Elongase Activity Indices

Desaturase and elongase activity indices were calculated using the product/precursor ratio of the percentages of individual FAs according to the following notation: 16:1n-7/16:0 = Δ9-desaturase, 18:1n-9/18:0 = Δ9-desaturase, 20:4n-6/20:3n-6 and 20:5n-3/20:4n-3 = Δ5-desaturase and 18:0/16:0 = elongase [25]. Additionally, we measured a conversion efficiency of 18:2n-6 to 20:4n-6 (20:4n-6/18:2n-6) and a conversion efficiency of 18:3n-3 to 20:5n-3 (20:5n-3/18:3n-3).

2.5. Statistical Analysis

The tables and figures show the averages of three to six biological replicates and their standard errors. Statistical processing of experimental data was carried out using the statistical analysis package in Microsoft Office Excel 2010 and STATISTICA-9 software (Stat Soft Inc., USA). The normality of the distribution of the data obtained was checked using the Kolmogorov–Smirnov one-sample test for normality D_{K-S}.

3. Results

The contents of total lipids in oat leaves of both the control and the treatment gradually increased as they grew and developed (Table 1). With the decrease in the average daily air temperature from 9 to 1 and −3 °C (periods of the first and second hardening phases), the content of total lipids in oat leaves increased by a factor of 1.2 compared with the control plants of the same stage of development (*t*-test Student's = 3.34) (Table 1).

Table 1. Contents of total lipids (TL, mg/g dry weight ± standard error) in the leaves of the annual cereal *Avena sativa* sown on May 31 and July 15, 2014 at different stages of development and growing at different temperatures.

Date	t, °C *		Stages of Development	TL, mg/g DW
	min	Average		
Control (sown on May 31, 2014)				
07.07	14	18	Stem elongation	99 ± 4
11.07	13	21	Stem elongation	114 ± 4
14.07	17	23	Ear emergence	127 ± 5
25.07	16	21	Dough development	129 ± 5
Treatment (sown on July 15, 2014)				
25.07	16	21	Germination	73 ± 3
11.09	1	9	Stem elongation, ear emergence	128 ± 5
25.09	−4	1	Dough development (cold hardening phase I)	154 ± 6
30.09	−7	−3	Dough development (cold hardening phase II)	155 ± 6

*—daily air temperature.

In the summertime (June–July), the perennial smooth brome grown without mowing demonstrated lower absolute content of total lipids at all stages of development, i.e., below 60 mg/g dry weight, compared to the aftergrass (Table 2).

Cool-season cereals growing after mowing, which were hardened by low positive temperatures, i.e., when the average daily air temperature reached 1 °C, showed the amount of total lipids 2.4 times higher (Student's *t*-test = 14.93) compared to the control plants in the same stage of development (Table 2).

Table 2. Total contents of lipids (TLs, mg/g dry weight ± standard error) in the leaves of the perennial cereal *Bromopsis inermis* growing at different temperatures, at different stages of development.

Date	t, °C *		Stages of Development	TLs, mg/g DW
	min	Average		
Control—grass without mowing			
06.06	3	12	Tillering	26 ± 2
16.06	12	16	Stem elongation	30 ± 2
11.07	13	21	Ear emergence	44 ± 2
25.07	16	21	Dough development	57 ± 3
Treatment—grass after mowing (July 15, 2014)			
25.07	16	21	Aftergrass	93 ± 3
18.08	7	16	Stem elongation	89 ± 3
11.09	1	9	Ear emergence	124 ± 5
25.09	−4	1	Dough development (cold hardening phase I)	134 ± 4
30.09	−7	−3	Dough development (cold hardening phase II)	137 ± 4

*—daily air temperature.

The following phospholipids (PLs) were found in the cereal plants: phosphatidylcholine (PC), phosphatidylinositol (PI), phosphatidylethanolamine (PE), phosphatidylglycerol (PG), phosphatidic acid (PA), and diphosphatidylglycerol (DPG). The dominant PLs were PC and PG (Figures 1 and 2).

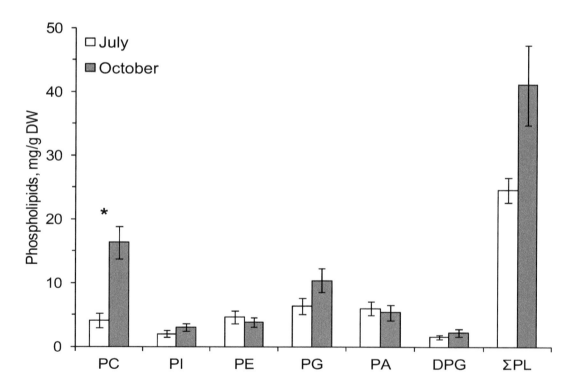

Figure 1. The contents (mg/g dry weight, standard error) of total phospholipids (ΣPL), phosphatidylcholine—PC, phosphatidylinositol—PI, phosphatidylethanolamine—PE, phosphatidylglycerol—PG, phosphatidic acid—PA and diphosphatidylglycerol—DPG in the leaves of *Avena sativa* on 25.07.2014 (July) and 3.10.2014 (October). *—Significant differences according to Student's *t*-test.

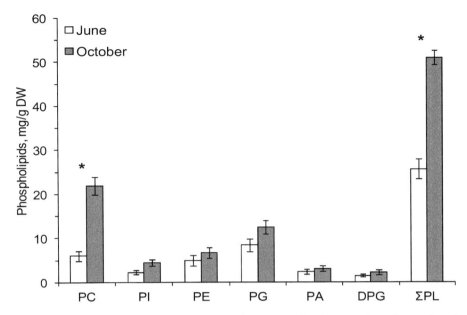

Figure 2. The contents (mg/g dry weight, standard error) of total phospholipids (ΣPL), phosphatidylcholine—PC, phosphatidylinositol—PI, phosphatidylethanolamine—PE, phosphatidylglycerol—PG, phosphatidic acid—PA and diphosphatidylglycerol—DPG in the leaves of *Bromopsis inermis* on 16.06.2014 (June) and 3.10.2014 (October). *—Significant differences according to Student's *t*-test.

In autumn, at the onset of low positive temperatures, the amount of PC increased in common oats by a factor of 4 and in the smooth brome by a factor of 3.7 compared with the content of these phospholipids in summer (Figures 1 and 2). The content of membrane PLs in the leaves of the smooth brome hardened by low positive temperatures significantly increased compared to summer values (Figure 2).

Sixteen fatty acids were identified in all samples of the smooth brome. The quantitatively and qualitatively prominent FAs are shown in Table 3. Among FAs, 18:3n-3, 16:0 and 18:2n-6 dominated, their total content reaching 85–90%. The total content of FAs in the leaves of the brome in the autumn period was significantly (1.8 times) higher than in the summer period (Table 3).

Table 3. Contents of fatty acids (mg/g of dry weight and % of total FA ± standard error) in the leaves of the perennial cereal *Bromopsis inermis* before mowing—07.07.2013 (July) and after mowing—25.09.2013 (September), and values of Student's *t*-test (*t*).

Fatty Acids	July *	September	*t*	July	September	*t*
	mg/g	mg/g		%	%	
14:0	0.10 ± 0.06	0.14 ± 0.01	0.62	0.6 ± 0.3	0.45 ± 0.02	−0.57
16:0	3.5 ± 0.4	4.9 ± 0.3	**2.83**	20 ± 1	15.8 ± 0.3	**−2.80**
18:0	0.5 ± 0.1	0.6 ± 0.1	1.12	2.7 ± 0.3	1.8 ± 0.1	−2.56
20:0	0.19 ± 0.01	0.25 ± 0.03	1.76	1.10 ± 0.04	0.8 ± 0.1	**−3.66**
22:0	0.23 ± 0.03	0.28 ± 0.02	1.57	1.3 ± 0.1	0.91 ± 0.04	**−3.33**
16:1n-9+n-7	0.1 ± 0.1	0.27 ± 0.03	**2.91**	0.3 ± 0.3	0.9 ± 0.2	1.60
16:1n-5	0.36 ± 0.03	0.6 ± 0.1	**4.90**	2.1 ± 0.1	2.02 ± 0.01	−0.43
18:1n-9	0.6 ± 0.3	0.6 ± 0.1	−0.28	3.6 ± 1.6	1.81 ± 0.04	−1.11
18:2n-6	2.0 ± 0.2	4.1 ± 0.4	**4.87**	11.5 ± 0.4	13.0 ± 0.3	**3.12**
18:3n-3	9.7 ± 1.2	19 ± 1	**5.37**	55 ± 1	61 ± 1	1.52
SFAs	4.7 ± 0.5	6.4 ± 0.5	2.54	27 ± 2	21 ± 1	**−2.79**
MUFAs	1.2 ± 0.3	1.6 ± 0.2	1.15	6.8 ± 1.7	5.2 ± 0.2	−0.90
PUFAs	12 ± 1	23 ± 2	**5.45**	67 ± 4	74.2 ± 0.4	2.07
ΣFAs	18 ± 2	31 ± 2	**4.93**	-	-	-

* the average air temperature in July = 10.9 °C, the average air temperature in September = −0.6°C; ΣFAs—total fatty acids, SFAs—saturated fatty acids, MUFAs—monounsaturated fatty acids, PUFAs—polyunsaturated fatty acids, bold font—significant differences according to Student's *t*-test.

The content of polyunsaturated fatty acids (PUFAs) in the smooth brome leaves significantly increased, while the content of total saturated fatty acids (SFAs) did not change with the decrease in air temperatures (Table 3). The content of 16:0, 16:1 isomers, 18:2n-6, and 18:3n-3 in the leaves of brome in autumn was significantly higher than in summer (Table 3).

The percentage of SFAs in brome leaves was lower in September compared with July. The decrease in SFAs was due to a decrease in the percentage of 16:0, 20:0, and 22:0 (Table 3). The percentage of PUFAs did not change with the decrease in air temperatures while the percentage of 18:2n-6 significantly increased (Table 3).

Fifty-three FAs were identified in the samples of liver, muscle and subcutaneous adipose tissues of the Yakutian horses. The percentages of important and quantitatively significant FAs are shown in Figure 3. The percentage of SFAs in the liver of the animals was significantly higher than in the muscle and adipose tissues (Figure 3a). Among the SFAs in the liver, 18:0 dominated. Its percentage was about 4 and 6 times higher than that in the muscle and adipose tissues, respectively. Shorter-chain SFAs, such as 14:0 and 16:0, dominated in the muscle and adipose tissues, and their percentages were significantly higher than in the liver (Figure 3a). The percentages of monounsaturated FAs (MUFAs) in the muscle and adipose tissues of the horses were twice higher than in the liver (Figure 3b). Among MUFAs, 18:1n-9 dominated in all types of tissues. Nevertheless, its percentage in the liver was significantly lower than in the other tissues (Figure 3b). The percentage of PUFAs was significantly higher in the animal liver than in the muscle and adipose tissues (Figure 3c). Among PUFAs in the liver, omega-6 PUFA, namely 18:2n-6, dominated. Its percentage was more than twice higher than in the muscle and adipose tissues. In contrast, the muscle and adipose tissues were dominated by omega-3 PUFA, namely 18:3n-3. Its percentage was more than twice as high as the percentage of this FA in the liver (Figure 3c). In total, 70% of all FAs in the muscle and adipose tissues were represented by 18:1n-9, 16:0, 18:3n-3, and 18:2n-6 and in the liver by 18:2n-6, 18:0, 16:0, and 18:1n-9 (Figure 3). No trans-FAs were found in the FA tissue of the horses, and the percentage of branched FAs was less than 1% of the total FAs. The percentages of many FAs were similar in the muscle and adipose tissues of the horses. In adipose tissues, however, the percentages of 18:3n-3 and short-chain SFA (12:0 and 14:0) were significantly higher than in muscles, but almost all long-chain PUFAs, including physiologically important arachidonic (ARA, 20:4n-6), eicosapentaenoic (EPA, 20:5n-3) and docosahexaenoic (DHA, 22:6n-3) acids, were absent (Figure 3). The percentages of EPA and DHA in the liver and muscle tissues were insignificant and ranged from 0.1% to 0.3% of the total FAs.

The contents of physiologically important EPA and DHA in the muscle tissue and liver of the horses were similar (Table 4). The n-6/n-3 ratio in the muscle tissue was about 7 times lower than in the liver, but did not differ significantly from that in the adipose tissue. The total content of FAs in the adipose tissue was 30 times higher than that in the muscle tissue and in the liver (Table 4).

Table 4. Content of EPA+DHA and total fatty acids (mg/100g and mg/g of wet weight, respectively) and the ratio of total omega-6 and omega-3 PUFAs in the muscle, liver, and subcutaneous adipose tissue of Yakutian horses. Means in lines labeled with the same letters are not significantly different at $p < 0.05$ after Tukey's HSD *post hoc* test (normal distribution, standard errors are given) or Kruskal–Wallis test with multiple comparisons of mean ranks (non-normal distribution standard errors are omitted).

	Muscle	Liver	SCfat
EPA+DHA, mg/100 g ww	11 ± 1 [A]	11 ±1 [A]	-
Total FA, mg/g ww	31 [A]	28 [A]	854 [B]
n-6/n-3	1.1 ± 0.2 [A]	5.5 ± 1.2 [B]	0.44 ± 0.03 [A]

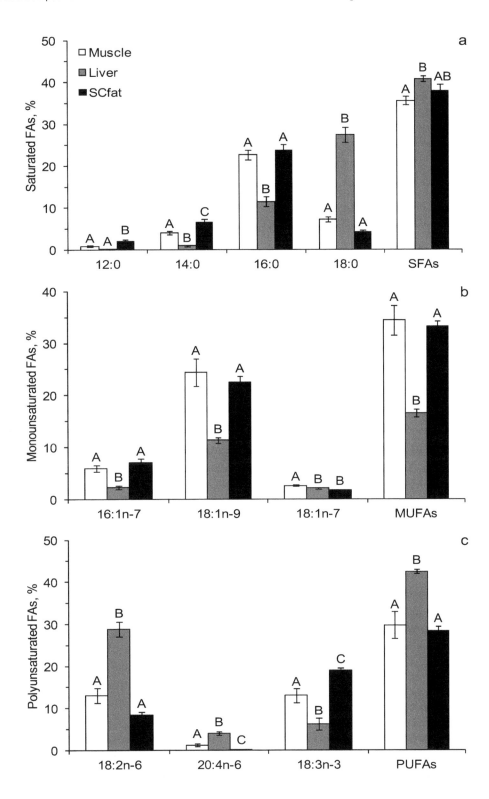

Figure 3. Contents of the prominent saturated fatty acids (**a**), monounsaturated fatty acids (**b**) and polyunsaturated fatty acids (**c**) (% of total FAs, standard error) in liver, muscle and subcutaneous adipose tissues of the Yakutian horses. Means for the same FAs labeled with the same letters are not significantly different at $p < 0.05$ after Tukey's HSD *post hoc* test.

The desaturase and elongase activity, estimated by an indirect method (by product/precursor ratio), were significantly different between the adipose tissue and the liver (Table 5).

Table 5. Calculated desaturase and elongase activity indices.

Product/Precursor Ratio	Liver	SCfat	t
18:0/16:0	2.5 ± 0.3	0.19 ± 0.02	**7.48**
16:1n-7/16:0	0.20 ± 0.02	0.3 ± 0.1	**−3.85**
18:1n-9/18:0	0.42 ± 0.04	5.3 ± 0.4	**−12.81**
20:4n-6/20:3n-6	7.9 ± 0.7	-	-
20:5n-3/20:4n-3	8.8 ± 0.8	0.5 ± 0.1	**10.19**
20:5n-3/18:3n-3	0.06 ± 0.03	0.0006 ± 0.0002	**2.34**
20:4n-6/18:2n-6	0.14 ± 0.01	0.003 ± 0.001	**11.80**

Bold font—significant differences according to Student's *t*-test.

The conversion efficiencies of 16:0 and 18:0 to 16:1n-7 and 18:1n-9, respectively, were higher in the adipose tissue and the conversion efficiencies of 16:0 to 18:0, 20:4n-3 to 20:5n-3, 18:3n-3 to 20:5n-3, and 18:2n-6 to 20:4n-6 were higher in the liver (Table 5). The conversion efficiency of 18:2n-6 to 20:4n-6 was higher than the conversion efficiency of 18:3n-3 to 20:5n-3 both in the liver and in the adipose tissue (Student's *t*-test = 2.75 and t = 2.66, respectively). The conversion efficiency of 20:4n-3 to 20:5n-3 and 20:3n-6 to 20:4n-6 in the liver did not differ significantly (Student's *t*-test = 1.09).

4. Discussion

Lower ambient temperatures significantly affect the 'liquidity' of plant membranes, reducing their fluidity. This leads to the increased expression of the genes responsible for FA desaturation [26]. The increased fraction of unsaturated FAs in plants with the temperature decrease stabilizes membrane fluidity and restores physiological activities of the associated enzyme and electron transport systems, photosynthesis in particular [27–29]. Affected by low temperatures, the genes that encode the synthesis of desaturases involved in the formation of 18:2n-6 and 18:3n-3 are activated in the plants [26,30]. The increase in total lipids, phospholipids, and total FAs that we detected in the cereals showed that these substances along with sugars, proteins, antioxidants, and carotenoids [1,31] are involved in the cold adaptation of cool-season plants in the cryolithozone of Central Yakutia. In the same way as *B. inermis* studied in our work, other herbaceous plants (*Avena sativa*, *Elytrigia repens*, *Equisetum variegatum*, and *Equisetum scirpoides*) accumulated significantly more FAs in their vegetative organs during the period of winter cold adaptation than in summer [17,32–34].

During cold adaptation of plants, the contents of phospholipids and polyunsaturated fatty acids increase in their tissues [35–37]. However, in contrast to most published data, we found a significant increase in the content of phosphatidylcholine but not in the other phospholipids. This finding probably shows the key role of phosphatidylcholine in temperature adaptation of the cereals.

Along with *Bromopsis inermis* and *Avena sativa*, the ability to cryopreserve green mass was found in many other plants in Central Yakutia, for example, cereals—*Arctophila fulva*, *Deschampsia borealis*, *Puccinellia jacutica*, *Poa alpigena*, hydrophytic sedges—*Carex rhynchophysa*, *C. atherodes*, *C. vesicata*, *C. enervis*, and most cotton grasses—*Eriophorum scheuchzeri*, *E. vaginatum*, *E. russeolumsubsp. leiocarpum*, *E. angustifolium* [6]. In the pre-winter period of fat accumulation, herbivores actively consume cool-season and winter-green parts of these fodder plants with the high contents of nutrients [2].

The main consumer of green cryo-fodder in plant ecosystems of cryolithozone in Central and North-Eastern Yakutia is the Yakutian horse. In autumn, from August to the beginning of October, the horses feed on green cryo-fodder. In favorable years, the accumulation of fat on green cryo-fodder by the Yakutian horses lasts up to mid-November [38,39].

The tissues of the Yakutian horses and their fodder were rich in two PUFAs, namely 18:3n-3 and 18:2n-6. These FAs are essential for the majority of animals [40–42]. Vertebrates can synthesize physiologically important long-chain PUFAs—20:5n-3, 22:6n-3, and 20:4n-6—from their dietary precursors 18:3n-3 and 18:2n-6, respectively, but the rate of synthesis is generally ineffective [40,43]. Suagee et al. found that mesenteric adipose tissue in horses had a high lipogenic capacity followed by subcutaneous adipose tissue and then liver [44]. Very low percentages of 20:5n-3 and 22:6n-3 (the average value = 0.4% of total FAs) as well as a low 20:5n-3/18:3n-3 ratio in the tissues probably indicated a low conversion efficiency of omega-3 PUFAs in the Yakutian horses. The conversion of omega-6 PUFAs seemed to be more efficient than conversion of omega-3 PUFAs, at least in the liver. The literature data and our results suggest that dietary sources of 20:5n-3, 22:6n-3 and 20:4n-6 were absent from the diet of the Yakutian horses [17]. Thus, we suppose that these long-chain PUFAs were synthesized in horses' tissues. According to our data, the efficiency of elongation of 16:0 to 18:0 was significantly higher in the liver while the conversion efficiency of SFAs to MUFAs was higher in the subcutaneous adipose tissue. Similar trends in the conversion efficiency of SFAs to MUFAs and elongation of 16:0 to 18:0 in liver and subcutaneous adipose tissue were reported by Adolph et al. for Warmblood horses [25].

Different contents of C18, C20, and C22 PUFAs in the tissues of the horses may indicate different functions of these PUFAs. Unlike Warmblood horses, the subcutaneous adipose tissue of the Yakutian horses was rich in PUFAs, especially in the omega-3 family [25]. Mordovskaya et al. and Slobodchikova et al. also noted enrichment of the Yakutian horses' adipose tissue with 18:3n-3 [45,46]. High percentages of 18:3n-3 in the subcutaneous fat of the Yakutian horses may increase fluidity (liquidity) of adipose tissue during the winter period of extremely low temperatures. This may be a reason for high mitochondrial activity in adipocytes, which increases energy production at low ambient temperatures. The beneficial effects of omega-3 PUFAs on the thermogenic function of adipocytes have recently been demonstrated [47–49]. Thus, we suppose that 18:3n-3 served as an energy-related component in the horses. By contrast, omega-6 PUFAs, namely 18:2n-6 and 20:4n-6, which were accumulated in liver and muscle tissue, likely served as important structural components or precursors of lipid mediators. Similar results were reported in a study of the FA composition of different lipid classes in Iberian horses [50]. In the muscle tissue, 18:2n-6 and 20:4n-6 accumulated in polar lipids, apparently performing a building function, and 18:3n-3, on the contrary, accumulated in neutral lipids, performing an energy function [50].

The PUFA content in the horse muscle and adipose tissues varies greatly depending on the diet, breed, and age of the animals [51–53]. For example, the content of 18:2n-6 in horse muscles varied from 12% to 32%, and the content of 18:3n-3 varied from 0.43% to 23.9% [53]. Along with individual fatty acids, the total contents of SFAs, MUFAs and PUFAs can also vary greatly in the horse muscle tissue: 34.2–47.8%, 16.4–50.2% and 15.6–46%, respectively [53]. The muscle tissue of the horses we studied contained equal proportions of SFAs, MUFAs, and PUFAs, which corresponded to the minimum SFA values and the average MUFA and PUFA values available in the literature. Similar to the Yakutian horses, horses that were fed on the native grass pasture had the same percentages of SFAs, MUFAs, and PUFAs, as well as high percentages of 18:3n-3, in subcutaneous adipose tissue [54]. The total FA content of the subcutaneous adipose tissue of the Yakutian horse corresponded to the high values known for horses, varying between 457 and 904 mg/g wet weight [53,54]. Obviously, nutrition has a significant effect on the variability of FA percentages in horses. Horses eating fresh plant food, but living in a mild climate, had similar contents and distribution of FAs, including 18:2n-6 and 18:3n-3, to those in the Yakutian horses. High contents of lipids and FAs such as 18:3n-3 and 18:2n-6 in green cryo-fodder probably help the Yakutian horses successfully survive the extreme temperatures of Central Yakutia. However, the results obtained only indirectly indicate this and do not allow us to clarify the subject.

In contrast to many farm animals, horses are able to efficiently assimilate PUFAs from plant food owing to the structure of their gastrointestinal tract, activity of microorganisms, and the presence of specific pancreatic lipases related with protein 2 (PLRP2) [53,55]. Thus, horse meat is considered as a useful dietary product, i.e., a source of essential PUFAs, namely 18:2n-6 and 18:3n-3 [56–58], and can be potentially enriched with long-chain omega-3 PUFAs, 20:5n-3 and 22:6n-3. However, our data suggest that meat, subcutaneous fat, and liver of the Yakutian horses are not rich in 20:5n-3 and 22:6n-3 if their food does not contain these PUFAs. This may indicate limitation in PUFA synthesis in the horses' tissues. The contents of 20:5n-3 and 22:6n-3 (% and mg/g wet weight) in the liver of the Yakutian horses and other horse breeds were significantly lower than in the liver of cows, pigs, and chickens [59]. The contents of 20:5n-3 and 22:6n-3 (mg/g wet weight) in the meat of horses were the same as in beef and higher than in pork [53,60]. In general, because of the high content of 18:3n-3 and the optimal ratio of n-6 to n-3 PUFAs, the Yakutian horse meat is a more valuable and health food product compared to beef, pork, and chicken, which is consistent with the data of many authors [45,60–62].

5. Conclusions

The cereal plants studied (*B. inermis* and *A. sativa*) accumulate lipids, phosphatidylcholine and fatty acids, in particular, during the period of natural cold hardening in extremely cold climates of the permafrost zone. Cereals enriched with nutrients are the basis for the Yakutian horse feeding during pre-winter fat accumulation. The muscle and adipose tissues and liver of the horses contained high percentages of 18:2n-6 and 18:3n-3, which were abundant in the cereals studied in this work. A likely reason for the diverse distribution of these FAs in tissues is that these FAs perform different functions in the animals. 18:2n-6 is probably used as a precursor in the synthesis of physiologically valuable 20:4n-6, while 18:3n-3 mainly performs an energy-related function. Such a high content of 18:2n-6 and 18:3n-3 in the tissues of horses of the Yakutian breed apparently helps animals successfully survive the extreme temperatures of Central Yakutia, although more research is needed. Additionally, the Yakutian horse meat has proved to be a valuable dietary product due to its low n-6/n-3 ratio.

Author Contributions: Conceptualization, supervision, K.A.P.; Data analysis, O.N.M. and K.A.P; Investigation, K.A.P., O.N.M., V.V.N. and K.N.S.; Methodology, L.V.D., O.N.M. and V.V.N.; Resources, K.A.P. and O.N.M.; Writing—review & editing, O.N.M. and K.A.P. All authors have read and agreed to the published version of the manuscript.

Acknowledgments: We thank the anonymous reviewers for their comments. We are also grateful to Elena Krasova for linguistic check and improvements.

References

1. Petrov, K.A.; Sofronova, V.E.; Chepalov, V.A.; Perk, A.A.; Maksimov, T.K. Sezonnye izmeniniia soderzhaniia fotosinteticheskikh pigmentov u mnogoletnikh travianistykh rastenii kriolitozony [Seasonal changes in the content of photosynthetic pigments in perennial grasses of the permafrost zone]. *Fiz.rast. [Russ. J. Plant Physiol.]* **2010**, *57*, 192–199.

2. Petrov, K.A. *Kriorezistentnost' Rastenii: Ekologo-fiziologicheskie I Biokhemicheskie Aspekty [Cryoresistance of Plants: Ecological, Physiological and Biochemical Aspects]*; SB RAS Publishing House: Novosibirsk, Russia, 2016; 276p.

3. Chirkova, T.V. *Fiziologicheskie Osnovy Ustoichivosti Rastenii [Physiological Basis of Plant Resistance]*; Publishing House of Saint-Petersburg State University: Saint-Petersburg, Russia, 2002; 244p.

4. Zakharova, V.I.; Kuznetsova, L.V.; Ivanova, E.I.; Vasilyeva-Kralina, I.I.; Gabyshev, V.A.; Egorova, A.A.; Zolotov, V.I.; Ivanova, A.P.; Ignatov, M.S.; Ignatova, E.A.; et al. *Raznoobrazie Rastitel'nogo Mira Iakutii [The Diversity of Plant Life in Yakutia]*; SB RAS Publishing House: Novosibirsk, Russia, 2005; 328p.

5. Aleksandrova, V.D.; Andreev, V.N.; Vakhtina, T.V.; Dydina, R.A.; Karev, G.I.; Petrovskii, V.V.; Shamurin, V.F. *Kormovaya Kharakteristika Rastenii Krainego Severa [Fodder Characteristics of the Plants of the Far North]*; Nauka Publishing House: Moscow, Russia, 1964; 484p.

6. Andreev, V.N.; Belyaeva, N.V.; Galaktionova, T.F.; Govorov, P.M.; Egorov, A.D.; Kurilyuk, T.T.; Myarikyanov, M.I.; Permyakova, A.A.; Perfilieva, V.I.; Petrov, A.M.; et al. *Tebenevochnye Pastbishcha Vostoka Iakutii [Winter-Grazing Pastures of the North-East of Yakutia]*; Yakutsk Book Publishing House: Yakutsk, Russia, 1974; 246p.

7. Egorov, A.D.; Potapov, V.Y.; Romanov, P.A. *Zonal'no-biokhimicheskie Osobennosti Kormovykh Rastenii iakutii i Nekotorye Problemy Razvitiia Zhivotnovodstva [Zonal-biochemical Characteristics of Fodder Plants in Yakutia and Some Problems in the Development of Animal Husbandry]*; Yakutsk Book Publishing House: Yakutsk, Russia, 1962; 51p.

8. Potapov, V.Y. *Uglevody i Lignin v Kormovykh Travakh Iakutii [Carbohydrates and Lignin in the Fodder Grasses of Yakutia]*; Nauka Publishing House: Moscow, Russia, 1967; 173p.

9. Pakendorf, B.; Novgorodov, I.N.; Osakovskij, V.L.; Danilova, A.P.; Protod'jakonov, A.P.; Stoneking, M. Investigating the effects of prehistoric migrations in Siberia: Genetic variation and the origins of Yakuts. *Hum. Genet.* **2006**, *120*, 334–353. [CrossRef] [PubMed]

10. Crubézy, E.; Amory, S.; Keyser, C.; Bouakaze, C.; Bodner, V.; Gibert, V.; Röck, F.; Parson, W.; Alexeev, A.; Ludes, B. Human evolution in Siberia: From frozen bodies to ancient DNA. *BMC Evol. Biol.* **2010**, *10*, 1–25. [CrossRef] [PubMed]

11. Keyser, C.; Hollard, C.; Gonzalez, A.; Fausser, J.; Rivals, E.; Alexeev, A.; Riberon, A.; Crubézy, E.; Ludes, B. The ancient Yakuts: A population genetic enigma. *Philos. Trans. R. SocLond. B Biol. Sci.* **2015**, *370*, 20130385. [CrossRef] [PubMed]

12. Guryev, I.P. *K voprosu o Proiskhozhdenii Iakutskoi Loshadi. Teriologicheskie Issledovaniia v Iakutii [On the Issue of the Yakutian Horse Origin. Theriological Research in Yakutia]*; Publishing House of Yakutsk Branch of SB AS USSR: Yakutsk, Russia, 1983; pp. 50–57.

13. Knyazev, S.P. Analiz geneticheskikh markerov aborigennykh iakutskikh loshadei v sviazi s filogeniei i domestikatsiei loshadei [Analysis of genetic markers of indigenous Yakutian horses in connection with the phylogeny and domestication of horses]. In *Proceedings of the International Conference "Animal Molecular and Genetic Markers"*; Agrarnaya Nauka: Kiev, Ukraine, 1996; pp. 31–32.

14. Tikhonov, V.N. Populiatsionno-geneticheskie parametry aborigennykh iakutskikh loshadei v sviazi s fiologeniei sovremennykh porod domashnie loshadi *Equus caballus* L. [Population and genetic parameters of indigenous Yakutian horses in connection with the physiology of modern domestic horse *Equus caballus* L.]. *Genetika [Genetics]* **1998**, *34*, 796–809. [PubMed]

15. Librado, P.; Sarkissian, C.; Ermini, L.; Schubert, M.; Jonsson, H.; Albrechtsen, A.; Fumagalli, M.; Yang, M.; Gamba, C.; Seguin-Orlando, A.; et al. Tracking the origins of Yakutian horses and the genetic basis for their fast adaptation to subarctic environments. *Proc. Natl. Acad. Sci. USA* **2015**, *112*, E6889–E6897. [CrossRef]

16. Petrov, K.A.; Chepalov, V.A.; Sofronova, V.E.; Ilyin, A.N.; Ivanov, R.V. Ekologo-fiziologicheskie i biokhimicheskie osnovy formirovaniia zelenogo kriokorma v Iakutii [Environmental, physiological and biochemical basis of green cryo-fodder formation in Yakutia]. *Sel'skokhoziaistvennaia Bioloiia [Agric. Biol.]* **2017**, *52*, 1129–1138.

17. Petrov, K.A.; Dudareva, L.V.; Nokhsorov, V.V.; Perk, A.A.; Chepalov, V.A.; Sophronova, V.E.; Voinikov, V.K.; Zulfugarov, I.S.; Lee, C.-H. The role of plant fatty acids in regulation of the adaptation of organisms to the cold climate in cryolithic zone of Yakutia. *J. Life Sci.* **2016**, *26*, 519–530. [CrossRef]

18. Alekseev, N.D.; Neustroev, M.P.; Ivanov, R.V. *Biologicheskie Osnovy Povysheniya Produktivnosti Loshadej [The Biological Basis for Increasing Horse Productivity]*; Gnu Yaniiskh so Raskhn: Yakutsk, Russia, 2006; 280p.

19. Bligh, E.C.; Dyer, W.J. A rapid method of total lipid extraction and purification. *Can. J. Biochem. Physiol.* **1959**, *37*, 911–917. [CrossRef]

20. Christie, W.W. Preparation of ester derivatives of fatty acids for chromatographic analysis. *Adv. Lipid Methodol.* **1993**, *2*, 69–111.

21. Vaskovsky, V.E.; Kostetsky, E.Y.; Vasendin, J.M. Universal reagent for determination of phosphorea in lipids. *J. Chromatogr.* **1975**, *114*, 129–141. [CrossRef]

22. Wagner, H.; Horhammer, L.; Walff, P. Dunnschtchromatographic von Phosphatiden und Glykolipiden. *Biochem. Z.* **1961**, *334*, 129–141.

23. Kates, M. *Techniques of Lipidology: Isolation, Analysis and Identification of Lipids*; American Elsevier: Amsterdam, The Netherlands, 1972; pp. 269–610.

24. Makhutova, O.N.; Borisova, E.V.; Shulepina, S.P.; Kolmakova, A.A.; Sushchik, N.N. Fatty acid composition and content in chironomid species at various life stages dominating in a saline Siberian lake. *Contemp. Probl. Ecol.* **2017**, *10*, 230–239. [CrossRef]

25. Adolph, S.; Schedlbauer, C.; Blaue, D.; Schöniger, A.; Gittel, C.; Brehm, W.; Fuhrmann, H.; Vervuert, I. Lipid classes in adipose tissues and liver differ between Shetland ponies and Warmblood horses. *PLoS ONE* **2019**, *14*, e0207568. [CrossRef]

26. Murata, N.; Los, D.A. Membrane fluidity and temperature perception. *Plant Physiol.* **1997**, *115*, 875–879. [CrossRef]

27. Tocher, D.R.; Leaver, M.J.; Hodson, P.A. Recent advances in the biochemistry and molecular biology of fatty

acyl desaturase. *Prog. Lipid Res.* **1998**, *37*, 73–117. [CrossRef]

28. Murata, N.; Los, D.A. Genome-wide analysis of gene expression characterizes histidine kinase Hik33 as an important component of the cold-signal transduction in cyanobacteria. *Physiol. Plant* **2006**, *57*, 235–247.

29. Guschina, I.A.; Harwood, J.L. Algal lipids and effect of the environment on their biochemistry. In *Lipids in Aquatic Ecosystems*; Arts, M.T., Kainz, M., Brett, M.T., Eds.; Springer: New York, NY, USA, 2009; pp. 1–24.

30. Trunova, T.I. *Rastenie i Nizkotemperaturnyi Stress [The Plant and the Stress Caused by Low Temperatures]*; Nauka: Moscow, Russia, 2007; pp. 1–54.

31. Sofronova, V.E.; Chepalov, V.A.; Petrov, K.A.; Dymova, O.V.; Golovko, T.K. Fond zelenykh i zheltykh pigmentov iarovogo ovsa, kul'tiviruemogo dlia polucheniia kriokorma v usloviiakh Tsentral'noi Iakutii [Green and yellow pigments of spring oats cultivated as cryo-fodder in the conditions of Central Yakutia]. *Agrarny Vestnik Urala [Ural Agric. Bull.]* **2019**, *4*, 72–77.

32. Petrov, K.A.; Perk, A.A.; Chepalov, V.A.; Chapter, I.V. Linoleic and Other Fatty Acids, Cryoresistance, and Fodder Value of Yakutian Plants. In *Linoleic Acids. Sources, Biochemical Properties and Health Effects*; Onakpoya, I., Ed.; Nova Science Publishers: New York, NY, USA, 2012; pp. 83–96.

33. Dudareva, L.V.; Rudikovskaya, E.G.; Nokhsorov, V.V.; Petrov, K.A. Fatty- acid profiles of aerial parts of three horsetail species growing in Central and Northern Yakutia. *Chem. Nat. Compd.* **2015**, *51*, 220–223. [CrossRef]

34. Nokhsorov, V.V. Adaptivnye izmeneniia sostava i soderzhaniia lipidov rastenii kriolitozony Iakutii pri gipotermii [Adaptive changes in the composition and content of lipids in the plants of the cryolithic zone in Yakutia in the hypothermal conditions]. Ph.D. Thesis, Siberian Institute of Physiology and Biochemistry of Plants Sb RAS, Irkutsk, Russia, 2017.

35. Welti, R.; Li, W.; Li, M.; Sang, Y.; Biesiada, H.; Zhou, H.-E.; Rajashekar, C.B.; Williams, T.D.; Wang, X. Profiling membrane lipids in plant stress responses—Role of phospholipase Dα in freezing-induced lipid changes in Arabidopsis. *J. Biol. Chem.* **2002**, *277*, 31994–32002. [CrossRef]

36. Wang, S.Y.; Lin, H.-S. Effect of plant growth temperature on membrane lipids in strawberry (Fragaria × ananassa Duch.). *Sci. Hortic.* **2006**, *108*, 35–42. [CrossRef]

37. Vereshchagin, A.G. *Lipidy v Zhizni Rastenii [Lipids in Plant Life]*; Nauka: Moscow, Russia, 2007; pp. 1–78.

38. Gabyshev, M.F. *The Yakut Horse*; Yakutsk Book Publishers: Yakutsk, Russia, 1957; p. 238.

39. Aleekseev, N.D. Novoe o proiskhozhdenii loshadei iakutskoi porody (biologicheskie aspekty [New in the origin of the Yakutian horses (biological aspects)]. *Nauka i Obrazovanie [Sci. Educ.]* **2005**, *2*, 114–118.

40. Gladyshev, M.I.; Sushchik, N.N.; Makhutova, O.N. Production of EPA and DHA in aquatic ecosystems and their transfer to the land. *Prostaglandins Other Lipid Mediat.* **2013**, *107*, 117–126. [CrossRef] [PubMed]

41. Twining, C.W.; Brenna, J.T.; Hairston, N.G., Jr.; Flecker, A.S. Highly unsaturated fatty acids in nature: What we know and what we need to learn. *Oikos* **2016**, *125*, 749–760. [CrossRef]

42. Malcicka, M.; Visser, B.; Ellers, J. An evolutionary perspective on linoleic acid synthesis in animals. *Evol. Biol.* **2018**, *45*, 15–26. [CrossRef]

43. Tocher, D.R.; Dick, J.R.; MacGlaughlin, P.; Bell, J.G. Effect of diets enriched in Δ6 desaturated fatty acids (18:3n-6 and18:4n-3), on growth, fatty acid composition and highly unsaturated fatty acid synthesis in two populations of Arctic charr (*Salvelinus alpines* L.). *Comp. Biochem. Physiol. Part B* **2006**, *144*, 245–253. [CrossRef]

44. Suagee, J.K.; Corl, B.A.; Crisman, M.V.; Wearn, J.G.; McCutcheon, L.J.; Geor, R.J. De novo fatty acid synthesis and NADPH generation in equine adipose and liver tissue. *Comp. Biochem. Physiol.* **2010**, *155*, 322–326. [CrossRef]

45. Mordovskaya, V.I.; Krivoshapkin, V.G.; Pogozheva, A.V.; Baiko, V.G. Fatty acid composition of adipose tissue lipids in horses of Yakut breed. *Vopr. Pitan.* **2005**, *74*, 17–23.

46. Slobodchikova, M.N.; Ivanov, R.V.; Stepanov, K.M.; Pustovoy, V.F.; Osipov, V.G.; Mironov, S.M. Lipid fatty acid composition of the fat tissue of the Yakut horse. *HorseBreed. Eques. Sport.* **2011**, *6*, 28–30.

47. Ghandour, R.A.; Colson, C.; Giroud, M.; Maurer, S.; Rekima, S.; Ailhaud, G.; Klingenspor, M.; Amri, E.-Z.; Pisani, D.F. Impact of dietary ω3 polyunsaturated fatty acid supplementation on brown and brite adipocyte function. *J. Lipid Res.* **2018**, *59*, 452–461. [CrossRef]

48. Colson, C.; Ghandour, R.A.; Dufies, O.; Rekima, S.; Loubat, A.; Munro, P.; Boyer, L.; Pisani, D.F. Diet supplementation in ω3 polyunsaturated fatty acid favors an anti-inflammatory basal environment in mouse adipose tissue. *Nutrients* **2019**, *11*, 438. [CrossRef] [PubMed]

49. Pisani, D.F.; Ailhaud, G. Involvement of polyunsaturated fatty acids in the control of energy storage and expenditure. *Oilseeds Fats Crop. Lipids* **2019**, *26*, 37. [CrossRef]

50. Belaunzaran, X.; Lavín, P.; Mantecón, A.R.; Kramer, J.K.G.; Aldai, N. Effect of slaughter age and feeding system on the neutral and polar lipid composition of horse meat. *Animal* **2018**, *12*, 417–425. [CrossRef] [PubMed]

51. Tonial, I.B.; Aguiar, A.C.; Oliveira, C.C.; Bonnafé, E.G.; Visentainer, J.V.; de Souza, N.E. Fatty acid and cholesterol content, chemical composition and sensory evaluation of horsemeat. *S. Afr. J. Anim. Sci.* **2009**, *39*, 328–332. [CrossRef]

52. Lorenzo, J.M.; Sarriés, M.V.; Tateo, A.; Polidori, P.; Franco, D.; Lanza, M. Carcass characteristics, meat quality and nutritional value of horsemeat: A review. *Meat Sci.* **2014**, *96*, 1478–1488. [CrossRef]

53. Belaunzaran, X.; Bessa, R.J.B.; Lavín, P.; Mantecón, A.R.; Kramer, J.K.G.; Aldai, N. Horse-meat for human consumption—Current research and future opportunities. *Meat Sci.* **2015**, *108*, 74–81. [CrossRef]

54. Ferjak, E.N.; Cavinder, C.A.; Sukumaran, A.T.; Burnett, D.D.; Lemley, C.O.; Dinh, T.T.N. Fatty acid composition of mesenteric, cardiac, abdominal, intermuscular, and subcutaneous adipose tissues from horses of three body condition scores. *Livest. Sci.* **2019**, *223*, 116–123. [CrossRef]

55. De Caro, J.; Eydoux, C.; Chérif, S.; Lebrun, R.; Gargouri, Y.; Carrière, F.; De Caro, A. Occurrence of pancreatic lipase-related protein-2 in various species and its relationship with herbivore diet. *Comp. Biochem. Physiol. Part B* **2008**, *150*, 1–9. [CrossRef]

56. Juárez, M.; Polvillo, O.; Gómez, M.D.; Alcalde, M.J.; Romero, F.; Valera, M. Breed effect on carcass and meat quality of foals slaughtered at 24 months of age. *Meat Sci.* **2009**, *83*, 224–228. [CrossRef]

57. Guil-Guerrero, J.L.; Rincón-Cervera, M.A.; Venegas-Venegas, C.E.; Ramos-Bueno, R.P.; Suárez, M.D. Highly bioavailable α-linolenic acid from the subcutaneous fat of the Palaeolithic Relict "Galician horse". *Int. Food Res. J.* **2013**, *20*, 3249–3258.

58. Belaunzaran, X.; Lavín, P.; Barron, L.J.R.; Mantecon, A.R.; Kramer, J.K.G.; Aldai, N. An assessment of the fatty acid composition of horse-meat available at the retail level in northern Spain. *Meat Sci.* **2017**, *124*, 39–47. [CrossRef] [PubMed]

59. Gladyshev, M.I.; Makhutova, O.N.; Gubanenko, G.A.; Rechkina, E.A.; Kalachova, G.S.; Sushchik, N.N. Livers of terrestrial production animals as a source of longchain polyunsaturated fatty acids for humans: An alternative to fish? *Eur. J. Lipid Sci. Technol.* **2015**, *117*, 1417–1421. [CrossRef]

60. Del Bò, C.; Simonetti, P.; Gardana, C.; Riso, P.; Lucchini, G.; Ciappellano, S. Horse meat consumption affects iron status, lipid profile and fatty acid composition of red blood cells in healthy volunteers. *Int. J. Food Sci. Nutr.* **2013**, *64*, 147–154. [PubMed]

61. Lee, C.-E.; Seong, P.-N.; Oh, W.-Y.; Ko, M.-S.; Kim, K.-I.; Jeong, J.-H. Nutritional characteristics of horsemeat in comparison with those of beef and pork. *Nutr. Res. Pract.* **2007**, *1*, 70–73. [CrossRef]

62. Lorenzo, J.M.; Munekata, P.E.S.; Campagnol, P.C.B.; Zhu, Z.; Alpas, H.; Barba, F.J.; Tomasevic, I. Technological aspects of horse meat products—A review. *Food Res. Int.* **2017**, *102*, 176–183. [CrossRef]

13

Fatty Acids of Marine Mollusks: Impact of Diet, Bacterial Symbiosis and Biosynthetic Potential

Natalia V. Zhukova [1,2]

[1] National Scientific Center of Marine Biology, Far East Branch of the Russian Academy of Sciences, Vladivostok 690041, Russia; nzhukova35@list.ru

[2] School of Biomedicine, Far Eastern Federal University, Vladivostok 690950, Russia

Abstract: The n-3 and n-6 polyunsaturated fatty acid (PUFA) families are essential for important physiological processes. Their major source are marine ecosystems. The fatty acids (FAs) from phytoplankton, which are the primary producer of organic matter and PUFAs, are transferred into consumers via food webs. Mollusk FAs have attracted the attention of researchers that has been driven by their critical roles in aquatic ecology and their importance as sources of essential PUFAs. The main objective of this review is to focus on the most important factors and causes determining the biodiversity of the mollusk FAs, with an emphasis on the key relationship of these FAs with the food spectrum and trophic preference. The marker FAs of trophic sources are also of particular interest. The discovery of new symbioses involving invertebrates and bacteria, which are responsible for nutrition of the host, deserves special attention. The present paper also highlights recent research into the molecular and biochemical mechanisms of PUFA biosynthesis in marine mollusks. The biosynthetic capacities of marine mollusks require a well-grounded evaluation.

Keywords: fatty acids; mollusks; symbiotic bacteria; biosynthesis

---

1. Introduction

The importance of fatty acids (FAs) in marine environments commonly focus on polyunsaturated fatty acids (PUFAs), eicosapentaenoic acid (EPA, 20:5n-3), docosahexaenoic acid (DHA, 22:6n-3) and, to a lesser extent, arachidonic acid (ARA, 20:4n-6), which are vitally important not only to human health but also to health and survival of marine and terrestrial organisms. They are derived from two metabolically distinct n-3 and n-6 FA families. The metabolic precursor of EPA and DHA is α-linolenic acid (ALA, 18:3n-3), whereas linoleic acid (LA, 18:2n-6) is the metabolic precursor of ARA. It is common knowledge that animals and humans cannot synthesize both n-3 and n-6 PUFAs de novo. Nevertheless, they are required for normal development, growth and optimal health. They can be produced endogenously by humans, but the rate of their biosynthesis is too low to satisfy the physiological requirements. Thus, n-3 and n-6 PUFAs are considered as essential for important physiological processes and must be supplied in the diet. The beneficial effects of n-3 and n-6 PUFA supplementation in diets have been well established both for humans and for marine animals.

The major sources of n-3 PUFAs are aquatic food webs [1–3]. They play a key role in biological processes and are among the most important molecules transferred via the plant–animal interface in aquatic food webs. According to generally accepted views, PUFAs are produced de novo mainly by unicellular phytoplankton and seaweeds and further transferred from primary producers to consumers on the following trophic levels of the marine food chains [4]. The most physiologically important EPA and DHA are accumulated within aquatic ecosystems, as they are transferred to animals that can be consumed by humans. Numerous studies have shown the relationship of the FA composition of consumers and food consumed, and, therefore, FA can be used as efficient and useful biomarkers for the study of trophic interactions between organisms in aquatic ecosystems [5,6].

However, information about the endogenous mechanisms of marine invertebrates responsible for synthesis of n-3 and n-6 PUFAs is still being accumulated. Recent researches have shown the potential of some marine mollusks for endogenous synthesis of long chain PUFAs (LC-PUFAs) [7–9]. Based on the transcriptome and genome sequences, as well as various publicly available databases, a number of novel fatty acyl desaturases (*Fad*) and elongations of very long-chain fatty acid (*Elovl*) genes have been identified from the major orders of the phylum Mollusca, suggesting that many mollusks possess most of the required enzymes for the synthesis of long chain LC-PUFAs [10]. The question whether these findings of the desaturase sequences in invertebrate species really cast doubt on the idea that the organic matter is transferred along the food chains, and thus the existence of trophic links between primary producers and consumers and the relationship of the FA composition of animals and the FA composition of food, are currently under discussion [8].

Mollusk FA have attracted the attention of researchers that has been driven by their critical roles in aquatic ecology and in trophic food webs, as well as by their importance as sources of essential FAs with important impacts on human health [11]. Among marine animals, mollusks are especially important as a source of PUFAs (after fish). Many members of the phylum Mollusca, commonly known as clams and snails, are traditional seafood items in human diets, and rich in essential PUFAs. The edible mollusks are commercially harvested and cultured [12]. Marine bivalve mollusks are highly appreciated, partly because of their positive effects on human health arising from their constituents—highly valued n-3 LC-PUFA—and so their consumption is increasing every year [13]. The mollusks represent different trophic levels, trophic groups, and differ by various dietary habits. To date, extensive data has been accumulated on mollusk FAs. The great diversity of mollusks is accompanied by their wide chemodiversity because of their trophic preferences and defense modes, as well as the biosynthetic capacities that influence their chemical composition.

The main objective of this review is to focus on the most important factors and causes determining the biodiversity of the mollusk FAs, with an emphasis on the key relationship of these FAs with the trophic sources and the food spectrum, rather than to make a complete description of the FA composition of the known mollusk species. The marker FAs of the trophic sources are also of particular interest. The discovery of new symbioses involving invertebrates and bacteria, which are responsible for nutrition of the host, deserves special attention. The present paper also highlights recent research into the molecular and biochemical mechanisms of PUFA biosynthesis in marine mollusks. The biosynthetic capacities of marine mollusks require a well-grounded assessment.

2. Importance of Essential Polyunsaturated Fatty Acids for Human Health

FAs are involved in several biochemical pathways and, being an important source of energy and components of cell membranes, are responsible for determining their structure, functions and cell signaling [14]. They ensure fluidity of the lipid bilayer, selective permeability and flexibility of cellular membranes, and are responsible for the mobility and function of embedded proteins and membrane-associated enzymatic activities [15].

Many biological actions of PUFAs are mediated via bioactive lipid mediators produced by fatty acid oxygenases and serve as endogenous mediators of cell signaling and gene expression that regulate inflammatory and immune responses, platelet aggregation, blood pressure and neurotransmission [16]. They support the physiological functions as homeostatic mediator [17]. PUFAs n-6 and n-3 are precursors of signaling molecules with opposing effects. ARA is converted to prostaglandins, leukotrienes and lipoxins, whose effect is predominantly pro-inflammatory. In contrast, EPA- and DHA-derived eicosanoids have chiefly an anti-inflammatory effect. PUFAs n-3 exhibit the most potent anti-inflammatory effects that helps to control inflammation underlying many chronic diseases, including atherosclerosis, coronary heart disease, diabetes, rheumatoid arthritis, cancer and mental health [18]. A large number of epidemiological studies and clinical trials suggest a beneficial relationship between n-3 PUFA consumption and reduced inflammatory symptoms. So, EPA and DHA are capable of partly inhibiting inflammation reactions, including leukocyte chemotaxis, adhesion molecule expression, leucocyte–endothelial adhesive interactions, production of inflammatory cytokines, and T cell reactivity [19]. Low intake of dietary EPA and DHA is associated with increased inflammatory processes, general cardiovascular health and risk of the development of Alzheimer's disease, as well as with poor fetal development, including neuronal, retinal and immune function [11,20,21]. Low maternal DHA intake may also cause increased risk of early preterm birth and asthma in children [22,23].

Many beneficial cardiovascular effects have been ascribed to PUFAs, including hypolipidemic, antithrombotic, antihypertensive, anti-inflammatory and antiarrhythmic properties, as well as the reduction of blood pressure [24]. The effectiveness of n-3 PUFAs for the prevention of cardiovascular diseases (CVD) is based on multiple molecular mechanisms, including membrane modification [25,26] where n-3 PUFAs are incorporated into lipid bilayer and affect membrane fluidity, formation of lipid micro-domains and also mechanisms such as attenuation of ion channels, regulation of pro-inflammatory gene expression and production of lipid mediators [27,28]. The use of n-3 PUFAs is recommended for ameliorating the CVD risk factors [11,29].

DHA, the dominant n-3 FA in the brain and retina, plays an important role in neural function, exhibits neuroprotective properties and represents a potential remedy against a variety of neurodegenerative and neurological disorders [30,31]. The potentially beneficial effect of DHA in preventing or ameliorating age-related cognitive decline has been revealed in a clinical study [30]. The n-3 LC-PUFAs exert positive effects on memory functions in healthy elderly adults [21] and support the neurological development of the infant brain [32,33]. Consumption of n-3 LC-PUFAs, particularly DHA, may enhance cognitive performance relating to learning, cognitive development, memory and rate of fulfilling cognitive tasks [34]. EPA and DHA play a critical role in neuronal cell functions and neurotransmission, as well as in inflammatory and immune reactions that are involved in neuropsychiatric disease states. Most experimental and epidemiological studies show the beneficial effect of n-3 PUFAs in various neurological and psychiatric disorders [35]. A diet supplemented with n-3 PUFAs exerts positive effects on brain structure and function in healthy elderly adults [36].

Several studies have confirmed that n-3 PUFAs possess a potential for prevention and therapy of several types of cancers and, moreover, they can improve the efficacy and tolerability of chemotherapy [37,38]. According to other studies, n-6 PUFAs, vice versa, induce progression in certain types of cancer [38]. Epidemiological and experimental studies have found a relationship between a PUFA-supplemented diet and the development of some types of cancer, including colon and colorectal carcinoma, breast cancer, prostate cancer, as well as lung cancer and neuroblastoma [38]. The promising effect of n-3 PUFAs on certain types of cancer is explained by their ability to modulate membrane-associated signal transductions and gene expression involved in cancer pathogenesis, as well as to suppress systemic inflammation [39].

Dietary intake of these essential components, as substances with therapeutic action, may maintain health, prevent the development of many diseases and mitigate a number of pathological conditions. Supplementation of PUFAs at a rate of at least 1 g per day, either in capsules or by marine products, demonstrated a protective effect against cardiovascular disorders, hyper-triglyceridemia, hyperlipidemia, metabolic syndrome or type 2 diabetes [29].

3. Primary Producers of Polyunsaturated Fatty Acids in Marine Ecosystems

3.1. Microalgae

Each algal class is characterized by a specific FA profile. The occurrence of certain compounds can be used as an FA signature for different algal classes. Chemotaxonomic differences in FA may be useful in the estimation of the input of specific microalgae in the tracing of these components on marine food webs.

Members of Bacillariophyceae are abundant in aquatic habitats and are considered as the most important primary producers of n-3 LC-PUFAs in marine food chains. Diatoms frequently dominate in seasonal phytoplankton blooms and, accordingly, these algae are the most studied classes of microalgae in terms of their lipids and FAs. The FAs reported for different species of Bacillariophyceae are typical for diatoms. The most abundant FAs are 20:5n-3 (it averages at 20–40% of total FA), 16:1n-7, 16:0, 14:0 and C16 PUFAs, 16:2n-4, 16:3n-4 and 16:4n-1, which account for about 80% of total FAs [40–44]. Hence, reliable markers of Bacillariophyceae have a high percentage of EPA, the predominance of 16:1n-7 over 16:0 and the presence of 16:2n-4, 16:3n-4 and 16:4n-1 along with low amounts of C18 PUFAs and DHA.

Dinophyceae species are major contributors to marine food webs and are second to diatoms as primary producers of organic matter in the oceans. They are especially abundant in coastal waters worldwide, where their exuberant growth, named algal bloom, is often observed. They are known as the main supplier of n-3 LC-PUFAs to marine animals. The more prominent FAs found in dinoflagellates are 16:0, 18:4n-3 (2.3–15.3% of total FA), 18:5n-3 (6.4–43.1%), 20:5n-3 (ranged from 1.8 to 20.9% in different species), and 22:6n-3, DHA (9.5–26.3%) [42,45–47]. Summing up the information on the FAs of this algal class, the high contents of 18:4n-3 and 22:6n-3 have generally been considered as useful signature compounds of dinoflagellates.

Green algae are classified into two classes, Chlorophyceae and Prasinophyceae, with their FA composition varying considerably. The most abundant FAs of the class Chlorophyceae are C18 and C16 PUFA isomers n-3 and n-6, of which, for example, 18:3n-3 reaches 43% of the total FAs [40,42,48]. The distinctive C16 PUFA isomers, 16:2n-6, 16:3n-3 and 16:4n-3, can be used in ecological studies as signature lipids to estimate abundance of these algae in phytoplankton or their input in the diet of invertebrates or transfer of these compounds into food webs. In general, the specific features of green algae are high concentrations of C16 PUFAs consisting of 16:2n-6, 16:3n-3 and 16:4n-3, and C18 PUFAs, such as 18:2n-6 and 18:3n-3, which are essential FAs and the precursors of metabolically distinct families of n-3 and n-6 PUFAs.

Eustigmatophyceae species contribute significantly to the organic matter of coastal waters in the Northern and Southern Hemispheres. Their FAs are dominated by three components, 16:0, 16:1n-7 and 20:5n-3, which together account for about 75% of the total FAs. In addition, an appreciable percentage of 20:4n-6 is detected (4–8.8%), whereas C18 PUFAs are present as minor components [42,49].

Cryptophyceae species are small marine or freshwater flagellates, which are abundant in some seasons and, hence, play an important role as food for invertebrates. A common characteristic of many cryptomonads is a very high proportion of n-3 PUFAs (up to 60–81.1% of total FAs) [42,50,51]. Among them, 18:4n-3 and 18:3n-3 are the most pronounced (together making up 40–50% of total FAs), but a high concentration of 20:5n-3 is also common (13–26%) [42,50,51]. Thus, the high percentage of 16:0, 18:4n-3, 18:3n-3 and 20:5n-3, along with a very low abundance of C16 PUFAs, is typical of most cryptomonads, which are considered as a highly valuable food source rich in n-3 PUFAs in aquatic ecosystems.

The class Prymnesiophyceae is divided into four orders, which have essential differences in lipid composition [42,52,53]. In general, the members of this class, similarly to diatoms, contain 14:0, 16:0, 16:1n-7 and 20:5n-3 as main components, but their distinguishing feature is the abundance of 18:4n-3 and 22:6n-3.

The FA profile of members of the Rhodophyceae is dominated by three major FAs, 16:0, 20:4n-6 and 20:5n-3, which together account for about 80% of the total FA [42,50]. It is worth noting that only red microalgae show a significant concentration of 20:4n-6 (up to 28%), which is a relatively rare or minor component in other classes.

Thus, the taxonomic differences in the FA composition between microalgae classes are obvious and each class is characterized by its specific FA profile. An FA analysis of microalgae has revealed signature compounds that may be useful to evaluate them as sources of different PUFAs. Uncommon FAs or groups of FAs may serve useful biochemical indicators in ecological studies. Chemotaxonomic differences, particularly those in terms of FAs, may be used for assessing the input of specific microalgae in the diet of animals.

3.2. Heterotrophic Protists

Another important source of PUFAs for marine mollusks is heterotrophic protists, zooflagellates and ciliates, constituting the links in the food web named the "microbial loop". Among marine heterotrophic nanoplankton, flagellates are the dominant group in terms of abundance, biomass and diversity [54], while flagellates, in turn, are consumed by ciliates in the food chain. Heterotrophic protists, flagellates and ciliates, similarly to microalgae, are responsible for the production of LC-PUFAs, which are essential for organisms at higher trophic levels in marine ecosystems. The marine ciliate *Parauronema acutum* is reported to contain a significant level of PUFAs: 18:4n-3 (9% of total FAs), 20:5n-3 (10%) and 22:6n-3 (5%) [55]. A similar pattern exists for marine free-living heterotrophic flagellates [56]. The ability of zooflagellates and ciliates to efficiently produce n-3 PUFAs, 20:5n-3, 22:6n-3 and 20:4n-6, was proven experimentally [56,57]. Thus, zooflagellates and ciliates that constitute links of the microbial loop can be a source of PUFAs for suspension- and deposit-feeding mollusks in marine ecosystems [56–58].

4. Biochemical Markers for Identification of Mollusk Feeding Patterns

Due to the great structural diversity of FAs and their substantial taxonomic specificity, the identification of characteristic FA patterns at different trophic levels allows estimation of relationships between primary producers and consumers of different trophic levels of a food web [5,59]. The current trend in lipid biochemistry is the use of FAs as biochemical markers for determination of animals' food sources and trophic relationships between species in aquatic communities [6,59–61]. The specificity of the FA composition of algae and microorganisms, which serve as food for consumers, are well documented (references for Section 3), and many of these FAs are transferred from prey to predators without modification [5,6,59]. This approach is based on the limited ability of animals to synthesize FAs, much of them animals receive from consumed food, particularly PUFAs, which can only be biosynthesized by microalgae and protozoa and become an essential dietary component for higher trophic levels. Potential food sources, such as diatoms, dinoflagellates, zooplankton and bacteria, have a distinctive FA composition with unique FAs or a specific FA ratio used as dietary tracers of mollusks (Table 1). For this reason, FAs are considered as biochemical markers, a very efficient and useful tool to provide information on the food spectrum and diversity of food sources for marine organisms and for studying food chains in marine ecosystems.

Table 1. Fatty acids as biomarkers of food sources for mollusks.

Fatty Acid Markers	Food Source	References
20:5n-3, 16:1n-7/16:0 > 1, 14:0, 16:2n-4, 16:3n-4, 16:4n-1	Diatoms	[41,42]
18:4n-3, 22:6n-3	Dinoflagellates	[46]
18:2n-6, 20:4n-6, 22:6n-3	Heterotrophic flagellates	[56,57]
22:6n-3, 18:1n-9	Animal material Meiobenthos	[59,62]
15:0, 15:1, *iso*-15:0, *anteiso*-15:0, *iso*-16:0, 17:0, *iso*-17:0, *anteiso*-17:0	Heterotrophic bacteria	[63–66]
16:0, 18:0, 22:0	Detritus	[67,68]
18:1n-9, 18:2n-6, 18:3n-3, 18:4n-3, 20:4n-6, 20:5n-3	Brown algae	[69]
Very long-chain FAs: *iso*-5,9-25:2; 25:2Δ5,9; 26:2Δ5,9; 27:2Δ5,9; 26:3Δ5,9,19; 26:3Δ5,9,17; 27:3Δ5,9,19	Sponges	[70–72]
Tetracosapolyenoic acids: 24:5n-6, 24:6n-3	Soft corals	[73]

5. Fatty Acids of Marine Mollusks

Mollusks are extremely widely represented in the oceans, both in number of species and in density of populations. Of the seven classes of this phylum, Gastropoda, Bivalvia and Cephalopoda account for more than 95% of the mollusk species and are a major marine fishery resource. Plenty of information on the lipids and FAs of these classes, their commercial importance and, particularly, on their nutritional value as sources of n-3 PUFAs has been accumulated to date. In her review, Joseph emphasizes the important influence of environmental and biological factors on FA for members of this phylum [74]. Currently, new data are collected, which make it possible to review the features of the mollusks' FAs and the impact of different factors on FA profiles. In this Section, we focus on the different diets of members of these classes, determining the principal differences in FAs between their species.

5.1. Gastropoda

The diet of gastropods, which are represented by the greatest number of species, differs according to the trophic group considered. According to the type of food, gastropods are generally divided into two groups: herbivorous and predators [75]. Their dietary specialization and trophic relationships are reflected in the FA composition of the species. Their trophic habits and food preferences influence the composition of their FAs, which can differ fundamentally for species with different diets (Figure 1).

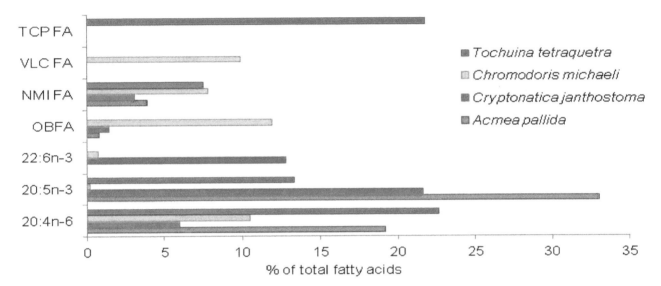

Figure 1. Distribution of fatty acids in gastropods with different types of feeding: herbivorous and carnivorous. Results are expressed as the mean [73,76,77]. TCP FA, tetracosapolyenoic fatty acid; VLC FA, very long chain fatty acid; NMI, non-methylene-interrupted; OBFA, odd-chain and branched fatty acids.

Evidently, the most primitive type of gastropod feeding involves browsing and grazing of algae from rocks. High percentages of 16:1n-7 and 20:5n-3, typical of diatoms, have been found in the pelagic pteropod *Limacina helicina* that inhabits Arctic and Antarctic waters, indicating a strong evidence of diatom ingestion [78]. The limpet *Acmaea pallida* feeds most frequently on brown algae, while *Lottia dorsuosa* feeding on filamentous and unicellular algae, scraping them from the surface of stones. Consequently, FAs of algal origin found in snails, such as 18:3n-3, 18:4n-3, 20:4n-6 and 20:5n-3, reflects a herbivorous feeding strategy (Table 2) [76]. Two intertidal grazers, *Patella aspera* and *P. candei*, also exhibit high levels of EPA and ARA [79]. Meanwhile, in lipids of carnivores, *Cryptonatica janthostoma* and *Nucella heyseana*, 22:6n-3 is dominant, as a result of their animal diet (Table 2). These species are known as consumers of mollusks, mainly bivalves [75]. The FA composition of limpets and snails is characterized generally by predominance of 20:5n-3 and 22:6n-3, which constitute usually 25%–35% of total FAs, being a rich source of n-3 PUFAs.

In contrast, sea slugs dot not have this specific feature; these two marine PUFAs are minor components and in sum do not exceed 1–3% of total FA. FA profiles of nudibranchs differ principally from those of other mollusks in the abundance of numerous very long chain FAs (VLC FAs) specific for marine sponges [70–72] or by the high portion of tetracosapolyenoic acids (TCP FAs), produced by octocorals (Figure 1) [73]. The opisthobranchs, including sea slugs, are predators on sessile animals, such as sponges, corals, bryazoans and ascidians. The majority of nudibranchs are predators on sponges, and the occurrence of VLC FAs with double bonds at Δ5, 9, including 5,9–24:2, 5,9–25:2, 5,9–26:2 and *iso*-5,9–25:2, are certainly a result of feeding on sponges [77,80]. TCP FAs, 24:5n-6 and 24:6n-3, found in high proportions (each is more than 10% of total FAs) in the tritonid nudibranch *Tochuina tetraquetra*, originate undoubtedly from soft corals of the subclass Alcyonaria, which it feeds on [73]. FAs of the nudibranch *Armina maculate*, which feeds on a pennatulacean commonly named "Sea Pen", *Veretillum cynomorium*, constituted predominantly 16:0, 18:0, 20:4n-6 and 20:5n-3 (62% of total FA); thus, evidencing a similarity with the FA profile of "Sea Pen" represented by the same major compounds, whereas FAs of the cephalaspidean *Aglaja tricolorata*, presumably feeding on foraminiferans from sandy bottoms, is rich in EPA and DHA (27% of total FAs) [81]. Moreover, the studied nudibranchs exhibit one more specific feature: their lipids are rich in n-6 PUFA, and their level is much higher than that of n-3 PUFAs. Dorid nudibranchs, besides 20:4n-6, contain also 22:4n-6 and 18:2n-6 [77]. High values of n-6 relative to n-3 PUFAs are unusual for marine organisms and are reported mostly for snails grazing on brown algae, being rich in 20:4n-6 (Table 1) [74,76].

Table 2. Fatty acid composition of gastropod mollusks from the East Pacific (% of total FAs) [76].

Fatty Acids	Acmea pallida	Lottia dorsuosa	Ischnochiton hakodadensis	Cryptonatica janthostoma
14:0	0.4	4.6	4.5	3.4
15:0	0.3	1.0	0.5	0.6
16:0	5.9	13.9	13.2	6.4
16:1	0.8	5.8	3.8	2.8
17:0	–	0.5	–	–
16:3n-4	0.8	2.9	0.4	1.7
17:1n-8	0.4	–	0.4	0.8
18:0	6.3	4.5	–	8.8
18:1	13.2	15.9	15.7	3.6
18:2n-6	–	4.9	2.0	2.6
18:3n-6	–	–	0.5	0.2
18:3n-3	1.1	–	4.4	0.6
20:1	10.5	9.2	3.9	7.3
18:4n-3	1.1	2.2	1.7	0.5
20:2NMI	1.5	1.2	–	8.1
20:3n-6	1.6	0.9	3.5	4.2
20:4n-6	19.2	15.5	6.8	6.0
22:2NMI	3.9	4.2	3.6	3.1
20:5n-3	33.0	11.8	13.3	21.6
22:4n-6	–	0.7	4.2	1.1
22:5n-6	–	–	0.8	0.5
22:5n-3	–	0.9	4.4	2.6
22:6n-3	–	–	0.8	12.8

5.2. Bivalvia

Most mollusks from the phylum Bivalvia are known to be suspension-feeders, their diet consisting mainly of plankton from the water column, protists from the near-bottom water layer and deposit-feeders collecting food from the surface of bottom sediments. Thus, planktonic and benthic microalgae, zooplankton, protozoans, including heterotrophic flagellates and ciliates, and also bacteria from detritus are the main components of diet of filter-feeding bivalves [75]. This feeding mode and, consequently, the diet primarily impact the composition of the mollusk FAs (Table 3), which exhibit an abundance of EPA, DHA, and quite often, ARA [61,74].

Table 3. Fatty acid composition of bivalve mollusks from the East Pacific (% of total FA) [76].

Fatty Acids	Arcidae			Mytilidae			Ostreidae	Cardiidae	Veneridae				Mactridae		Pectinidae	
	Scapharca broughtoni	Arca boucardi	Anadara maculosa	Mytilus edulis	Crenomitilus grayanus	Modiolus difficilus	Crassostrea gigas	Clinocardium californiense	Callista brevisiphonata	Saxidomus purpuratus	Protothaca jedoensis	Mercenaria stimpsoni	Spisula voyi	Mactra chinensis	Patinopecten yessoensis	Chlamys swifti
14:0	0.6	0.6	2.7	2.1	3.9	2.7	3.0	2.0	0.6	1.8	2.8	7.8	6.9	3.1	3.0	4.2
15:0	1.1	0.4	0.3	0.7	0.5	0.5	0.8	0.9	0.5	0.4	0.4	0.6	0.6	0.6	0.5	0.4
16:0	10.2	9.2	13.3	14.8	16.6	14.9	14.9	12.9	13.5	11.0	10.9	14.8	12.4	15.0	11.2	13.7
16:1n-7	1.5	3.2	2.4	5.0	8.3	5.5	4.9	4.9	2.4	3.3	6.6	9.2	5.6	9.6	5.0	4.9
16:3n-4	3.1	1.2	4.2	0.8	1.5	2.0	2.1	1.8	1.0	0.1	2.4	1.4	1.9	1.8	1.1	0.3
17:1n-8	2.2	1.3	0.2	0.2	0.6	1.4	1.0	2.4	2.0	0.6	0.8	0.6	0.8	0.8	08	0.9
18:0	10.6	6.6	13.5	3.5	2.8	5.8	3.6	5.5	0.5	5.8	4.0	6.4	6.8	4.0	6.1	5.3
18:1n-7	5.7	2.8	4.7	3.6	5.7	6.6	12.1	7.1	0.3	4.2	6.8	3.2	4.9	6.0	6.1	8.2
18:2n-6	2.2	1.6	3.8	1.7	2.2	2.1	2.2	1.2	0.5	0.8	1.3	1.4	1.3	0.4	9.0	1.6
18:3n-6	0.1	–	0.5	0.1	–	–	0.3	0.6	0.4	0.1	0.2	0.6	0.2	0.3	0.2	0.5
18:3n-3	1.0	2.1	1.5	1.4	1.5	1.5	1.8	1.5	0.2	0.7	0.3	0.7	0.8	0.4	0.4	0.9
20:1	10.8	12.5	8.9	12.3	5.6	7.4	6.8	3.2	12.5	17.3	5.8	3.6	6.5	6.8	3.9	5.5
18:4n-3	0.6	1.4	0.8	1.8	2.7	2.1	3.1	1.2	1.3	2.7	1.0	4.5	3.7	2.7	2.6	4.5
20:2NMI	0.1	0.7	0.1	0.8	3.9	1.2	1.3	3.4	0.2	0.3	0.2	1.9	1.0	–	0.9	0.2
20:3n-6	0.1	–		0.8	1.7	–	–	1.2	0.5	0.7	–	1.1	1.5	1.4	0.1	0.7
20:4n-6	5.8	6.4	7.8	3.9	2.5	3.3	2.1	4.3	3.6	2.9	3.5	1.5	2.7	2.1	3.0	3.8
22:2NMI	20.7	12.8	12.7	4.6	4.6	3.8	4.7	6.1	6.0	0.7	6.5	1.9	1.7	0.6	0.7	0.6
20:5n-3	6.1	10.3	4.0	14.5	16.2	22.9	16.7	13.4	18.3	22.3	14.4	24.5	17.3	21.2	19.7	20.2
22:3n-6	1.2	1.3	0.4	1.5	0.8	1.3	1.0	1.4	1.6	1.5	1.6	1.6	1.4	1.4	0.9	1.1
22:4n-6	0.9	0.8	1.3	0.1	0.2	0.4	0.1	2.5	1.1	2.2	1.4	0.4	0.8	0.8	0.1	0.2
22:5n-6	1.3	1.4	2.3	0.7	0.4	0.6	0.4	1.1	1.6	1.1	1.1	0.3	0.9	0.9	0.6	0.6
22:5n-3	1.0	1.3	0.7	1.1	0.8	1.1	0.1	3.0	2.0	3.3	1.5	0.1	2.0	2.0	0.5	0.9
22:6n-3	14.2	22.4	13.1	23.0	15.3	12.0	16.0	16.7	19.5	15.2	24.2	12.3	17.4	17.4	18.5	21.2

Variations in the trophic environment and also the food selectivity of the species result in the dominance of the FA biomarkers of diatoms or dinoflagellates, zooplankton or detritus, or a combination of these sources. The DHA to EPA ratio reflects the proportion of zooplankton, diatoms and dinoflagellates in the bivalve's diet [6,82,83]. DHA often dominates in FAs of zooplankton and dinoflagellates [6,46,56,57], whereas EPA originates from diatoms [40–42]. A high concentration of 16:1n-7 and 20:5n-3, as well as a higher EPA/DHA ratio, suggests the importance of diatoms in the diet of the mollusks, whereas an elevated level of 18:2n-6, 20:4n-6 and DHA indicates the important contribution of microheterotrophs (flagellates and ciliates) in the diet. A higher proportion of odd-chain and branched FAs (OBFAs) is the evidence of the presence of bacteria in the diet of bivalves [84,85].

FA composition of the different taxa of marine bivalves from temperate waters of the East Pacific shows that their characteristic feature is a high abundance of n-3 PUFAs (Table 3). The concentration of both EPA and DHA reaches 25%, and ARA extends to about 8% of total FAs. FA composition varies from species to species, but n-3 PUFA are usually dominant. Furthermore, the high content of EPA and DHA shown in Table 3 is similar to the values obtained for the other species from different regions, for example, *Crassostrea angulata*, *Mytilus edulis*, *C. edule* and *Venerupis pullastra* from the coastal and estuarine systems of Portugal [86]; the oyster *Crassostrea virginica* [87] and sea scallop *Placopecten magellanicus* from the coast of Canada [88]; and the pod razor clam *Ensis siliqua* [89]. PUFAs, especially EPA (19–22% of total FAs) and DHA (20–32% of total FAs) were found to account for the majority of total FAs in tissues of the scallops *Patinopecten yessoensis* and *Chlamys farreri*, which provides an opportunity to use them as a potentially health-promoting food for human consumption [90]. Previous studies also reported the dominance of these PUFAs in tissues of *P. yessoensis* [84] and *Pecten maximus* [91].

In addition, spatio-temporal intraspecific variations in mollusk FAs are observed. So, FAs of *Pecten maximus* showed strong differences between individuals from shallow and deep-water habitats. This trend was driven by the content of marker FAs of diatoms, which are abundant near coasts. Scallops from deeper habitats are characterized by higher contents of flagellate FA markers compared with scallops from shallow habitats that emphasize the variability of the FA content according to the diet of this species along its distribution range [91]. FA biomarkers (Table 1) explain the spatial and temporal heterogeneity in nutrient sources for mollusks. The pattern of spatial and temporal variations of the biomarker FAs in the bivalve *Spondylus crassisquama* [83] and *Mytilus galloprovincialis* [92] revealed the nature and origins of food sources for these bivalves. Species-specific feeding adaptations to environmental variability of two bivalves, the clam *Callista chione* and the cockle *Glycymeris bimaculate*, from two shallow sites of the coastal oligotrophic Mediterranean Sea are revealed. The species demonstrate the differences in FAs mainly due to EPA and DHA percentage during the seasons. FA markers revealed a mixed diet where *Callista chione* fed more upon fresh material (diatoms and zooplankton) than *Glycymeris bimaculate*, which relied largely on bacteria-derived detritus [85].

Pinna nobilis, endemic to the Mediterranean Sea, is known to ingest different food items depending on its shell size. As a result, small-sized *P. nobilis* are associated with a detrital food chain characterize by saturated FAs (38%) and OBFAs (9.9%), while the diet of large- and medium-sized individuals have a greater proportion of PUFAs (EPA from 13% to 22% and DHA from 13 to 44% of total FAs). Thus, FA composition of the species reflects a lower contribution by markers of detritus and an increasing contribution of phytoplankton and zooplankton with increasing shell size [93].

5.3. Cephalopoda

Compared to data on lipids of gastropods and bivalves, information on cephalopods is not as abundant. Nevertheless, it is evident that their FA composition, similarly to that in gastropods and bivalves, is dietary dependent [94,95]. They inhabit pelagic ecosystems and are active predators preying on a variety of fish and invertebrates, such as crustaceans and mollusks. Their diet varies between species and is affected by gender, size, sexual maturity and season of year [96]. Cephalopods are generally known to be consumers of higher trophic levels, or top predators, actively accumulating n-3 PUFAs, EPA and, in particular, DHA in their tissues, which, are transferred up food chains from primary producers and ingested with their food [94]. They are excellent sources of n-3 PUFAs, especially EPA and DHA. An FA analysis of the most commonly consumed cephalopods, such as common cuttlefish *Sepia officinalis*, European squid *Loligo vulgaris*, common octopus *Octopus vulgaris* and musky octopus *Eledone moschata*, showed the dominance of DHA (21–39% of total FA), EPA (8–17%), ARA (1.5–12%), as well as saturated 16:0 (16–25%) and 18:0 (4–10%) during the seasons [97].

The FA composition of the mantle and digestive gland differed markedly between the squid species. The digestive gland is rich in monounsaturated FAs whereas the mantle contains high concentrations of PUFAs, particularly DHA (about 40% of total FAs) (Figures 2 and 3). These findings imply that the squid, as a top predator, actively concentrates EPA and, in particular, DHA in the tissues from the diets. Published data show a similarity in FA of mantle tissue between various species from the different geographic regions (Figure 3), including *Nototodarus gouldi*, inhabiting the tropical and temperate waters of Australia and New Zealand [98], and *Moroteuthis ingens*, an endemic species to the Southern Ocean, having a circumpolar distribution in the sub-Antarctic [99]. Meanwhile, FAs of the digestive gland of squids differ significantly between species (Figure 3), largely reflecting the variety of diet consumed, e.g., [95]. An FA analysis, frequently applied in dietary studies of cephalopods, indicates that the digestive gland is an accurate source of dietary tracers [95], thus revealing a recent history of dietary intake [94,99].

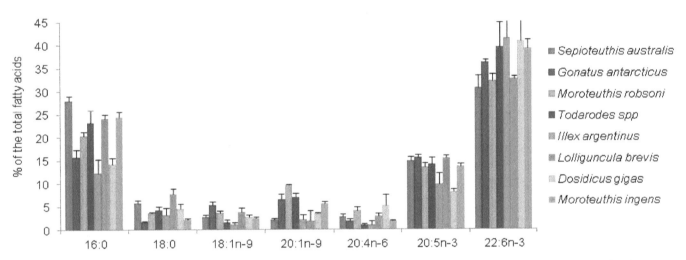

Figure 2. Major fatty acids (% of total FAs) in the mantle of squids. Values are mean ± standard deviation (SD) [94,95,99–101].

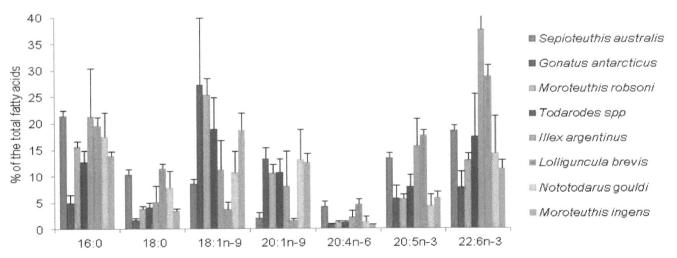

Figure 3. Major fatty acids (% of the total FAs) in the digestive gland of squids. Values are mean ± SD [94,95,98,99,101].

A combination of stomach content and FA signature analyses provided clear evidence of seasonal shifts in prey composition of the arrow squid *Nototodarus gouldi* and suggested temporal variations in its diet. Additionally, FA analyses show dietary differences related with gender, size and maturity of females. According to these relationships, the diet of *N. gouldi* is closely associated with prey size, abundance, availability and, possibly, to life-history stages [98]. The spatial variations in diet are believed to result in the differences in FA profiles of the digestive gland of the onychoteuthid squid, *Moroteuthis ingens*, from four different areas of the Southern Ocean. The FA analysis indicates that crustaceans are an important prey for smaller squid, whereas fish constitute a major portion of prey of larger squid [94]. Moreover, on the example of the jumbo squid, *Dosidicus gigas*, it was shown that an FA analysis can trace the geographic origin of squid individuals [100].

The FA biomarker concept has proven to be useful in the study of energy sources for reproduction in the squid *Illex argentines*. It was found that the FA composition of ovaries shows a more pronounced correlation with that of digestive glands than with the mantle, an energy reserved organ that reflects the dietary intake. The similarity in FA composition between the ovaries and the mantle during the early maturation and spawning period indicates that during these two periods, the somatic energy reserve is involved in reproductive growth. Thus, the potential implication of FAs is useful to provide insights into the breeding strategies among cephalopods [101].

Octopus species consume mainly mollusks, crustaceans, fishes and, sometimes, smaller species of *Octopus* as supplementary dietary components. Significant differences between FAs of the tissues of common octopus *Octopus vulgaris* are evident [102]. Among PUFAs, ARA, EPA and DHA are present at high concentrations in all tissues. ARA is more abundant in the digestive gland compared to muscles (11.4 and 7.9%, respectively), DHA dominates muscle tissues (20.7 and 14.0%) and the percentage of EPA is similar in these tissues (about 15% of total FAs). In contrast to squids, monoenoic FAs are not the main components in octopus, but saturated FAs, 16:0 and 18:0, are prominent in all tissues. Furthermore, the muscles contain more 16:0 compared with the digestive gland (20.1 and 12.4%, respectively) and 18:0 content is similar in these tissues (about 12%) [102].

Cuttlefish is an inhabitant of the seafloor that ambushes small animals such as crabs, shrimps, fishes and small mollusks. Feeding experiments have demonstrated that the FA profile of the digestive gland of the cuttlefish, *Sepia officinalis*, reflects the FA profile of its prey. Cuttlefish that had been fed fish showed comparatively high levels of fish-derived signatures, and this dietary dependence was also found for cuttlefish fed on crustaceans [103]. The proportions of the specific prey FAs are mirrored on the animal FAs. The major FAs in the cuttlefish mantle are DHA, EPA, 16:0 and 18:0 [97].

Thus, cephalopods, being top predators, actively accumulate n-3 PUFAs, EPA and, in particular, DHA in their tissues, coming from their prey, and therefore are valuable marine sources containing high levels of DHA, EPA and a noticeable level of ARA.

6. Contribution of Symbionts to the Fatty Acid Pool of Mollusks

Symbiotic associations between mollusks and microorganisms are widespread; they result in unique ecological strategies and increased metabolic diversity of the partners (Table 4). Symbiotic microbes typically supply nutrients to host animals that provide the microbes with shelter.

Table 4. Symbiotic microbes in marine mollusks.

Type of Nutrition	Symbionts	Function	Host	References
Chemotrophic	Bacteria	Nutritional	Bivalves and gastropods	[104]
Phototrophic	Zooxanthellae	Nutritional	Giant clam *Tridacna squamosa*, Gastropod *Strombus gigas*	[105]
	Algal chloroplasts	Nutritional	Sea slug *Elysia chlorotica*	[106]
	Chlorella	Nutritional	Clams, e.g., *Anodonta*	[107]
Heterotrophic	Bacteria	Nutritional	Bivalve shipworm *Bankia setacea*	[108]
		Light production	Squid *Euprymna scolopes*	[109]
		Chemical defense	Sacoglossan *Elysia rufescens*	[110]

Mollusks inhabit a variety of marine ecosystems. In environments characterized by poor nutrient contents, alternative strategies for nutrition have evolved. For example, some marine invertebrates, including mollusks living near deep-sea hydrothermal vents, cold seeps, on whales, wood falls on the deep-sea floor and shallow-water coastal sediments, derive their nutrition from chemoautotrophic microbes housed in their tissues and specialized structures [104]. The occurrence of symbiotic microbes with invertebrates that fix carbon dioxide autotrophically and synthesize organic compounds that are passed on to the host, play a critical role in establishing the lipid composition of the animals. Bacteria are known to produce various odd and branched FAs (OBFAs) named "bacterial acids" (Table 1). Additionally, *cis*-vaccenic acid, 18:1n-7, is biosynthesized by the anaerobic pathway unique for bacteria [111]. These FAs are widely offered as an indicator of the bacterial input in marine environment. Accordingly, elevated concentrations of the specific bacterial FAs, such as OBFAs, 16:1n-7 and 18:1n-7, coupled with a considerable reduction in n-3 and n-6 PUFAs produced by algae in lipids of the animals suggest a contribution of bacteria to the mollusk nutrition.

FAs have been used as a biomarker to reveal symbiotic relationships between bacteria and the bivalve mollusks *Solemya velum* [112], *Pillucina picidium* [113] and *Axinopsida orbiculata* from a shallow-water hydrothermal vent ecosystem of Kraternaya Bay [114], as well as the nudibranch *Dendrodoris nigra* [115]. These animals exhibit a high percentage of monoenoic FAs (about 40% of total FAs) mainly due to 18:1n-7, low concentrations of n-3 and n-6 PUFAs and an increased level of dienoic NMI FAs (Figure 4). In contrast, lipids of filter-feeding mollusks are dominated by 20:5n-3 and 22:6n-3, accounting for one-third of total FAs (Figure 4) [113]. A gastropod species, *Ifremeria nautilei*, from the deep-sea hydrothermal vent systems of the West Pacific, harbors two types of bacterial symbionts: a high abundance of sulphide-oxidizing and a low abundance of methane-oxidizing bacteria. It results in the dominance of 18:1n-7 (about 25% of the total FAs), 16:1n-7 (20–40%), and 16:0 (up to 15%) in its lipids and a low content of EPA and DHA [116].

Unlike bivalves from hydrothermal vents, deep-sea species living near cold seeps contain neither n-3 nor n-6 PUFAs. Two cold-seep bathymodiolin mussels, *Bathymodiolus japonicus* and *B. platifrons*, contain n-4 and n-7 PUFAs (25–27% of total FAs), including 18:3n-7,10,13; 18:4n-4,7,10,13; 20:3n-7,10,13; and 20:4n-4,7,10,13 with the main 16:1n-7 and 18:1n-7 (up to 25% in sum) [117], because they host methane-oxidizing bacteria and survive independently of photosynthetic products. A unique FA composition was reported for the cold-seep vesicomyid clam, *Calyptogena phaseoliformis*, which houses sulfur-oxidizing bacteria. The major FAs found in this clam belong to the novel n-4 and n-1 NMI PUFAs, such as 20:3n-4,7,15; 21:3n-4,7,16, and 20:4n-1,4,7,15, with significant levels of 20:2n-7,15 and 21:2n-7,16 as n-7 NMI FAs [118]. Similar traits exhibit another species of vesicomyid clams, *Phreagena* (synonym *Calyptogena*) *soyoae* and *Archivesica gigas*, harboring symbiotic sulphide-oxidizing chemoautotrophic bacteria in their gills. They are common in deep-sea chemosynthesis-based communities in the North Pacific. An FA analysis confirmed the lack of n-3 and n-6 PUFAs in their composition and revealed a high percentage of n-7, n-4 and n-1 NMI PUFAs. A comparison of FA compositions of various organs of the clams showed that the content of these NMI FAs in gills was much lower than that of other organs, it suggests that the biosynthesis of n-7, n-4 and n-1 NMI PUFAs occurs in tissues of vesicomyid clams [119].

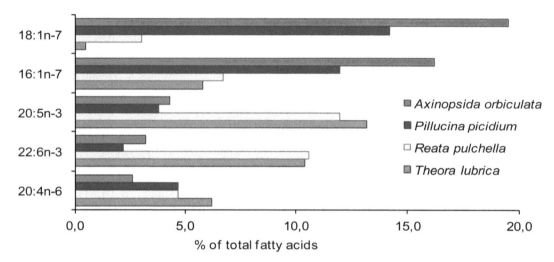

Figure 4. Distribution of the most remarkable of marker fatty acids of bacterial symbionts in the bivalve mollusks *Axinopsida orbiculata* [114], *Pillucina picidium* [113], containing sulfate-reducing symbiotic bacteria, and the symbiont-free bivalves, *Reata pulchaella* and *Theara lubrica* [113].

Thus, in contrast to shallow-water filter-feeding or grazing mollusks, which contain photosynthetic n-3 and n-6 PUFAs as the main components, mollusks with symbiotic chemoautotrophic bacteria show significant modification of FA composition: A decline or complete lack of essential n-3 and n-6 PUFAs and appearance of significant amounts and a variety of NMI FAs, which can probably by synthesized by mollusks using the bacterial FAs as precursors.

The specific FA composition of symbiont-containing species may give a hint of the character of symbiosis. Photosynthetic symbionts, such as dinoflagellates from the genus *Symbiodinium*, settle on corals and giant clams *Tridacna* and supply photosynthetically fixed carbon to their hosts, which contribute to their lipid composition [120,121]. Some of these compounds, such as 18:4n-3, 18:3n-3, 18:5n-3, 22:6n-3 and 16:0, being biomarkers of symbiotic dinoflagellates [46,122], are detected in host organisms. Metabolic interactions consist in exchange of nutrients between the host and its symbionts [120], providing them with a competitive advantage in tropical waters poor in nutrients.

By comparing FAs of the herbivorous limpet *Acmea pallida* and nudibranch species, it is obvious that a striking feature of the nudibranchs is the unusually high level of OBFAs specific for bacteria (Figure 5). It is evident that the share of total OBFAs, predominantly 15:0, 17:0, 17:1n-8 and *iso-* and *anteiso-*C15, C16, C17, C18, and C19, in *A. pallida* is 0.7% of total FAs, whereas in the *D. nigra* it reaches 18.6%, in *Chromodoris* sp. 15.8% and in *Phyllidia coelestis* exceeds 30% [80]. They are normally minor metabolites in marine invertebrates, but the level of these bacterial acids recorded from nudibranchs proved to be extraordinary. A high level of bacterial FAs in nudibranchs may serve, in our opinion, as an indicator that the symbiotic bacteria provide the host with nutrients. Indeed, transmission electron microscopy (TEM) confirmed the presence of rod-shaped Gram-negative symbiotic bacteria in the cytoplasm of epithelial cells and the glycocalyx layer covering the epithelium of the notum and the mantle of *D. nigra* [115]. Moreover, some bacterial OBFAs, such as 17:1n-8 and 19:1n-8, evidently, serve as potential precursors for the biosynthesis of odd-chain PUFA, such as 21:2Δ7,13 identified in nudibranchs [80].

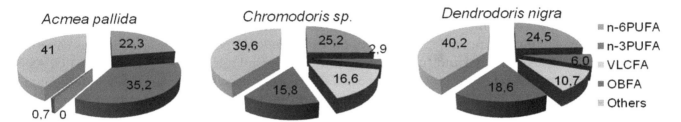

Figure 5. Distribution of fatty acids in the carnivorous nudibranch *Chromodoris* sp. [77], in *Dendrodoris nigra* [115] feeding on sponges and in the herbivorous limpet *Acmea pallida* [76] feeding on brown algae. *D. nigra* is known to harbor symbiotic intracellular bacteria [115].

7. Biosynthesis of Fatty Acids in Mollusks

Biosynthetic pathways of PUFAs are described in detail in many articles and reviews. In brief, monoenoic FAs, such as 18:1n-9 and 16:1n-7, are produced through the action of $\Delta 9$ fatty acyl desaturases. This activity is ubiquitous and found in all living organisms. Oleic acid, 18:1n-9, can be further desaturated through the action of n-6 or $\Delta 12$ desaturase to form, first, LA; then n-3 or $\Delta 15$ desaturase converts LA into ALA. Until recently it was believed that only plants are able to produce de novo LA and ALA, which are essential for animals [1,4,8]. The enzymes involved in LC-PUFA biosynthesis, namely, $\Delta 8$, $\Delta 6$, $\Delta 5$ and $\Delta 4$ desaturases, necessary for the production of EPA and DHA from 18:3n-3, have been described in algae [123].

Hitherto, there was a concept that the mollusks, as well as other marine invertebrates, are not able to synthesize n-3 and n-6 PUFAs de novo to satisfy physiological needs, and PUFAs in the marine invertebrates are derived exclusively from phyto- and zooplankton. This opinion was confirmed by experimental data on the incorporation of ^{14}C-acetate in FAs of the yellow clam *Mesodesma macroides* [124,125]. Similarly, in the experiments with juvenile oysters, *C. gigas* fed algae that had previously been cultured with labeled ^{14}C-acetate confirmed that dietary FAs are incorporated directly into oyster lipids, mostly in a unaltered form, and only less than 1% of the ^{14}C-label was found in 20:5n-3 and 22:6n-3 [126]. Recently, it was shown experimentally that the majority of radioactivity from ^{14}C-FAs incorporates into lipids of *Octopus vulgaris* paralarvae [127] and *Sepia officinalis* hatchlings [7], found as unmodified FAs with elongation being the only metabolism detected, and no desaturation activity towards the FAs was recorded. Moreover, in the study of the FA biosynthesis by bivalves, *Scapharca broughtoni*, *Callista brevisiphonata* and *M. edulis*, the active incorporation of ^{14}C-acetate was found in saturated, monoenoic and dienoic NMI FAs, whereas the radioactive label in LC-PUFAs n-3 and n-6 families, as well as in their precursors, LA and ALA, was not detected. It has been found that NMI dienoic acids, 20:2Δ5, 11, 20:2Δ5, 13 and 22:2Δ7,13, 22:2Δ7,15, are the only PUFAs that mollusks are able to synthesize [128,129]. These FAs with isolated double bonds were suggested to be derived as a result of the action of $\Delta 5$desaturase and elongations. To produce these dienoic FAs, $\Delta 5$desaturase mediates the insertion of the double bond in 20:1n-7 and 20:1n-9, common for invertebrates [128]. Indeed, a fatty acyl desaturase (*Fad*) gene with $\Delta 5$ activity has been characterized both molecularly and functionally from the octopus *Octopus vulgaris* [102], the gastropod *Patella vulgata* [8], the abalone *Haliotis discus* [130], the noble scallop *Chlamys nobilis* [131] and the bivalve *Sinonovacula constricta* [132].

Thus, returning to the issue of the biosynthesis of FAs in mollusks, it should be notde that relying on the experimental data on radioactive acetate or FA incorporation it is proved that marine mollusks are not capable to synthesize de novo LA and ALA, and their long-chain homologues, EPA and DHA, which are essential acids and PUFAs, must be considered essential dietary nutrients.

With the development of genetic methods, interest in the issue of the capability of marine invertebrates to biosynthesize n-3 and n-6 PUFAs has risen and many noteworthy findings in this field have been made recently. The availability of gene sequences databases of multitude species of invertebrates contributes to understanding of the biochemical mechanisms of PUFA biosynthesis in marine invertebrates at a molecular level [8,9,133,134]. Using molecular genetics approaches, the *Fad* and elongation of very long-chain FA (*Elovl*) genes have been characterized both molecularly and functionally, namely, isolated, identified, and expressed in the yeast heterologous hosts. It has been proven that multiple invertebrates, including representative of mollusks, the gastropod *Patella vulgata*, possess the endogenous capability to produce n-3 PUFAs de novo and further biosynthesize physiologically essential n-3 LC-PUFAs [8]. Among aquatic invertebrates, the biosynthesis of LC-PUFAs has been more extensively investigated in marine mollusks [10,102,132,133]. Cloning and characterization of functional diversity of *Fad* and *Elovl* involved in the PUFA biosynthetic pathway was carried out in the cephalopods *Octopus vulgaris* and *Sepia officinalis*, abalone *Haliotis discus* and bivalves, *Chlamys nobilis* and *C. angulata*, and the achievements were reviewed [9]. The capability of LC-PUFA biosynthesis of particular species has been established to depend upon the complement of the key enzymes required, *Fad* and *Elovl* [9].

Thus, the presence of the *Fad* and *Elovl* genes coding the critical enzymes participating in the PUFA biosynthesis in invertebrates can be considered as the biosynthetic potential of mollusks to endogenously produce PUFAs.

8. Dietary Source of PUFAs Versus Own Biosynthetic Capability of Mollusks

The use of genetic methods contributed to a significant advance in the search of genes of the enzymes involved in the PUFA biosynthesis. Does the discovery of desaturase genes in a number of metazoans that enable them to endogenously produce PUFAs, actually mean that we should revise the concept of microalgae as primary producers of PUFAs and the subsequent transport of organic matter to invertebrates up the food chains and dietary origin of the PUFAs? It is worth clarification whether data on the presence of *Fad* and *Elovl* genes in mollusks can contradict the concept of the dietary dependence of fatty acids of animals on the fatty acid composition of their food.

The number of studied mollusk species, which possess genes encoding enzymes important in the LC-PUFA biosynthesis, is quite limited. Furthermore, the distribution of ωx desaturase genes within particular taxonomic groups is non-uniform, and the ability of the production of LC-PUFAs has been established to vary greatly between different mollusk species and heavily depend on the complementation of the desaturase and elongase genes, as well as on their enzymatic activity [9]. For example, ωx desaturase genes are found in the freshwater mussel *Elliptio complanate* and the common octopus *O. vulgaris* [102,134], but not in the marine oyster *C. gigas* nor the mussel *M. galloprovincialis* [8]. The presence of different types of biosynthetic enzymes, as well as their different enzymatic specificity, suggests that the abilities of mollusks for PUFA biosynthesis vary among species. A similar conclusion follows from the studies of biosynthetic capacity of mollusks using radio-labelled precursors [7,126–129].

The mechanism of gene expression is known to be complex and depend on various factors. Genes usually interact with and respond to the organism's environment. *Fad* and *Elovl* genes are usually identified through a search for available sequence databases, and through an analysis of the distribution of ωx desaturase genes across the Phylum and functional characterization of the enzymes using a yeast heterologous expression system [8]. Marine ecosystems are highly rich in n-3 PUFAs produced by planktonic microalgae and, therefore, it can be assumed that the abundance of PUFA in the diet of mollusks may be among the environmental factors that can determine ωx desaturase gene expression. Probably, the genetically incorporated mechanism of PUFA n-3 synthesis is not implemented in mollusks, at least not to the full extent, since the amount of these vital components coming from food is sufficient for animals to provide their physiological and biochemical requirements.

However, in conditions of a PUFA deficiency in animals living in extreme environmental conditions, for example, in deep-sea ecosystems, there is no compensation for this deficiency by endogenous synthesis of PUFAs. As a result, a low content of PUFAs in mollusks or their absence is observed with the simultaneous compensatory increase in the number of NMI FAs [116,135–139]. There is ample evidence of the increase in NMI FAs with a shortage of PUFA [71,76,119,140]. A similar pattern is observed with the symbiosis of bivalves and chemoautotrophic bacteria, leading to a decrease in the PUFA and an increase in the NMI FA levels in the host [113,119,141]. This emphasizes the importance of understanding the biosynthetic capabilities of invertebrates, as well as the importance of combining dietary and biosynthetic approaches to understanding the origin of mollusk fatty acids.

Thus, some mollusk species, similarly to the majority of other invertebrates, possess genes encoding desaturases and elongases involved in pathways of biosynthetic pathways of LC-PUFAs [8,9,132,133], although both their potential and functions remain to be clarified. However, the most species of marine mollusks are apparently not capable to carry out the synthesis of these FA to a sufficient extent to satisfy the physiological requirements. The dependence of the FA composition of the mollusks on food consumed shows that n-3 PUFAs should be considered essential FAs, since their endogenous production appears to be limited.

9. Variations in Fatty Acids in Response to Environmental Factors

Numerous studies explore the influence of biotic and abiotic factors on their FA composition. The majority of studies focus on the seasonal fluctuations in the lipid and FA compositions that are found to directly relate with the reproductive cycle [142] or correlate with changes in the mollusk diet which followed the seston dynamics during the seasons [82,143]. Other studies assess the relationship between FAs and water temperature fluctuations [144,145]. Meanwhile, the species differ in their sensitivity to environmental factors. Since the importance of nutrient quality in terms of FA composition has already been addressed above, in this section the emphasis is on the environmental factors, such as bottom sediments, water salinity, temperature and water pollution.

For benthic animals, the structure and composition of bottom sediments are the important characteristics of their habitat, being one of the key factors that determine the trophic potential of benthic epifauna. In muddy areas, a benthic community is composed mainly of diatoms, heterotrophic nanoflagellates and oligotrich ciliates, whereas in sandy areas, heterotrophic nanoflagellates, euglenoid flagellates, oligotrich ciliates and scuticociliates are dominant among protists [146]. For example, the FA compositions of the scallop *P. yessoensis* from two habitats, muddy and sandy, show pronounced differences pointing to different food availabilities. An elevated content of diatom biomarkers, 20:5n-3, 16:1n-7 and C16 PUFAs, indicates that scallops from a muddy bottom are more reliant on diatom sources. Scallops inhabiting a sandy site have higher amounts of biomarkers of flagellates, ciliates and invertebrate larvae, such as 18:2n-6, 20:4n-6 and 22:6n-3, compared to individuals from the muddy site. This suggests that a scallop's diet depends on food availability [84].

Water salinity is one of the most influential environmental factors especially in estuarine systems and intertidal zone, where its variations cause major physiological and biochemical stress for aquatic organisms. Various species show different tolerance to water salinity. Under salinity stress, the bivalves *Cerastoderma edule* and *Scrobicularia plana* reduce food consumption and physiological pathways; a decrease in PUFA content is observed in *C. edule*. They can store the FAs which are of high physiological importance by reducing their activity and energy consumption [147]. The authors attribute the observed variations in the FA composition, particularly the contents of n-3 PUFAs, NMI FAs and n-6 PUFAs in the littoral mussel *M. edulis* from two different habitats, presumably to the necessity to survive the frequent fluctuations in such environmental factors as water temperature and salinity [148]. The observed modifications in the membrane lipids of the mussel gills lead to a change in the physical state of the membranes, their fluidity and permeability, the functions of ion channels, enzymes and receptors, which ensures the normal functioning of the organism under fluctuations in sea water salinity [148].

Intraspecific variations in FAs are also found to occur in response to water temperature variations. Negative relationship was observed between the acclimation temperature and the unsaturation index of membrane lipids in the oyster *C. virginica*, according to the homeoviscous adaptation theory. These temperature-related changes are mainly due to the variations in EPA content in fast-growing oysters, and in DHA and EPA contents in slow-growing animals [145]. The blue mussel *M. edulis* and the oyster *C. virginica* showed an increased 20:4n-6 level in their tissues as temperature rose, suggesting an increased availability of this FA for eicosanoid biosynthesis during stress response [144].

Contaminants, including the wide variety of pesticides and heavy metal pollutants, increase in marine ecosystems as the results of development of industrial and agricultural activities, being a stress factor for marine organisms. Such contaminants, like heavy metals, were reported to influence feeding, growth, reproduction, cardiac activity, and maturation of bivalve mollusks [149]. These physiological changes lead to changes in the lipid and FA metabolism, while the EPA level is reduced due to exposure to metals and organic pollutants [150,151]. A stress response, manifested as a decrease in PUFA and NMI FA content, was observed in the mussel *M. galloprovinciales* exposed to cadmium and copper [149], as well as to polycyclic aromatic hydrocarbon contamination [152]. A lower value of EPA compared to the control was recorded from the bivalves *Cerastoderma edule* and *Scrobicularia plana* exposed to copper sulphate [153], *Mizuhopecten yessoensis* exposed to cadmium [150] and *Scrobicularia plana* from a habitat contaminated by dioxin and polycyclic aromatic hydrocarbons [154], which can be considered a possible biochemical and physiological consequence for these animals.

Thus, the influence of some factors is not as evident; it is explained by the masking effect of a more significant contribution of food to the FA composition of the organisms. It should be noted that despite the variations in the FA composition of the filter-feeding mollusks in response to biotic and abiotic factors, their specific features, such as EPA and DHA dominance, are retained unchanged.

10. Conclusions

PUFAs, especially EPA and DHA, are fundamental to the health and survival of marine and terrestrial organisms. Mollusk FAs play critical roles in aquatic ecology and trophic food webs, and also play an important role as sources of essential PUFAs, significantly contributing to human health. This review highlights that the extensive taxonomic biodiversity of mollusks accompanies a wide chemical diversity, since the trophic preferences, biosynthetic abilities and physiological requirements of mollusks effect their chemical composition. The review provides evidence of a trophic transfer of FAs from various food sources to marine mollusks, which further emphasizes the nutritional contribution of the FA composition of animals. The variation in FA distribution and abundance between mollusks of different taxonomic and trophic groups is estimated. Mollusks differ in their feeding strategy, divided into the following trophic groups: filter-feeding, gathering, carnivorous, and the symbiont contribution. Some mollusks give shelter to microbes that provide them with nutrients, and these enable animals to settle in nutrient poor environments. Mollusks, which rely completely on bacterial endosymbionts in their diet, have specific FA profiles rich in "bacterial FAs" and poor in PUFAs. In addition, the biosynthetic potential of mollusks influences the FA composition. Mollusks are capable of producing numerous NMI FAs, which are especially important in low nutrient environments. Based on experimental data on incorporation of radioactive acetate or FA into marine mollusks, it is proved that they are not able to synthesize de novo LA and ALA and their long-chain homologues, EPA and DHA. However, it has recently been shown that some mollusk species possess genes encoding desaturases and elongases involved in pathways of biosynthetic pathways of LC-PUFAs. The biosynthetic capacities of marine mollusks require a well-grounded evaluation.

Acknowledgments: I thank the anonymous reviewers for their comments.

References

1.   Parrish, C.C. Essential fatty acids in aquatic food webs. In *Lipids in Aquatic Ecosystems*; Arts, M.T., Brett, M.T., Kainz, M.J., Eds.; Springer: New York, NY, USA, 2009; pp. 309–326. ISBN 978-0-387-88607-7. [CrossRef]

2.   Gladyshev, M.I.; Sushchik, N.N.; Kalachova, G.S.; Makhutova, O.N. Stable isotope composition of fatty acids in organisms of different trophic levels in the Yenisei River. *PLoS ONE* **2012**, *7*, e34059. [CrossRef] [PubMed]

3.   Colombo, S.M.; Wacker, A.; Parrish, C.C.; Kainz, M.J.; Arts, M.T. A fundamental dichotomy in long-chain

polyunsaturated fatty acid abundance between and within marine and terrestrial ecosystems. *Environ. Rev.* **2017**, *25*, 163–174. [CrossRef]

4. Müller-Navarra, D.C.; Brett, M.T.; Liston, A.M.; Goldman, C.R. A highly unsaturated fatty acid predicts carbon transfer between primary producers and consumers. *Nature* **2000**, *403*, 74–77. [CrossRef] [PubMed]

5. Sargent, J.H.; Whittle, K.J. Lipids and hydrocarbons in the marine food web. In *Analysis of Marine Ecosystems*; Longhurst, A., Ed.; Academic Press: New York, NY, USA, 1981; pp. 491–533.

6. Dalsgaard, J.; John, M.S.; Kattner, G.; Muller-Navarra, D.; Hagen, W. Fatty acid trophic markers in the pelagic marine environment. *Adv. Mar. Biol.* **2003**, *46*, 241–251.

7. Reis, D.B.; Rodríguez, C.; Acosta, N.G.; Almansa, E.; Tocher, D.R.; Andrade, J.P.; Sykes, A.V. In vivo metabolism of unsaturated fatty acids in *Sepia officinalis* hatchlings. *Aquaculture* **2016**, *450*, 67–73. [CrossRef]

8. Kabeya, N.; Fonseca, M.M.; Ferrier, D.E.K.; Navarro, J.C.; Bay, L.K.; Francis, D.S.; Tocher, D.R.; Castro, L.F.C.; Monroig, Ó. Genes for *de novo* biosynthesis of omega-3 polyunsaturated fatty acids are widespread in animals. *Sci. Adv.* **2018**, *4*, eaar6849. [CrossRef]

9. Swanson, D.; Block, R.; Mousa, S.A. Omega-3 fatty acids EPA and DHA: Health benefits throughout life. *Adv. Nutr.* **2012**, *3*, 1–7. [CrossRef]

10. Monroig, Ó.; Kabeya, N. Desaturases and elongases involved in polyunsaturated fatty acid biosynthesis in aquatic invertebrates: A comprehensive review. *Fish. Sci.* **2018**, *84*, 911–928. [CrossRef]

11. Calder, P.C. Very long-chain n-3 fatty acids and human health: Fact, fiction and the future. *Proc. Nutr. Soc.* **2018**, *77*, 52–72. [CrossRef]

12. Surm, J.M.; Prentis, P.J.; Pavasovic, A. Comparative analysis and distribution of omega-3 LCPUFA biosynthesis genes in marine molluscs. *PLoS ONE* **2015**, *10*, e0136301. [CrossRef]

13. Khan, B.M.; Liu, Y. Marine mollusks: Food with benefits. *Compr. Rev. Food Sci. Food Saf.* **2019**, *18*, 548–564. [CrossRef]

14. Ibarguren, M.; López, D.J.; Escribá, P.V. The effect of natural and synthetic fatty acids on membrane structure, microdomain organization, cellular functions and human health. *Biochim. Biophys. Acta* **2014**, *1838*, 1518–1528. [CrossRef] [PubMed]

15. Ishihara, T.; Yoshida, M.; Arita, M. Omega-3 fatty acid-derived mediators that control inflammation and tissue homeostasis. *Int. Immunol.* **2019**, *31*, 559–567. [CrossRef] [PubMed]

16. Maulucci, G.; Cohen, O.; Daniel, B.; Sansone, A.; Petropoulou, P.I.; Filou, S.; Spyridonidis, A.; Pani, G.; De Spirito, M.; Chatgilialoglu, C.; et al. Fatty acid-related modulations of membrane fluidity in cells: Detection and implications. *Free Radic. Res.* **2016**, *50*, S40–S50. [CrossRef] [PubMed]

17. Calder, P.C. Omega-3 polyunsaturated fatty acids and inflammatory processes: Nutrition or pharmacology? *Br. J. Clin. Pharmacol.* **2013**, *75*, 645–662. [CrossRef]

18. Das, U.N.; Ramos, E.J.; Meguid, M.M. Metabolic alterations during inflammation and its modulation by central actions of omega-3 fatty acids. *Curr. Opin. Clin. Nutr. Metab. Care* **2003**, *6*, 413–419. [CrossRef]

19. Klingenberg, R.; Hansson, G.K. Treating inflammation in atherosclerotic cardiovascular disease: Emerging therapies. *Eur. Heart J.* **2009**, *30*, 2838–2844. [CrossRef]

20. Miles, E.A.; Calder, P.C. Can early omega-3 fatty acid exposure reduce risk of childhood allergic disease? *Nutrients* **2017**, *21*, 784. [CrossRef]

21. Schubert, R.; Kitz, R.; Beermann, C.; Rose, M.A.; Baer, P.C.; Zielen, S.; Boehles, H. Influence of low-dose polyunsaturated fatty acids supplementation on the inflammatory response of healthy adults. *Nutrition* **2007**, *23*, 724–730. [CrossRef]

22. Külzow, N.; Witte, A.V.; Kerti, L.; Grittner, U.; Schuchardt, J.P.; Hahn, A.; Flöel, A. Impact of omega-3 fatty acid supplementation on memory functions in healthy older adults. *J. Alzheimers Dis.* **2016**, *51*, 713–725. [CrossRef]

23. Yelland, L.N.; Gajewski, B.J.; Colombo, J.; Gibson, R.A.; Makrides, M.; Carlson, S.E. Predicting the effect of maternal docosahexaenoic acid (DHA) supplementation to reduce early preterm birth in Australia and the United States using results of within country randomized controlled trials. *Prostaglandins Leukot. Essent. Fatty Acids* **2016**, *112*, 44–49. [CrossRef] [PubMed]

24. Endo, J.; Arita, M. Cardioprotective mechanism of omega-3 polyunsaturated fatty acids. *J. Cardiol.* **2016**, *67*, 22–27. [CrossRef] [PubMed]

25. Zhukova, N.V.; Novgorodtseva, T.P. Lipid composition of erythrocytes at cardiovascular and hepatobiliary diseases. In *Lipids: Categories, Biological Functions and Metabolism, Nutrition and Health*; Gilmore, P.L., Ed.;

Nova Science Publishers Inc.: New York, NY, USA, 2010; pp. 1–45. ISBN 978-1-61668-464-8.

26. Novgorodtseva, T.P.; Denisenko, Y.K.; Zhukova, N.V.; Antonyuk, M.V.; Knyshova, V.V.; Gvozdenko, T.A. Modification of the fatty acid composition of the erythrocyte membrane in patients with chronic respiratory diseases. *Lipids Health Dis.* **2013**, *12*, 117. [CrossRef] [PubMed]

27. Simopoulos, A.P. The importance of the omega-6/omega-3 fatty acid ratio in cardiovascular disease and other chronic diseases. *Exp. Biol. Med.* **2008**, *233*, 674–688. [CrossRef] [PubMed]

28. Dyall, S.C. Long-chain omega-3 fatty acids and the brain: A review of the independent and shared effects of EPA, DPA and DHA. *Front. Aging Neurosci.* **2015**, *7*, 52. [CrossRef]

29. Novgorodtseva, T.H.; Karaman, Y.K.; Zhukova, V.V.; Lobanova, T.G.; Antonyuk, M.V.; Kantur, T.A. Composition of fatty acids in plasma and erythrocytes and eicosanoids level in patients with metabolic syndrome. *Lipids Health Dis.* **2011**, *10*, 82. [CrossRef]

30. Echeverria, F.; Valenzuela, R.; Catalina Hernandez-Rodas, M.; Valenzuela, A. Docosahexaenoic acid (DHA), a fundamental fatty acid for the brain: New dietary sources. *Prostaglandins Leukot. Essent. Fatty Acids* **2017**, *124*, 1–10. [CrossRef]

31. Rangel-Huerta, O.D.; Gil, A. Omega 3 fatty acids in cardiovascular disease risk factors: An updated systematic review of randomised clinical trials. *Clin Nutr.* **2018**, *37*, 72–77. [CrossRef]

32. Reimers, A.; Ljung, H. The emerging role of omega-3 fatty acids as a therapeutic option in neuropsychiatric disorders. *Ther. Adv. Psychopharm.* **2019**, *9*, 1–18. [CrossRef]

33. Innis, S.M. Impact of maternal diet on human milk composition and neurological development of infants. *Am. J. Clin. Nutr.* **2014**, *99*, 734S–741S. [CrossRef]

34. Campoy, C.; Escolano-Margarit, V.; Anjos, T.; Szajewska, H.; Uauy, R. Omega 3 fatty acids on child growth, visual acuity and neurodevelopment. *Br. J. Nutr.* **2012**, *107*, S85–S106. [CrossRef] [PubMed]

35. Stonehouse, W.; Conlon, C.A.; Podd, J.; Hill, S.R.; Minihane, A.M.; Haskell, C.; Kennedy, D. DHA supplementation improved both memory and reaction time in healthy young adults: A randomized controlled trial. *Am. J. Clin. Nutr.* **2013**, *97*, 1134–1143. [CrossRef] [PubMed]

36. Witte, A.V.; Kerti, L.; Hermannstädter, H.M.; Fiebach, J.B.; Schreiber, S.J.; Schuchardt, J.P.; Hahn, A.; Flöel, A. Long-chain omega-3 fatty acids improve brain function and structure in older adults. *Cereb Cortex.* **2014**, *24*, 3059–3068. [CrossRef] [PubMed]

37. Larsson, S.C.; Kumlin, M.; Ingelman-Sundberg, M.; Wolk, A. Dietary long-chain n-3 fatty acids for the prevention of cancer: A review of potential mechanisms. *Am. J. Clin. Nutr.* **2004**, *79*, 935–945. [CrossRef]

38. Huerta-Yépez, S.; Tirado-Rodriguez, A.B.; Hankinson, O. Role of diets rich in omega-3 and omega-6 in the development of cancer. *Bol. Med. Hosp. Infant Mex.* **2016**, *73*, 446–456. [CrossRef]

39. Nabavi, S.F.; Bilotto, S.; Russo, G.L.; Orhan, I.E.; Habtemariam, S.; Daglia, M.; Devi, K.P.; Loizzo, M.R.; Tundis, R.; Nabavi, S.M. Omega-3 polyunsaturated fatty acids and cancer: Lessons learned from clinical trials. *Cancer Metastasis Rev.* **2015**, *34*, 359–380. [CrossRef]

40. Volkman, J.K.; Jeffrey, S.W.; Nichols, P.D.; Rogers, G.I.; Garland, C.D. Fatty acid and lipid composition of 10 species of microalgae used in mariculture. *J. Exp. Mar. Biol. Ecol.* **1989**, *128*, 219–240. [CrossRef]

41. Yongmanitchai, W.; Ward, O.P. Growth of and omega-3 fatty acid production by *Phaeodactylum tricornutum* under different culture conditions. *Appl. Environ. Microbiol.* **1991**, *57*, 419–425.

42. Dunstan, G.A.; Volkman, J.K.; Barrtt, S.M.; Leroi, J.-M.; Jeffrey, S.W. Essential polyunsaturated fatty acids from 14 species of diatom (Bacillariophyceae). *Phytochemistry* **1994**, *35*, 155–161. [CrossRef]

43. Patel, A.; Matsakas, L.; Hruzova, K.; Rova, U.; Christakopoulos, P. Biosynthesis of nutraceutical fatty acids by the oleaginous marine microalgae *Phaeodactylum tricornutum* utilizing hydrolysates from organosolv-pretreated birch and spruce biomass. *Mar. Drugs* **2019**, *17*, 119. [CrossRef]

44. Nichols, P.D.; Jones, G.J.; De Leeuw, J.W.; Johns, R.B. The fatty acid and sterol composition of two marine dinoflagellates. *Phytochemistry* **1984**, *23*, 1043–1047. [CrossRef]

45. Mansour, M.P.; Volkman, J.K.; Jackson, A.E.; Blackburn, S.I. The fatty acid and sterol composition of five marine dinoflagellates. *J. Phycol.* **1999**, *35*, 710–720. [CrossRef]

46. Jónasdóttir, S.H. Fatty acid profiles and production in marine phytoplankton. *Mar. Drugs* **2019**, *17*, 151. [CrossRef] [PubMed]

47. Dunstan, G.A.; Volkman, J.K.; Jeffrey, S.W.; Barrtt, S.M. Biochemical composition of microalgae from the green algal classes Chlorophyceae and Prasinophyceae. 2. Lipid classes and fatty acids. *J. Exp. Mar. Biol. Ecol.* **1992**, *161*, 115–134. [CrossRef]

48. Volkman, J.K.; Brown, M.R.; Dunstan, G.A.; Jeffrey, S.W. The biochemical composition of marine microalgae from the class Eustigmatophyceae. *J. Phycol.* **1993**, *29*, 69–78. [CrossRef]

49. Dunstan, G.A.; Brown, M.R.; Volkman, J.K. Cryptophyceaea and Rodophyceae; chemotaxonomy, phylogeny, and application. *Phytochemistry* **2005**, *66*, 2557–2570. [CrossRef]

50. Peltomaa, E.; Johnson, M.D.; Taipale, S.J. Marine cryptophytes are great sources of EPA and DHA. *Mar. Drugs* **2018**, *16*, 3. [CrossRef]

51. Volkman, J.K.; Dunstan, G.A.; Jeffrey, S.W.; Kearney, P.S. Fatty acids from microalgae of the genus *Pavlova*. *Phytochemistry* **1991**, *30*, 1855–1859. [CrossRef]

52. Alonzo, L.; Grima, E.M.; Pérez, J.A.S.; Sánchez, J.L.G.; Camacho, F.G. Fatty acid variation among different isolates of a single strain of *Isochrysis galbana*. *Phytochemistry* **1992**, *31*, 3901–3904. [CrossRef]

53. Fenchel, T. *Ecology of Protozoa*; Science Technical Publishers: Madison, WL, USA, 1987; pp. 1–197. ISBN 978-3-662-06819-9. [CrossRef]

54. Sul, D.; Erwin, J.A. The membrane lipids of the marine ciliated protozoan *Parauronema acutum*. *Biochim. Biophys. Acta* **1997**, *1345*, 162–171. [CrossRef]

55. Zhukova, N.V.; Kharlamenko, V.I. Sources of essential fatty acids in the marine microbial loop. *Aquat. Microb. Ecol.* **1999**, *17*, 153–157. [CrossRef]

56. Bec, A.; Martin-Creuzburg, D.; Von Elert, E. Fatty acid composition of the heterotrophic nanoflagellate *Paraphysomonas* sp.: Influence of diet and *de novo* biosynthesis. *Aquat. Biol.* **2010**, *9*, 107–112. [CrossRef]

57. Zhukova, N.V.; Aizdaicher, N.A. Fatty acid composition of 15 species of marine microalgae. *Phytochemistry* **1995**, *39*, 351–356. [CrossRef]

58. Desvilettes, C.; Bec, A. Formation and transfer of fatty acids in aquatic microbial food webs—role of heterotrophic protists. In *Lipids in Aquatic Ecosystems*; Arts, M.T., Brett, M., Kainz, M., Eds.; Springer: New York, NY, USA, 2009; pp. 25–42.

59. Kelly, J.R.; Scheibling, R.E. Fatty acids as dietary traces in benthic food webs. *Mar. Ecol. Prog. Ser.* **2012**, *446*, 1–22. [CrossRef]

60. Makhutova, O.N.; Sushchik, N.N.; Gladyshev, M.I.; Ageev, A.V.; Pryanichnikova, E.G.; Kalachova, G.S. Is the fatty acid composition of freshwater zoobenthic invertebrates controlled by phylogenetic or trophic factors? *Lipids* **2011**, *46*, 709–721. [CrossRef]

61. Van der Heijden, L.H.; Graeve, M.; Asmus, R.; Rzeznik-Orignac, J.; Niquil, N.; Bernier, Q.; Guillou, G.; Asmus, H.; Lebreton, B. Trophic importance of microphytobenthos and bacteria to meiofauna in soft-bottom intertidal habitats: A combined trophic marker approach. *Mar. Environ. Res.* **2019**, *149*, 50–66. [CrossRef]

62. Graeve, M.; Kattner, G.; Piepenburg, D. Lipids in Arctic benthos: Does the fatty acid and alcohol composition reflect feeding and trophic interactions? *Polar Biol.* **1997**, *18*, 53–61. [CrossRef]

63. Perry, G.J.; Volkman, J.K.; Johns, R.B.; Bavor, H.J. Fatty acids of bacterial origin in contemporary marine sediments. *Geochim. Cosmochim. Acta* **1979**, *43*, 1715–1725. [CrossRef]

64. Gillan, F.T.; Hogg, R.W. A method for the estimation of bacterial biomass and community structure in mangrove-associated sediments. *J. Microbiol. Methods* **1984**, *2*, 275–293. [CrossRef]

65. Findlay, R.H.; Trexler, M.B.; Guckert, J.B.; White, D.C. Laboratory study of disturbance in marine sediments: Response of a microbial community. *Mar. Ecol. Prog. Ser.* **1990**, *62*, 121–133. [CrossRef]

66. Kaneda, T. Iso-fatty and anteiso-fatty acids in bacteria—biosynthesis, function, and taxonomic significance. *Microbiol. Rev.* **1991**, *55*, 288–302. [PubMed]

67. Suhr, S.B.; Pond, D.W.; Gooday, A.J.; Smith, C.R. Selective feeding by benthic foraminifera on phytodetritus on the western Antarctic Peninsula shelf: Evidence from fatty acid biomarker analysis. *Mar. Ecol. Prog. Ser.* **2003**, *262*, 153–162. [CrossRef]

68. Topping, J.N.; Murray, J.W.; Pond, D.W. Sewage effects on the food sources and diets of benthic foraminifera living in oxic sediment: A microcosm experiment. *J. Exp. Mar. Biol. Ecol.* **2006**, *329*, 239–250. [CrossRef]

69. Khotimchenko, S.V.; Vaskovsky, V.E.; Titlyanova, T.V. Fatty acids of marine algae from the Pacific coast of north California. *Bot. Mar.* **2002**, *45*, 17–22. [CrossRef]

70. Bergquist, P.R.; Lawson, M.P.; Lavis, A.; Cambie, R.C. Fatty acid composition and the classification of the Porifera. *Biochem. Syst. Ecol.* **1984**, *12*, 63–84. [CrossRef]

71. Barnathan, G. Non-methylene-interrupted fatty acids from marine invertebrates: Occurrence, characterization and biological properties. *Biochimie* **2009**, *91*, 671–678. [CrossRef] [PubMed]

72. Kornprobst, J.-M.; Barnathan, G. Demospongic acids revisited. *Mar. Drugs* **2010**, *8*, 2569–2577. [CrossRef]

73. Imbs, A.B. High level of tetracosapolyenoic fatty acids in the cold-water mollusk *Tochuina tetraquetra* is

a result of the nudibranch feeding on soft corals. *Polar Biol.* **2016**, *39*, 1511–1514. [CrossRef]

74. Joseph, J.D. Lipid composition of marine and estuarine invertebrates. Part II: Mollusca. *Prog. Lipid Res.* **1982**, *21*, 109–153. [CrossRef]

75. Tsikhon-Lukanina, E.A. *Trofologiya Vodnykh Mollyuskov (Trophology of Aquatic Mollusks)*; Nauka: Moscow, Russia, 1987; pp. 1–177.

76. Zhukova, N.V.; Svetashev, V.I. Non-methylene-interrupted dienoic fatty acids in mollusks from the Sea of Japan. *Comp. Biochem. Physiol.* **1986**, *83*, 643–646. [CrossRef]

77. Zhukova, N.V. Lipids and fatty acids of nudibranch mollusks: Potential sources of bioactive compounds. *Mar. Drugs* **2014**, *12*, 4578–4592. [CrossRef] [PubMed]

78. Kattner, G.; Hagen, W.; Graeve, M.; Albers, C. Exceptional lipids and fatty acids in the pteropod *Clione limacine* Gastropoda from both polar oceans. *Mar. Chem.* **1998**, *61*, 219–228. [CrossRef]

79. Fernandes, I.; Fernandes, T.; Cordeiro, N. Nutritional value and fatty acid profile of two wild edible limpets from the Madeira Archipelago. *Eur. Food Res. Technol.* **2019**, *245*, 895–905. [CrossRef]

80. Zhukova, N.V. Lipid classes and fatty acid composition of the tropical nudibranch mollusks *Chromodoris* sp. and *Phyllidia Coelestis*. *Lipids* **2007**, *42*, 1169–1175. [CrossRef]

81. Gomes, N.G.M.; Fernandes, F.; Madureira-Carvalho, Á.; Valentão, P.; Lobo-da-Cunha, A.; Calado, G.; Andrade, P.B. Profiling of heterobranchia sea slugs from Portuguese coastal waters as producers of anti-cancer and anti-inflammatory agents. *Molecules* **2018**, *23*, 1207. [CrossRef]

82. Lavaud, R.; Artigaud, S.; Le Grand, F.; Donval, A.; Soudant, P.; Flye-Sainte-Marie, J.; Strohmeier, T.; Strand, O.; Leynaert, A.; Beker, B.; et al. New insights into the seasonal feeding ecology of Pecten maximus using pigments, fatty acids and sterols analyses. *Mar. Ecol. Prog. Ser.* **2018**, *590*, 109–129. [CrossRef]

83. Mathieu-Resuge, M.; Kraffe, E.; Le Grand, F.; Boens, A.; Bideau, A.; Lluch-Cota, S.E.; Racotta, I.S.; Schaal, G. Trophic ecology of suspension-feeding bivalves inhabiting a north-eastern Pacific coastal lagoon: Comparison of different biomarkers. *Mar. Environ. Res.* **2019**, *145*, 155–163. [CrossRef]

84. Silina, A.V.; Zhukova, N.V. Growth variability and feeding of scallop *Patinopecten yessoensis* on different bottom sediments: Evidence from fatty acid analysis. *J. Exp. Mar. Biol. Ecol.* **2007**, *348*, 46–59. [CrossRef]

85. Purroya, A.; Najdekb, M.; Islac, E.; Župand, I.; Thébaulte, J.; Peharda, M. Bivalve trophic ecology in the Mediterranean: Spatio-temporal variations and feeding behavior. *Mar. Environ. Res.* **2018**, *142*, 234–249. [CrossRef]

86. Ruano, F.; Ramos, P.; Quaresma, M.; Bandarra, N.M.; Fonseca, I.P. Evolution of fatty acid profile and Condition Index in mollusc bivalves submitted to different depuration periods. *Rev. Port. Cienc. Vet.* **2012**, *107*, 75–84.

87. Chu, F.L.; Webb, K.L.; Chen, J. Seasonal changes of lipids and fatty acids in oyster tissues (*Crassostrea virginica*) an estuarine particulate matter. *Comp. Biochem. Physiol. A* **1990**, *95*, 385–391. [CrossRef]

88. Napolitano, G.E.; Ackman, R.G. Fatty acid dynamics in sea scallops *Placopecten magellanicus* (Gmelin 1791) from Georges Bank, Nova Scotia. *J. Shellfish. Res.* **1993**, *12*, 267–277.

89. Baptista, M.; Repolho, T.; Maulvault, A.L.; Lopes, V.M.; Narciso, L.; Marques, A.; Bandarra, N.; Rosa, R. Temporal dynamics of amino and fatty acid composition in the razor clam *Ensis siliqua* (Mollusca: Bivalvia). *Helgol. Mar. Res.* **2014**, *68*, 465–482. [CrossRef]

90. Hu, X.P.; Da An, Q.; Zhou, D.; Lu, T.; Yin, F.W.; Song, L.; Zhao, Q.; Zhang, J.H.; Qin, L.; Zhu, B.W.; et al. Lipid profiles in different parts of two species of scallops (*Chlamys farreri* and *Patinopecten yessoensis*). *Food Chem.* **2018**, *243*, 319–327. [CrossRef]

91. Nerot, C.; Meziane, T.; Schaal, G.; Grall, J.; Lorrain, A.; Paulet, Y.M.; Kraffe, E. Spatial changes in fatty acids signatures of the great scallop *Pecten maximus* across the Bay of Biscay continental shelf. *Cont. Shelf Res.* **2015**, *109*, 1–9. [CrossRef]

92. Fernández-Reiriz, M.-J.; Garrido, J.L.; Irisarri, J. Fatty acid composition in *Mytilus galloprovincialis* organs: Trophic interactions, sexual differences and differential anatomical distribution. *Mar. Ecol. Prog. Ser.* **2015**, *528*, 221–234. [CrossRef]

93. Najdek, M.; Blazina, M.; Ezgeta-Balic, D.; Peharda, M. Diets of fan shells (*Pinna nobilis*) of different sizes: Fatty acid profiling of digestive gland and adductor muscle. *Mar. Biol.* **2013**, *160*, 921–930. [CrossRef]

94. Phillips, K.L.; Nichols, P.D.; Jackson, G.D. Lipid and fatty acid composition of the mantle and digestive gland of four Southern Ocean squid species: Implications for food-web studies. *Antarct. Sci.* **2002**, *14*, 212–220. [CrossRef]

95. Stowasser, G.; Pierce, G.J.; Moffat, C.F.; Collins, M.A.; Forsythe, J.W. Experimental study on the effect of diet

on fatty acid and stable isotope profiles of the squid *Lolliguncula brevis*. *J. Exp. Mar. Biol. Ecol.* **2006**, *333*, 97–114. [CrossRef]

96. Villanueva, R.; Perricone, V.; Fiorito, G. Cephalopods as predators: A short journey among behavioral flexibilities, adaptions, and feeding habits. *Front. Physiol.* **2017**, *8*, 598. [CrossRef]

97. Ozogul, Y.; Duysak, O.; Ozogul, F.; Ozkutuk, A.S.; Tureli, C. Seasonal effects in the nutritional quality of the body structural tissue of cephalopods. *Food Chem.* **2008**, *108*, 847–852. [CrossRef]

98. Pethybridge, H.; Virtue, P.; Casper, R.; Yoshida, T.; Green, C.P.; Jackson, G.; Nichols, P.D. Seasonal variations in diet of arrow squid (*Nototodarus gouldi*): Stomach content and signature fatty acid analysis. *J. Mar. Biol. Assoc. UK* **2012**, *92*, 187–196. [CrossRef]

99. Phillips, K.L.; Nichols, P.D.; Jackson, G.D. Size-related dietary changes observed in the squid *Moroteuthis ingens* at the Falkland Islands: Stomach contents and fatty-acid analyses. *Polar Biol.* **2003**, *26*, 474–485. [CrossRef]

100. Gong, Y.; Li, Y.; Chen, X.; Chen, L. Potential use of stable isotope and fatty acid analyses for traceability of geographic origins of jumbo squid (*Dosidicus gigas*). *Rapid Commun. Mass Spectrom.* **2018**, *32*, 583–589. [CrossRef] [PubMed]

101. Lin, D.; Han, F.; Xuan, S.; Chen, X. Fatty acid composition and the evidence for mixed income–capital breeding in female Argentinean short-fin squid *Illex argentines*. *Mar. Biol.* **2019**, *166*, 90. [CrossRef]

102. Monroig, Ó.; Navarro, J.C.; Dick, J.R.; Alemany, F.; Tocher, D.R. Identification of a Δ5-like fatty acyl desaturase from the cephalopod *Octopus vulgaris* (Cuvier 1797) involved in the biosynthesis of essential fatty acids. *Mar. Biotechnol.* **2012**, *14*, 411–422. [CrossRef] [PubMed]

103. Fluckiger, M.; Jackson, G.D.; Nichols, P.D.; Wotherspoon, S. An experimental study of the effect of diet on the fatty acid profiles of the European Cuttlefish (*Sepia offcinalis*). *Mar. Biol.* **2008**, *154*, 363–372. [CrossRef]

104. Dubilier, N.; Bergin, C.; Lott, C. Symbiotic diversity in marine animals: The art of harnessing chemosynthesis. *Nat. Rev. Microbiol.* **2008**, *6*, 725–740. [CrossRef]

105. Yellowlees, D.; Rees, T.A.V.; Leggat, W. Metabolic interactions between algal symbionts and invertebrate hosts. *Plant Cell Environ.* **2008**, *31*, 679–694. [CrossRef]

106. Rumpho, M.E.; Pelletreau, K.N.; Moustafa, A.; Bhattacharya, D. The making of a photosynthetic animal. *J. Exp. Biol.* **2011**, *214*, 303–311. [CrossRef]

107. Venn, A.A.; Loram, J.E.; Douglas, A.E. Photosynthetic symbioses in animals. *J. Exp. Bot.* **2008**, *59*, 1069–1080. [CrossRef] [PubMed]

108. O'Connor, R.M.; Fung, J.M.; Sharp, K.H.; Benner, J.S.; McClung, C.; Cushing, S.; Lamkin, E.R.; Fomenkov, A.I.; Henrissat, B.; Londer, Y.Y.; et al. Gill bacteria enable a novel digestive strategy in a wood-feeding mollusk. *Proc. Natl. Acad. Sci. USA* **2014**, *111*, E5096–E5104. [CrossRef] [PubMed]

109. Visik, K.; Ruby, E.G. *Vibrio fischeri* and its host: It takes two to tango. *Curr. Opin. Microbiol.* **2007**, *9*, 632–638. [CrossRef] [PubMed]

110. Davis, J.; Fricke, W.F.; Hamann, M.T.; Esquenazi, E.; Dorrestein, P.C.; Hill, R.T. Characterization of the bacterial community of the chemically defended hawaiian Sacoglossan *Elysia rufescens*. *Appl. Environ. Microbiol.* **2013**, *79*, 7073–7081. [CrossRef] [PubMed]

111. Cronan, J.E.; Thomas, J. Bacterial fatty acid synthesis and its relationships with polyketide synthetic pathways. *Methods Enzymol.* **2009**, *459*, 395–433. [CrossRef] [PubMed]

112. Conway, N.; McDowell Capuzzo, J. Incorporation and utilization of bacterial lipids in the *Solemya velum* symbiosis. *Mar. Biol.* **1991**, *108*, 277–291. [CrossRef]

113. Saito, H. Unusual novel *n*-4 polyunsaturated fatty acids in cold-seep mussels (*Bathymodiolus japonicus* and *Bathymodiolus platifrons*), originating from symbiotic methanotrophic bacteria. *J. Chtomatogr.* **2008**, *1200*, 242–254. [CrossRef]

114. Zhukova, N.V.; Kharlamenko, V.I.; Svetashev, V.I.; Rodionov, I.A. Fatty-acids as markers of bacterial symbionts of marine bivalve mollusks. *J. Exp. Mar. Biol. Ecol.* **1992**, *162*, 253–263. [CrossRef]

115. Kharlamenko, V.I.; Zhukova, N.V.; Khotimchenko, S.V.; Svetashev, V.I.; Kamenev, G.M. Fatty acids as markers of food sources in a shallow-water hydrothermal ecosystem (Kraternaya Bight, Yankich Island, Kurile Islands). *Mar. Ecol. Prog. Ser.* **1995**, *120*, 231–241. [CrossRef]

116. Saito, H. Identification of novel n-4 series polyunsaturated fatty acids in a deep-sea clam, *Calyptogena phaseoliformis*. *J. Chromatogr. A* **2007**, *1163*, 247–259. [CrossRef]

117. Zhukova, N.V.; Eliseikina, M.G. Symbiotic bacteria in the nudibranch mollusk *Dendrodoris nigra*: Fatty acid composition and ultrastructure analysis. *Mar. Biol.* **2012** *159*, 1783–1794. [CrossRef]

118. Saito, H.; Hashimoto, J. Characteristics of the fatty acid composition of a deep-sea vent gastropod, *Ifremeria Naut. Lipids* **2010**, *45*, 537–548. [CrossRef] [PubMed]

119. Kharlamenko, V.I.; Kiyashko, S.I.; Sharina, S.N.; Ivin, V.V.; Krylova, E.M. An ecological study of two species of chemosymbiotrophic bivalve molluscs (Bivalvia: Vesicomyidae: Pliocardiinae) from the Deryugin Basin of the Sea of Okhotsk using analyses of the stable isotope ratios and fatty acid compositions. *Deep Sea Res. Part I* **2019**, *150*, 103058. [CrossRef]

120. Imbs, A.B.; Yakovleva, I.M.; Dautova, T.N.; Bui, L.H.; Jones, P. Diversity of fatty acid composition of symbiotic dinoflagellates in corals: Evidence for the transfer of host PUFAs to the symbionts. *Phytochemistry* **2014**, *101*, 76–82. [CrossRef]

121. Dubousquet, V.; Gros, E.; Berteaux-Lecellier, V.; Viguier, B.; Raharivelomanana, P.; Bertrand, C.; Lecellier, G.J. Changes in fatty acid composition in the giant clam *Tridacna maxima* in response to thermal stress. *Biol. Open* **2016**, *5*, 1400–1407. [CrossRef]

122. Zhukova, N.V.; Titlyanov, E.A. Fatty acid variations in symbiotic dinoflagellates from Okinawan corals. *Phytochemistry* **2003**, *62*, 191–195. [CrossRef]

123. Harwood, J.L.; Guschina, I.A. The versatility of algae and their lipid metabolism. *Biochimie* **2009**, *91*, 679–684. [CrossRef]

124. De Moreno, J.E.A.; Moreno, V.J.; Brenner, R.R. Lipid metabolism of the yellow clam, *Mesodesma macroides*: 2—Polyunsaturated fatty acid metabolism. *Lipids* **1976**, *11*, 561–566. [CrossRef]

125. De Moreno, J.E.A.; Moreno, V.J.; Brenner, R.R. Lipid metabolism of the yellow clam, *Mesodesma macroides*: 3—Saturated fatty acids and acetate metabolism. *Lipids* **1977**, *12*, 804–808. [CrossRef]

126. Waldock, M.J.; Holland, D.L. Fatty acid metabolism in young oyster, *Crassostrea gigas*: Polyunsaturated fatty acids. *Lipids* **1984**, *19*, 332–336. [CrossRef]

127. Reis, D.B.; Acosta, N.G.; Almansa, E.; Navarro, J.C.; Tocher, D.R.; Monroig, Ó.; Andrade, J.P.; Sykes, A.V.; Rodríguez, C. In vivo metabolism of unsaturated fatty acids in *Octopus vulgaris* hatchlings determined by incubation with 14C-labelled fatty acids added directly to seawater as protein complexes. *Aquaculture* **2014**, *431*, 28–33. [CrossRef]

128. Zhukova, N.V. The pathway of the biosynthesis of non-methylene-interrupted dienoic fatty acids in mollusks. *Comp. Biochem. Physiol. B* **1991**, *100*, 801–804. [CrossRef]

129. Zhukova, N.V. Biosynthesis of non-methylene-interrupted dienoic fatty acids for [C-14] acetate in molluscs. *Biochim. Et Biophys. Acta* **1986**, *878*, 131–133. [CrossRef]

130. Li, M.; Mai, K.; He, G.; Ai, Q.; Zhang, W.; Xu, W.; Wang, J.F.; Liufu, Z.; Zhang, Y.; Zhou, H. Characterization of two Δ5 fatty acyl desaturases in abalone (*Haliotis discus* hannai Ino). *Aquaculture* **2013**, *416–417*, 48–56. [CrossRef]

131. Liu, H.; Guo, Z.; Zheng, H.; Wang, S.; Wang, Y.; Liu, W.; Zhang, G. Functional characterization of a Δ5-likefatty acyl desaturase and its expression during early embryogenesis in the noble scallop *Chlamys nobilis* Reeve. *Mol. Biol. Rep.* **2014**, *41*, 7437–7445. [CrossRef] [PubMed]

132. Ran, Z.; Xu, J.; Liao, K.; Monroig, O.; Navarro, J.C.; Oboh, A.; Jin, M.; Zhou, Q.; Zhou, C.; Douglas, R.; et al. Biosynthesis of long-chain polyunsaturated fatty acids in the razor clam *Sinonovacula constricta*: Characterization of four fatty acyl elongases and a novel desaturase capacity. *Biochim. Et Biophys. Acta* **2019**, *1864*, 1083–1090. [CrossRef] [PubMed]

133. Monroig, Ó.; Tocher, D.R.; Navarro, J.C. Biosynthesis of polyunsaturated fatty acids in marine invertebrates: Recent advances in molecular mechanisms. *Mar. Drugs* **2013**, *11*, 3998–4018. [CrossRef]

134. Garrido, D.; Kabeya, N.; Hontoria, F.; Navarro, J.C.; Reis, D.B.; Martín, M.V.; Rodríguez, C.; Almansa, E.; Monroig, Ó. Methyl-end desaturases with Δ12 and ω3 regioselectivities enable the de novo PUFA biosynthesis in the cephalopod *Octopus vulgaris*. *Biochim. Et Biophys. Acta* **2019**, *1864*, 1134–1144. [CrossRef]

135. Ben-Mlih, F.; Marty, J.C.; Fiala-Medioni, A. Fatty acid composition in deep hydrothermal vent symbiotic bivalves. *J. Lipid Res.* **1992**, *33*, 1797–1806.

136. Zhang, H.; Liu, H.; Cheng, D.; Liu, H.; Zheng, H. Molecular cloning and functional characterisation of a polyunsaturated fatty acid elongase in a marine bivalve *Crassostrea angulata*. *J. Food Nutr. Res.* **2018**, *6*, 89–95. [CrossRef]

137. Allen, C.E.; Tyler, P.A.; Van Dover, C.L. Lipid composition of the hydrothermal vent clam *Calyptogena pacifica* (Mollusca: Bivalvia) as a trophic indicator. *J. Mar. Biol. Assoc. UK* **2001**, *81*, 817–821. [CrossRef]

138. Howell, K.L.; Pond, D.W.; Billett, D.S.M.; Tyler, P.A. Feeding ecology of deep- sea seastars (Echinodermata: Asteroidea): A fatty acid biomarker approach. *Mar. Ecol. Prog. Ser.* **2003**, *255*, 193–206. [CrossRef]

139. Phleger, C.F.; Nelson, M.M.; Groce, A.K.; Cary, S.C.; Coyne, K.J.; Nichols, P.D. Lipid composition of deep-sea hydrothermal vent tubeworm *Riftia pachyptila*, crabs *Munidopsis subsquamosa* and *Bythograea thermydron*, mussels *Bathymodiolus* sp. and limpets *Lepetodrilus* spp. *Comp. Biochem. Physiol. B* **2005**, *141*, 196–210. [CrossRef] [PubMed]

140. Klingensmith, J.S. Distribution of methylene and nonmethylene-interrupted dienoic fatty acids in polar lipids and triacylglycerols of selected tissues of the hardshell clam (*Mercinaria mercenaria*). *Lipids* **1982**, *17*, 976–981. [CrossRef] [PubMed]

141. Pranal, V.; Fiala-Medioni, A.; Guezennec, J. Fatty acid characteristics in two symbiotic gastropods from a deep hydrothermal vent of the West Pacific. *Mar. Ecol. Prog. Ser.* **1996**, *142*, 175–184. [CrossRef]

142. Hurtado, M.; Racotta, I.; Arcos, F.; Morales-Bojorquez, E.; Moal, J.; Soudant, P.; Palacios, E. Seasonal variations of biochemical, pigment, fatty acid, and sterol compositions in female *Crassostrea corteziensis* oysters in relation to the reproductive cycle. *Comp. Biochem. Physiol. B* **2012**, *163*, 172–183. [CrossRef] [PubMed]

143. Ezgeta-Balic, D.; Najdek, M.; Peharda, M.; Blazina, M. Seasonal fatty acid profile analysis to trace origin of food sources of four commercially important bivalves. *Aquaculture* **2012**, *334*, 89–100. [CrossRef]

144. Pernet, F.; Tremblay, R.; Comeau, L.; Guderley, H. Temperature adaptation in two bivalve species from different thermal habitats: Energetics and remodelling of membrane lipids. *J. Exp. Biol.* **2007**, *210*, 2999–3014. [CrossRef]

145. Pernet, F.; Tremblay, R.; Redjah, I.; Sevigny, J.-M.; Gionet, C. Physiological and biochemical traits correlate with differences in growth rate and temperature adaptation among groups of the eastern oyster *Crassostrea virginica*. *J. Exp. Biol.* **2008**, *211*, 969–977. [CrossRef]

146. Shimeta, J.; Amos, C.L.; Beaulien, S.E.; Katz, S.L. Resuspension of benthic protists at subtidal coastal sites with differing sediment composition. *Mar. Ecol. Prog. Ser.* **2003**, *259*, 103–115. [CrossRef]

147. Gonçalves, A.M.M.; Borroso, D.V.; Serafim, T.L.; Verdelhos, T.; Marques, J.C.; Gonçalves, F. The biochemical response of two commercial bivalve species to exposure to strong salinity changes illustrated by selected biomarkers. *Ecol. Indic.* **2017**, *77*, 59–66. [CrossRef]

148. Nemova, N.N.; Fokina, N.N.; Nefedova, Z.A. Modifications of gill lipid composition in littoral and cultured blue mussels *Mytilus edulis* L. under the influence of ambient salinity. *Polar Rec.* **2013**, *154*, 217–225. [CrossRef]

149. Fokina, N.N.; Ruokolainen, T.R.; Nemova, N.N.; Bakhmet, I.N. Changes of blue mussels *Mytilus edulis* L. lipid composition under cadmium and copper toxic effect. *Biol. Trace Elem. Res.* **2013**, *154*, 217–225. [CrossRef] [PubMed]

150. Chelomin, V.P.; Belcheva, N.N. Alterations of microsomal lipid-synthesis in gill cells of bivalve mollusk *Mizuhopecten-yessoensis* in response to cadmium accumulation. *Comp. Biochem. Phys. C* **1991**, *99*, 1–5. [CrossRef]

151. Gonçalves, F.; Mesquita, A.F.; Verdelhos, T.; Coutinho, J.A.P.; Marques, J.C.; Gonçalves, A.M.M. Fatty acids' profiles as indicators of stress induced by of a common herbicide on two marine bivalves species: *Cerastoderma edule* (Linnaeus, 1758) and *Scrobicularia plana* (da Costa, 1778). *Ecol. Indic.* **2016**, *63*, 209–218. [CrossRef]

152. Signa, G.; Di Leonardo, R.; Vaccaro, A.; Tramati, C.D.; Mazzola, A.; Vizzini, S. Lipid and fatty acid biomarkers as proxies for environmental contamination in caged mussels *Mytilus Galloprovincialis*. *Ecol. Idic.* **2015**, *57*, 384–394. [CrossRef]

153. Mesquita, A.F.; Gonçalves, F.; Verdelhos, T.; Marques, J.C.; Gonçalves, A.M.M. Fatty acids profiles modifications in the bivalves *Cerastoderma edule* and *Scrobicularia plana* in response to copper sulphate. *Ecol. Indic.* **2018**, *85*, 318–328. [CrossRef]

154. Perrat, E.; Couzinet-Mossion, A.; FossiTankoua, O.; Amiard-Triquet, C.; Wielgosz-Collinn, G. Variation of content of lipid classes, sterols and fatty acids in gonads and digestive glands of *Scrobicularia plana* in relation to environment pollution levels. *Ecotoxicol. Environ. Saf.* **2013**, *90*, 112–120. [CrossRef]

14

Origin of Carbon and Essential Fatty Acids in Higher Trophic Level Fish in Headwater Stream Food Webs

Megumu Fujibayashi [1,2,*], Yoshie Miura [1], Reina Suganuma [1], Shinji Takahashi [3], Takashi Sakamaki [4] and Naoyuki Miyata [1]

[1] Faculty of Bioresource Sciences, Akita Prefectural University, Kaidobata-Nishi 241-438, Shimoshinjo Nakano, Akita city, Akita Prefecture 010-0195, Japan

[2] Department of Urban and Environmental Engineering, Faculty of Engineering, Kyushu University, 744, Motooka, Nishiku, Fukuoka 819-0395, Japan

[3] Technical Division, School of Engineering, Tohoku University, 6-6-06 Aoba, Aoba, Sendai, Miyagi 980-8579, Japan

[4] Department of Civil and Environmental Engineering, School of Engineering, Tohoku University, 6-6-06 Aoba, Aoba, Sendai, Miyagi 980-8579, Japan

*   Correspondence:m.fujibayashi@civil.kyushu-u.ac.jp

Abstract: Dietary carbon sources in headwater stream food webs are divided into allochthonous and autochthonous organic matters. We hypothesized that: 1) the dietary allochthonous contribution for fish in headwater stream food webs positively relate with canopy cover; and 2) essential fatty acids originate from autochthonous organic matter regardless of canopy covers, because essential fatty acids, such as 20:5ω3 and 22:6ω3, are normally absent in allochthonous organic matters. We investigated predatory fish *Salvelinus leucomaenis* stomach contents in four headwater stream systems, which are located in subarctic region in northern Japan. In addition, stable carbon and nitrogen isotope ratios, fatty acid profile, and stable carbon isotope ratios of essential fatty acids were analyzed. Bulk stable carbon analysis showed the major contribution of autochthonous sources to assimilated carbon in *S. leucomaenis*. Surface baits in the stomach had intermediate stable carbon isotope ratios between autochthonous and allochthonous organic matter, indicating aquatic carbon was partly assimilated by surface baits. Stable carbon isotope ratios of essential fatty acids showed a positive relationship between autochthonous sources and *S. leucomaenis* across four study sites. This study demonstrated that the main supplier of dietary carbon and essential fatty acids was autochthonous organic matter even in headwater stream ecosystems under high canopy cover.

Keywords: fatty acids; dietary sources; allochthonous; *Salvelinus leucomaenis*

1. Introduction

Aquatic animals are supported by two basal organic carbon sources, autochthonous (aquatic primary producers) and allochthonous sources (fallen leaf litter and insects from surrounding terrestrial ecosystems) [1]. Contributions of these basal organic carbon sources to aquatic food webs depend on the proportion of microalgal and phytoplankton productivity, and the number of terrestrial subsidies [2]. For lower order headwater streams, as high canopy cover promotes abundant inputs of litter falls [3] and limited productivity of attached algae due to shading effects [4], the main carbon sources for aquatic consumers are predicted as allochthonous by the river continuum concept (RCC) [5]. However, food web studies in headwater streams have demonstrated that the dominant dietary contribution for macroinvertebrates is both autochthonous [6,7] and allochthonous [8,9], indicating that the predominant carbon sources for headwater stream food webs are unclear.

Previous studies on the dietary contribution of these basal organic carbon sources have mainly focused on quantitative contribution; however, studies focusing on dietary quality are relatively rare in headwater stream food webs [10,11]. For instance, essential fatty acids are known to be important nutrition for fish health [12–14]. In particular, the roles of 20:5ω3 and 22:6ω3 in fish growth, survival, and reproduction have been studied in many species [15–17]. These studies have demonstrated that dietary essential fatty acids could improve fish condition. Freshwater fish can synthesize 20:5ω3 and 22:6ω3 if 18:3ω3, which is precursor of these two essential fatty acids, is available from dietary sources [13]. However, the conversion efficiency in aquatic animals is generally very low. Consequently, direct intake of these fatty acids from dietary sources is required [18]. However, terrestrial organic matter contains only 18:3ω3 but not 20:5ω3 and 22:6ω3, indicating that terrestrial organic matter is nutritionally poor [18]. 20:5ω3 is present at high levels in diatoms [19], which are sometimes the dominant algae attached to the surface of substrates in headstream ecosystems [20]. This implies that autochthonous organic carbon (i.e., attached algae) may be the main essential fatty acid source for consumers, although other carbon components are derived from allochthonous inputs in headwater streams.

Stable isotope ratios of bulk carbon in consumers reflect those of assimilated diet with only minor fractionation (<1‰), which enable us to infer its dietary carbon sources [21]. In addition, stable carbon isotope ratios of autochthonous organic sources (e.g., attached algae) and allochthonous organic sources are distinguishable in many cases [8,22]. Accordingly, the contribution of allochthonous organic matter to stream food webs has been evaluated by bulk carbon stable isotope ratios [8]. Stable isotope ratios of bulk nitrogen ($\delta^{15}N$) have also been used in food web studies, because of its usefulness to evaluate the trophic position of animals owing to the substantial enrichment of approximately 3‰ relative to that of assimilated diet [23–25].

For tracing the dietary essential fatty acid origin, although their carbon stable isotope ratios would be helpful, information about isotopic fractionation of essential fatty acids between diet and consumers is limited. For instance, Budge et al. [26] demonstrated that there was no isotopic fractionation in several essential fatty acids between the diet and serum of fish in a feeding experiment with Atlantic Pollock. Moreover, an almost isotopically unchanged transfer of 18:2ω6 and 18:3ω3 between diet and zebrafish *Danio rerio* was observed in a 100-day feeding experiment with constant dietary sources [27]. Several aquatic food web studies have already assumed no fractionation of isotopic value of essential fatty acids [28,29]. On the contrary Gradyshev et al. [30] found gradual depletion of stable carbon isotope ratios of essential fatty acids, including 18:2ω6 and 18:3ω3, through higher trophic levels in a stream food chain, suggesting that fractionation was negative. Depleted fractionation in 18:3ω3 was also reported in zooplankton in a feeding experiment [31]. Depleted fractionation in other essential fatty acids have been also reported by a feeding experiments with Daphnia [32]. This can be explained by lighter compounds being assimilated preferentially. As above, the information on fractionation of fatty acids are not sufficient and they were conflicting (i.e., no or small fractionation and negative fractionation). Nielsen et al. [33] pointed out more information on the fractionation of fatty acids are required for diet tracing study. Thus, we did not apply a constant value of isotopic fractionation of essential fatty acids in a food chain in this study. We applied a correlation analysis of stable carbon isotope ratios of essential fatty acids between diet and consumer among the study sites. If organic sources consistently contribute to consumers and have wider isotopic differences, one would expect to detect a significant and positive relationship in stable carbon isotope ratios of essential fatty acids between consumers and assimilated food source regardless of isotopic fractionation [34–36].

Here, we tested the following two hypotheses that: 1) the dietary allochthonous contribution for fish in headwater stream food webs positively relates with canopy cover; and 2) essential fatty acids originate from autochthonous organic matter regardless of canopy covers. To test these hypotheses, we analyzed fatty acid compositions and bulk carbon and compound-specific isotope ratios in fish from four headstream ecosystems.

2. Materials and Methods

2.1. Sampling

We conducted field surveys in four headwater streams, located in subarctic area in the northern part of Japan from July to September 2016 (Table 1). Canopy cover was calculated from a hemispherical photography taken from the center of each stream using CanopOn2 program [37].

Table 1. Description of study sites in this study.

Study Site	GPS	Order	Sampling Date	Canopy (%)	Water Temperature (°C)
Babame	N39.8678°, E140.2552°	1	10 July	93.5	15.9
Hayakuchi	N40.4227°, E140.3470°	3	20 July	67.6	15.6
Kurikoma	N38.9169°, E140.7356°	1	2 September	91.1	14.2
Naruse	N 39.0716°, E 140.7187°	3	18 September	63.5	18.5

Salvelinus leucomaenis is a dominant predatory fish in these four study sites. The main dietary sources of *S. leucomaenis* are larvae and adults of aquatic insects and terrestrial insects [38,39]. *S. leucomaenis* was sampled by fishing. The total length and whole wet weight were measured and a muscle near the pelvic fins was dissected for further analyses. The stomach was preserved in 90% ethanol. For autochthonous organic sources analyses, epilithic biofilm, which was mainly composed of attached algae, was removed using a brush from several randomly selected stones. The collected epilithic biofilm was placed in a plastic sampling bottle with distilled water. The bottle containing algae was filtered through two glass filters (GFF; Whatman, Little Chalfont, UK) in the laboratory for fatty acid and bulk stable carbon and nitrogen isotope analyses. For allochthonous organic sources analyses, decomposed immersed leaf litter was sampled into a plastic bag. Both autochthonous and allochthonous organic sources were sampled in triplicate. The larvae of aquatic insects including Ephemeroptera, Trichoptera, and Plecoptera, which are potentially a direct dietary source for *S. leucomaenis*, were collected using D-frame nets (250 μm mesh) and sorted in the laboratory. Heptageniidae and Ephemerellidae were used for further analysis as they were dominant and commonly detected across all four study sites. All collected samples were transported to the laboratory in a cooler box. All samples were placed in a plastic bag separately and stored in a freezer at −20 °C until further analysis.

Additional sampling for analyses of bulk stable isotope ratios of *S. leucomaenis*, its stomach contents, epilithic biofilms, and leaf litter were conducted in the same location of Babame in July 2018. The same sampling procedure was applied except for stomach contents. The stomach of *S. leucomaenis* was placed in a plastic bottle with distilled water and moved to laboratory in a cooler. The stomach contents were identified and separated. *S. leucomaenis*, epilithic biofilms, and leaf litter were treated following the method mentioned above. All samples were preserved in a plastic bag and stored in a freezer at −20 °C until further analysis.

2.2. Analyses

The stomach contents of each *S. leucomaenis* individual were divided into four groups based on the morphological characteristic using a stereoscopic microscope: water baits (larvae of aquatic insects) and surface baits (adults of aquatic insects and terrestrial insects) according to the definition of Tsuda [38], terrestrial plants, and unknown. Each group was weighted and the contribution of each group was calculated.

Freeze-dried samples of *S. leucomaenis*, both organic sources, and aquatic insects were used for the 'one-step method' [40] for fatty acid analysis. For aquatic insects, two or three individuals were pooled as one sample, and prepared three samples in each study sites. Freeze-dried samples were moved to a 10 mL glass tube. For *S. leucomaenis*, aquatic insects, and leaf litter, approximately 50 mg of homogenized sample was used. For epilithic biofilms, one sheet of GFF was used in the analysis. One milliliter of an internal standard (0.1 mg of tricosanoic acid per 1 mL of hexane), 1 mL of hexane, and 0.8 mL of 14% boron trifluoride methanol were added to the 10 mL glass tube. Nitrogen gas was then used to fill the head space. The glass tube was placed in a 100 °C dry bath for 2 h, followed by cooling to room temperature, and 0.5 mL of hexane and 1 mL of ultrapure water were added. The glass tube was vigorously shaken manually and centrifuged for 3 min at 2,500 rpm. The upper layer of hexane, containing fatty acid methyl esters (FAMEs), was transferred to a 1.5 mL gas chromatography (GC) vial. Solid residues of *S. leucomaenis* were used for bulk carbon and nitrogen isotope ratio analysis.

One microliter of FAME solution was injected in a gas chromatograph (Trace GC, Thermo Fisher Scientific, Bremen, Germany) equipped with a capillary column (Select FAME, 100 m × 0.25 mm i.d.; Agilent Technologies, Santa Clara, CA, USA). The GC analysis was carried out under the analytical conditions described by Fujibayashi et al. [28]. Each fatty acid peak was identified by comparing their retention times with those of commercial authentic standard mixtures (Supelco, Inc., Bellefonte, PA, USA). The peak area was used for calculating the contribution of each fatty acid to total fatty acids.

After fatty acid analysis by GC, the remaining hexane sample was used to analyze the essential fatty acids stable carbon isotope ratio. FAMEs in the hexane solution were injected into a GC–isotope ratio mass spectroscopy instrument (Trace GC Ultra/Delta-V Advantage; Thermo Fisher Scientific, Bremen, Germany), which was equipped with a capillary column (SP2560, 100 m × 0.25 mm i.d.; Supelco, Inc., Bellefonte, PA, USA). The operating conditions were as described by Fujibayashi et al. [29]. Each essential fatty acid peak was identified as described above for the GC-FID analysis. Stable carbon isotope ratios of fatty acids were determined using the following formula:

$$\delta^{13}\text{C or N (‰)} = (\text{R}_{\text{sample}}/\text{R}_{\text{standard}} - 1) \times 1000 \qquad (1)$$

where R_{sample} is the $^{13}\text{C}/^{12}\text{C}$ ratio of the sample, and $\text{R}_{\text{standard}}$ is the $^{13}\text{C}/^{12}\text{C}$ ratio of the international isotopic standard (i.e., Vienna Pee Dee Belemnite).

Correction for the effect of additional carbon from boron trifluoride methanol on $\delta^{13}\text{C}$ was conducted according to Fujibayashi et al. [29]. The stable carbon isotopes of fatty acids in *S. leucomaenis*, epilithic biofilms, and terrestrial litter samples were analyzed.

Dried solid residues of *S. leucomaenis*, subsamples of freeze-dried terrestrial litter, GFFs (epilithic biofilms), and aquatic insects were used for bulk stable carbon and nitrogen isotope ratio analysis. Utilization of solid residues after a one-step method potentially changes the isotopic value. Therefore, the relationship between the stable isotope ratios of carbon and nitrogen in original samples and those in the corresponding dried solid residue after the one-step method were checked in advance with freshwater fish muscle samples (Supplementary file Figure S1). For nitrogen, while a significant positive relationship was detected, variation was relatively high. Thus, estimation of trophic position of *S. leucomaenis* using solid resides may include some extents of error. However, stable carbon isotope rations of solid residues well reflected that of the original samples, and the utilization of solid resides for stable carbon isotope analysis was applied in this study. For the sampling of aquatic insects, one individual was used for one sample, and we prepared three samples for each study site. All samples were weighed in microcapsules and injected into an elemental analyzer (Flash EA; Thermo Fisher Scientific, Bremen, Germany) linked to a mass spectrometer (Delta-V Advantage; Thermo Fisher Scientific, Bremen, Germany). Stable isotope ratios of bulk carbon and nitrogen were expressed as Equation (1); where R_{sample} is the $^{13}\text{C}/^{12}\text{C}$ or $^{15}\text{N}/^{14}\text{N}$ ratio of the sample, and $\text{R}_{\text{standard}}$ is the $^{13}\text{C}/^{12}\text{C}$ and the $^{15}\text{N}/^{14}\text{N}$ ratio of the international isotopic standard (Vienna Pee Dee Belemnite, and atmospheric N_2, respectively).

3. Results

Although canopy cover was high in both the first order rivers, Babame and Kurikoma, with 93.5% and 91.1%, respectively; the third-order rivers, Hayakuchi and Naruse, had relatively open canopy with 67.6% and 63.5%, respectively.

Ten and 11 individuals of *S. leucomaenis* were caught in Babame and Kurikoma by fishing; however, just one individual was caught in Hayakuchi and Naruse. There was a relatively high proportion, ranging from 31% to 68%, of unknown components in the stomach contents that could not be identified because of decomposition (Figure 1). Terrestrial plants were almost not detected in the stomach contents, while water and surface baits were dominant in the stomach contents in *S. leucomaenis*. There was no obvious relationship between water and surface bait contribution and canopy cover.

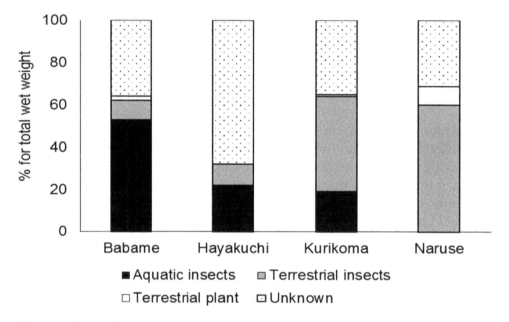

Figure 1. Stomach contents (wet weight %) of *S. leucomaenis* from four study sites.

Although stable carbon and nitrogen isotope ratios of *S. leucomaenis* varied among study sites, the trophic position of *S. leucomaenis* was always the highest (Figure 2). Aquatic insects were generally at a lower position than *S. leucomaenis* with similar carbon isotopic values. Leaf litter showed the most depleted isotopic value for both carbon and nitrogen in all study sites. The range of stable carbon isotope ratios of leaf litter was relatively narrow, from −31.5‰ in Hayakuchi to −29.8‰ in Kurikoma. For epilithic biofilms, stable carbon isotope ratios were more enriched than terrestrial litter and showed a wider range, from −27.2‰ in Kurikoma to −23.8‰ in Hayakuchi.

All essential fatty acids were detected in *S. leucomaenis* (Figure 3). The major essential fatty acid was 22:6ω3. The average contribution of 22:6ω3 was the highest in *S. leucomaenis* from Kurikoma. With respect to other essential fatty acids, the other omega-3 fatty acids, such as 18:3ω3 and 20:5ω3, presented a higher contribution than that of omega-6 fatty acids. In both aquatic insects, essential fatty acid distribution was similar, with no 22:6ω3. The major fatty acids in both aquatic insects were 18:3ω3 and 20:5ω3. This essential fatty acid pattern was similar among study sites. Epilithic biofilms mainly consisted of 18:3ω3 and 20:5ω3. The contribution of 20:5ω3 was relatively constant in all study sites, while 18:3ω3 contribution varied widely among study sites. Only small amounts of 22:6ω3 were detected from epilithic biofilms. Terrestrial litter only contained 18:2ω6 and 18:3ω3. Other C20 essential fatty acids were only detected at low percentages.

The stable carbon isotope ratios of 18:2ω6, 20:4ω6, and 20:5ω3 in *S. leucomaenis* and epilithic biofilms were almost the same across study sites (Supplementary file Figure S2); consequently, a significant or marginally positive relationship was detected between them (correlation analysis: 18:2ω6, n = 4, r = 0.98, *p* < 0.01; 20:5ω3, n = 4, r = 0.99, *p* < 0.001; 20:4ω6, n = 4, r = 0.94, *p* = 0.063) (Figure 4). Although there was no statistical significance, a positive trend was detected between the stable carbon isotope ratios of 18:3ω3 in epilithic biofilms and that in *S. leucomaenis*. The stable carbon isotope ratios of 18:3ω3 were generally lower in the epilithic biofilms than in *S. leucomaenis*

(Supplementary file Figure S2). Contrarily, the stable carbon isotope ratios of 18:2ω6 and 18:3ω3 in leaf litter were not positively correlated with those of in *S. leucomaenis*.

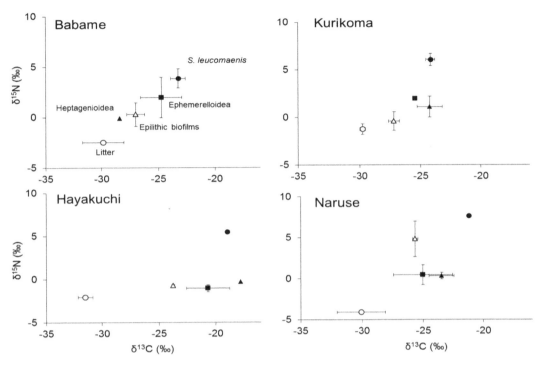

Figure 2. Stable isotope ratios biplot for bulk carbon and nitrogen in basal organic carbon sources and consumers in four study sites. Error bars represent standard deviation.

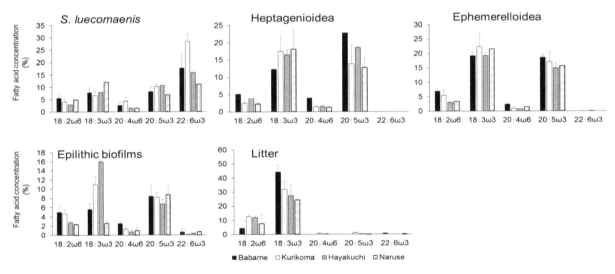

Figure 3. Contribution of essential fatty acids in basal organic carbon sources and consumers in the four study sites. Error bars represent standard deviation.

The bulk carbon and nitrogen isotope ratios of leaf litter in Babame in July 2018 showed values similar to those in July 2016, −29.3‰ for carbon, and −2.1‰ for nitrogen (Figure 5). The carbon and nitrogen stable isotope ratios of epilithic biofilms were −25.1‰ and 4.3‰, respectively. Terrestrial insects in the stomach of *S. leucomaenis* were between leaf litters and epilithic biofilms for carbon and nitrogen stable isotope ratios, with a mean value of −26.7‰ for carbon and 2.2‰ for nitrogen. The bulk stable carbon and nitrogen isotope ratios of *S. leucomaenis* were the most enriched among all samples, and close to those of epilithic biofilms.

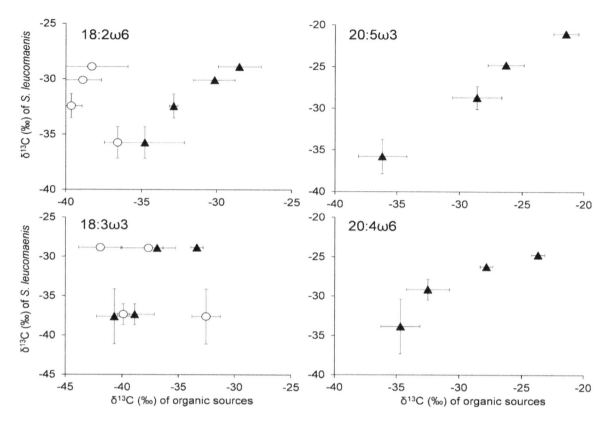

Figure 4. Relationship between stable isotope ratios of essential fatty acid in *S. leucomaenis* and basal organic sources from the four study sites. The black triangle and open circle represent autochthonous organic sources (epilithic biofilms) and allochthonous organic sources (leaf litters), respectively. Error bars represent standard deviation.

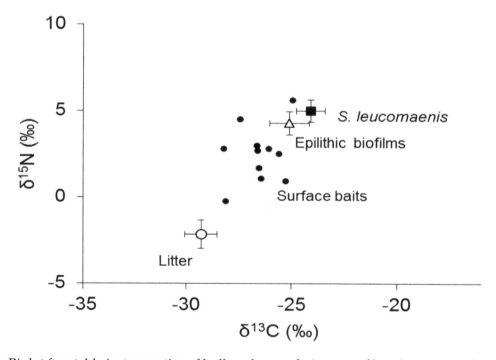

Figure 5. Biplot for stable isotope ratios of bulk carbon and nitrogen of basal organic carbon sources, *S. leucomaenis,* and surface baits from *S. leucomaenis* stomachs in Babame, 2018. Error bars represent standard deviation.

4. Discussion

4.1. Origin of Organic Sources

Canopy cover is an important factor for biogeochemical and biological processes in headwater streams [41]. High leaf litter input and limited primary production are expected in headwater streams. Therefore, a positive relationship between allochthonous contribution and canopy cover was expected [42]. For instance, dietary inputs of surface baits, such as emerged aquatic insects and terrestrial insects, can be expected to increase as the canopy cover increased. However, the stomach contents did not show the expected patterns. Furthermore, our results of bulk carbon stable isotope ratios demonstrated that allochthonous contribution was very rare in *S. leucomaenis*, even in the Babame and Kurikoma study sites, where the canopy cover was very high (>90%). The isotopic positions of *S. leucomaenis* were relatively close to aquatic insects for carbon and higher for nitrogen, indicating that aquatic insects were diet items for *S. leucomaenis*. There were isotopic differences of 2–3‰ in carbon between epilithic biofilms and aquatic insects. The differences seem to be relatively high if a dietary relationship was assumed between epilithic biofilms and aquatic insects, because 0–1‰ fractionations have generally been assumed [21]. These relatively high differences may be explained under some assumptions. First, epilithic biofilms are the mixture of various organic sources with not only algal species, but also terrestrial organic matter [43]. According to our fatty acid analysis of epilithic biofilms, 24:0, which is a fatty acid biomarker of higher plants [44], was detected at 1–2% (data not shown), indicating that the analyzed epilithic biofilms contained terrestrial organic matter. Therefore, the stable carbon isotope ratios are the average of all contained organic matter [43]. The stable carbon isotope ratios of leaf litter showed generally low values in our study sites, indicating pure attached algae stable isotope values were likely more enriched than the analyzed values. If aquatic grazers can selectively utilize specific preferred carbon sources from periphyton [45], the observed wider fractionation between aquatic insects and epilithic biofilms is explainable. This wider fractionation could also be explained by temporal variation of stable carbon isotope ratios in epilithic biofilms. Although the leaf litter stable carbon isotope ratios were relatively constant among the study sites, those of epilithic biofilms widely varied, even in the same study sites between 2016 and 2018 in Babame. It is known that algal stable carbon ratios varied along the gradient of some environmental factors such as isotopic value of dissolved inorganic carbon [8,46] and growth stage [46,47]. This potentially high variability made it difficult to infer the algae dietary contribution for consumers under one-time sampling of stable carbon isotope. Contrastingly, stable carbon isotope ratios of animals were considered to integrate relatively long times [22]. The stable carbon isotope ratios of aquatic insects were generally higher than those of leaf litter and relatively similar to those of attached algae. Thus, the main organic source for *S. leucomaenis* seems to be autochthonous, that is, epilithic biofilms transferred through aquatic insects, regardless of canopy cover in the study sites.

Major dietary contribution of autochthonous sources for headstream consumers has been reported for tropical [6], subtropical [48], and temperate regions [7]. For instance, Lewis et al. [49] showed a major contribution of autochthonous dietary input even under the dominant input of allochthonous organic matters. These contrary observations against the predictions of RCC can be attributed to poor food quality of terrestrial organic matters and high quality of algae [18,50]. However, as observed from the stomach contents analysis, there was a substantial contribution of fallen insects to *S. leucomaenis* diet, in accordance to previous studies on *S. leucomaenis* stomach content [51]. In this study, we could not identify each species. However, according to the stable carbon isotope ratio analysis of surface baits in stomach contents, the carbon sources of these surface baits must have been derived partly from aquatic algae. Some riparian insects spend larval life in aquatic ecosystems and have been known to play an important role transferring highly unsaturated fatty acids from aquatic to terrestrial ecosystems by emerging [52]. This indicated that parts of surface baits can also be a vector of autochthonous carbon going back to aquatic ecosystems. Several researchers pointed out that dietary utilization of riparian insects is one of the pathways to acquire autochthonous organic sources [18]. Therefore, to evaluate organic matter origin, assimilation-based methods (e.g., stable isotope and fatty acids) are required.

4.2. Origin of Essential Fatty Acids

It is well known that lipids have more negative $\delta^{13}C$ values than that of other biochemical compounds because lighter carbons tend to be used for conversion of pyruvate to acetyl coenzyme A in lipid synthesis [53]. Therefore, several studies have reported more depleted isotopic values in essential fatty acids than that of bulk carbon [27,30]. Similarly, stable isotope ratios of essential fatty acids were substantially depleted compared with that of bulk carbon in the current study. Furthermore, spatial difference was also observed in both bulk and essential fatty acids isotope. Enriched isotopic value was observed in Hayakuchi and Naruse Rivers where canopy cover was relatively low, indicating high availability of sunlight for photosynthesis. It is known that high growth rate makes algal isotopic value enriched due to the increase of contribution of heavy CO_2 [46]. The observed wider difference in isotopic ratios of epilithic biofilm can be reflected in the photosynthetic activity in each study site.

S. leucomaenis contained all essential fatty acids. 18:2ω6 and 18:3ω3 were detected in both organic carbon sources, leaf litters and epilithic biofilms. These fatty acids are not synthesized by animals [54], indicating the origin of these fatty acids in *S. leucomaenis* was either or both of them. We observed a significant positive relationship of stable carbon isotope ratios of 18:2ω6 between epilithic biofilms and *S. leucomaenis* indicating that this essential fatty acids in *S. leucomaenis* was mainly of autochthonous origin. For 18:3ω3, although a positive tendency was observed between epilithic biofilms and *S. leucomaenis*, this relationship was not statistically significant. However, since 18:3ω3 cannot be synthesized by *S. leucomaenis*, 18:3ω3 must come from either epilithic biofilms or leaf litter. Epilithic biofilms seem to be a probable candidate for the origin in 18:3ω3 for *S. leucomaenis* because isotopic values of leaf litter were highly depleted compared to that of *S. leucomaenis*, which cannot be explained by the previously reported isotopic fractionation, small [27] or depleted and inconsistent [32].

The origin of 20:5ω3 and 20:4ω6 seems to be epilithic biofilms or biosynthesis from their corresponding precursors, namely 18:3ω3 and 18:2ω6, respectively [13]. The stable carbon isotope ratios of both essential fatty acids showed positive relationships between *S. leucomaenis* and epilithic biofilm, indicating that these essential fatty acids were also of autochthonous origin. If we assume the origin of essential fatty acids to be epilithic biofilm, isotopic fractionation through two trophic levels, namely epilithic biofilm, aquatic insects, and *S. leucomaenis* for 18:2ω6, 20:5ω3, and 20:4ω6 was −0.9–0.4‰, 0–1.5‰, and −1.0–3.3‰, respectively. On the contrary, for 18:3ω3, the expected isotopic fractionation of epilithic biofilm to *S. leucomaenis* via aquatic insects was 1.6–8.0‰. Fujibayashi et al. [29] found no significant difference of isotopic value of 18:3ω3 between freshwater fish and blooming cyanobacteria. However, the mechanism of this inconsistent and variable fractionation for 18:3ω3 was not explainable in this study. Further research is required for isotopic fractionation of essential fatty acids in food chains.

While 22:6ω3 was the most abundant essential fatty in *S. leucomaenis*, both organic sources and aquatic insects did not contain 22:6ω3. However, we only analyzed the fatty acid content in two ephemeral groups. Moreover, the absence or very small contribution of 22:6ω3 in aquatic insects has been reported for a wide range of aquatic insect taxa [55–57]. As 22:6ω3 was less available from dietary sources, 22:6ω3 detected in *S. leucomaenis* must be biosynthesized from its precursor [11]. During elongation from 20:5ω3 to 22:6ω3, lighter carbon may be preferentially added. Consequently, 22:6ω3 isotopic value depleted compared with that of 20:5ω3 [27]. However, we found that almost the same or slightly enriched isotopic values in 22:6ω3 compared with that of 20:5ω3 (Supplementary file Figure S2). The same tendency was observed in several aquatic consumers, including fish, in Yenisei River [30]. Gladyshev et al. [30] pointed out the possibility that the acetate pool, which is required for elongation of fatty acid, is significantly enriched in ^{13}C compared with fatty acids. While further study is needed to comprehensively understand essential fatty acid dynamics in aquatic ecosystems, our results demonstrated that the main source of essential fatty acids in headstream food webs was autochthonous organic matter.

5. Conclusions

Dietary origin of total organic carbon and essential fatty acids for the predatory fish *S. leucomaenis* was investigated in the four headwater streams with the two hypotheses: (1) the dietary allochthonous contribution for fish in headwater stream food webs positively relate with canopy cover; and (2) essential fatty acids originate from autochthonous organic matter regardless of canopy cover. Our results indicated that autochthonous organic matters were the main dietary origin of not only essential fatty acids, but also total organic carbon regardless of canopy cover.

Author Contributions: Conceptualization, M.F., Y.M. and R.S.; Formal analysis, M.F.; Funding acquisition, M.F. and S.K.; Investigation, M.F., Y.M., R.S. and S.T.; Methodology, M.F., Y.M. and R.S.; Validation, M.F., Y. M. and R. S.; Visualization, M.F., Y. M. and R. S.; Writing – original draft, M.F.; Writing – review & editing, M.F., Y.M., R.S., S.T., T.S., N.M. and S.K.; Supervision, M.F.

References

1. Allan, J.D.; Castillo, M.M. *Stream ecology: Structure and function of running waters*; Springer: Dordrecht, The Netherlands, 2007.
2. Doi, H. Spatial patterns of autochthonous and allochthonous resources in aquatic food webs. *Popul. Ecol.* **2009**, *51*, 57–64. [CrossRef]
3. Minshall, G.W. Autotrophy in stream ecosystems. *BioScience* **1978**, *28*, 767–771. [CrossRef]
4. Lamberti, G.A.; Steinman, A.D. A comparison of primary production in stream ecosystems. *J. North Am. Benthol. Soc.* **1997**, *16*, 95–104. [CrossRef]
5. Vannote, R.L.; Minshall, G.W.; Cummins, K.W.; Sedell, J.R.; Cushing, C.E. The river continuum concept. *Can. J. Fish. Aquat. Sci.* **1980**, *37*, 130–137. [CrossRef]
6. Lau, D.C.P.; Leung, K.M.Y.; Dudgeon, D. Are autochthonous foods more important than allochthonous resources to benthic consumers in tropical headwater streams? *J. North. Am. Benthol. Soc.* **2009**, *28*, 426–439. [CrossRef]
7. Hayden, B.; McWilliam-Hughes, S.M.; Cunjak, R.A. Evidence for limited trophic transfer of allochthonous energy in temperate river food webs. *Freshw. Sci.* **2016**, *35*, 544–558. [CrossRef]
8. Doucett, R.R.; Power, G.; Barton, D.R.; Drimmie, R.J.; Cunjak, R.A. Stable isotope analysis of nutrient pathways leading to Atlantic salmon. *Can. J. Fish. Aquat. Sci.* **1996**, *53*, 2058–2066. [CrossRef]
9. Rosi-Marshall, E.J.; Wallace, J.B. Invertebrate food webs along a stream resource gradient. *Freshw. Biol.* **2002**, *47*, 129–141. [CrossRef]
10. Guo, F.; Bunn, S.E.; Brett, M.T.; Kainz, M.J. Polyunsaturated fatty acids in stream food webs—High dissimilarity among producers and consumers. *Freshw. Biol.* **2017**, *62*, 1325–1334. [CrossRef]
11. Guo, F.; Bunn, S.E.; Brett, M.T.; Fry, B.; Hager, H.; Ouyang, X.; Kainz, M.J. Feeding strategies for the acquisition of high-quality food sources in stream macroinvertebrates: Collecting, integrating, and mixed feeding. *Limnol. Oceanogr.* **2018**, *63*, 1964–1978. [CrossRef]
12. Watanabe, T.; Kitajima, C.; Fujita, S. Nutritional values of live organisms used in Japan for mass propagation of fish: A review. *Aquaculture* **1983**, *34*, 115–143. [CrossRef]
13. Glencross, B.D. Exploring the nutritional demand for essential fatty acids by aquaculture species. *Rev. Aquac.* **2009**, *1*, 71–124. [CrossRef]
14. Tocher, D.R. Fatty acid requirements in ontogeny of marine and freshwater fish. *Aquac. Res.* **2010**, *41*, 717–732. [CrossRef]
15. Smith, D.M.; Hunter, B.J.; Allan, G.L.; Roberts, D.C.K.; Booth, M.A.; Glencross, B.D. Essential fatty acids in the diet of silver perch (*Bidyanus bidyanus*): Effect of linolenic and linoleic acid on growth and survival. *Aquaculture* **2004**, *236*, 377–390. [CrossRef]
16. Sargent, J.R.; McEvoy, L.A.; Estevez, A.; Bell, J.G.; Bell, M.V.; Henderson, R.J.; Tocher, D.R. Lipid nutrition of marine fish during early development: Current status and future directions. *Aquaculture* **1999**, *179*, 217–229. [CrossRef]
17. Nowosad, J.; Kucharczyk, D.; Targońska, K. Enrichment of zebrafish *Danio rerio* (Hamilton, 1822) diet with polyunsaturated fatty acids improves fecundity and larvae quality. *Zebrafish* **2017**, *14*, 364–370. [CrossRef]

18. Brett, M.T.; Bunn, S.E.; Chandra, S.; Galloway, A.W.E.; Guo, F.; Kainz, M.J.; Kankaala, P.; Lau, D.C.P.; Moulton, T.P.; Power, M.E.; et al. How important are terrestrial organic carbon inputs for secondary production in freshwater ecosystems? *Freshw. Biol.* **2017**, *62*, 833–853. [CrossRef]

19. Taipale, S.; Strandberg, U.; Peltomaa, E.; Galloway, A.W.E.; Ojala, A.; Brett, M.T. Fatty acid composition as biomarkers of freshwater microalgae: Analysis of 37 strains of microalgae in 22 genera and in seven classes. *Aquat. Microb. Ecol.* **2013**, *71*, 165–178. [CrossRef]

20. Cattaneo, A.; Kerimian, T.; Roberge, M.; Marty, J. Periphyton distribution and abundance on substrata of different size along a gradient of stream trophy. *Hydrobiologia* **1997**, *354*, 101–110. [CrossRef]

21. DeNiro, M.J.; Epstein, S. Influence of diet on the distribution of carbon isotopes in animals. *Geochim. Cosmochim. Acta* **1978**, *42*, 495–506. [CrossRef]

22. Finlay, J.C. Stable-carbon-isotope ratios of river biota: Implications for energy flow in lotic food webs. *Ecology* **2001**, *82*, 1052–1064. [CrossRef]

23. Minagawa, M.; Wada, E. Stepwise enrichment of 15N along food chains: Further evidence and the relation between δ15N and animal age. *Geochim. Cosmochim. Acta* **1984**, *48*, 1135–1140. [CrossRef]

24. Zanden, M.; Vander, J.; Rasmussen, J.B. Variation in δ15N and δ13C trophic fractionation: Implications for aquatic food web studies. *Limnol. Oceanogr.* **2001**, *46*, 2061–2066. [CrossRef]

25. Vanderklift, M.A.; Ponsard, S. Sources of variation in consumerdiet δ15N enrichment: A meta-analysis. *Oecologia* **2003**, *136*, 169–182. [CrossRef] [PubMed]

26. Budge, S.M.; AuCoin, L.R.; Ziegler, S.E.; Lall, S.P. Fractionation of stable carbon isotopes of tissue fatty acids in Atlantic Pollock (*Pollachius virens*). *Ecosphere* **2016**, *7*, e01437. [CrossRef]

27. Fujibayashi, M.; Ogino, M.; Nishimura, O. Fractionation of the stable carbon isotope ratio of essential fatty acids in zebrafish *Danio rerio* and mud snails *Bellamya chinensis*. *Oecologia* **2016**, *180*, 589–600. [CrossRef] [PubMed]

28. Budge, S.M.; Wooller, M.J.; Springer, A.M.; Iverson, S.J.; McRoy, C.P.; Divoky, G.J. Tracing carbon flow in an arctic marine food web using fatty acid-stable isotope analysis. *Oecologia* **2008**, *157*, 117–129. [CrossRef] [PubMed]

29. Fujibayashi, M.; Okano, K.; Takada, Y.; Mizutani, H.; Uchida, N.; Nishimura, O.; Miyata, N. Transfer of cyanobacterial carbon to a higher trophic-level fish community in a eutrophic lake food web: Fatty acid and stable isotope analyses. *Oecologia* **2018**, *188*, 901–912. [CrossRef] [PubMed]

30. Gladyshev, M.I.; Sushchik, N.N.; Kalachova, G.S.; Makhutova, O.N. Stable isotope composition of fatty acids in organisms of different trophic levels in the Yenisei River. *PLoS One* **2012**, *7*, e34059. [CrossRef] [PubMed]

31. Bec, A.; Perga, M.E.; Koussoroplis, A.; Bardoux, G.; Desvilettes, C.; Bourdier, G.; Mariotti, A. Assessing the reliability of fatty acid-specific stable isotope analysis for trophic studies. *Methods Ecol. Evol.* **2011**, *2*, 651–659. [CrossRef]

32. Gladyshev, M.I.; Makhutova, O.N.; Kravchuk, E.S.; Anishchenko, O.V.; Sushchik, N.N. Stable isotope fractionation of fatty acids of Daphnia fed laboratory cultures of microalgae. *Limnologica* **2016**, *56*, 23–29. [CrossRef]

33. Nielsen, J.M.; Clare, E.L.; Hayden, B.; Brett, M.T.; Kratina, P. Diet tracing in ecology: Method comparison and selection. *Methods Ecol. Evol.* **2018**, *9*, 278–291. [CrossRef]

34. Sakamaki, T.; Richardson, J.S. Biogeochemical properties of fine particulate organic matter as an indicator of local and catchment impacts on forested streams. *J. Appl. Ecol.* **2011**, *48*, 1462–1471. [CrossRef]

35. Sakamaki, T.; Shum, J.Y.T.; Richardson, J.S. Watershed effects on chemicalproperties of sediment and primary consumption in estuarine tidal flats:importance of watershed size and food selectivity by macrobenthos. *Ecosystems* **2010**, *13*, 328–337. [CrossRef]

36. Fujibayashi, M.; Sakamaki, T.; Shin, W.; Nishimura, O. Food utilization of shell-attached algae contributes to the growth of host mud snail, *Bellamya chinensis*: Evidence from fatty acid biomarkers and carbon stable isotope analysis. *Limnologica* **2016**, *57*, 66–72. [CrossRef]

37. Takenaka, A. CanopOn 2: Hemispherical photography analysis program. Available online: http://takenaka-akio.org/etc/canopon2/index.html. (accessed on 30 June 2019).

38. Tsuda, M. Consideration on the contents of alimentary canals of Salvelinus leycomaenis pluvius and Oncorhynchus masou f. masou from Ina-gawa River System. *Jpn. J. Limnol.* **1967**, *28*, 51–55. [CrossRef]

39. Kato, F. Ecological and morphological notes on the charr, Salvelinus leucomaenis in the Nagara River and Ibi River systems. *SUISANZOSHOKU* **1992**, *40*, 145–152.

40. Abdulkadir, S.; Tsuchiya, T. One-step method for quantitative and qualitative analysis of fatty acids in marine animal samples. *J. Exp. Mar. Biol. Ecol* **2008**, *354*, 1–8. [CrossRef]

41. Sakamaki, T.; Richardson, J.S.; Arnott, S. Nonlinear variation of stream-forest linkage along a stream-size gradient: An assessment using biogeochemical proxies of in-stream fine particulate organic matter. *J. Appl. Ecol.* **2013**, *50*, 1019–1027. [CrossRef]

42. Collins, S.M.; Kohler, T.J.; Thomas, S.A.; Fetzer, W.W.; Flecker, A.S. The importance of terrestrial subsidies in stream food webs varies along a stream size gradient. *Oikos* **2016**, *125*, 674–685. [CrossRef]

43. Ishikawa, N.F.; Yamane, M.; Suga, H.; Ogawa, N.O.; Yokoyama, Y.; Ohkouchi, N. Chlorophyll a-specific $\Delta^{14}C$, $\delta^{13}C$ and $\delta^{15}N$ values in stream periphyton: Implications for aquatic food web studies. *Biogeosciences* **2015**, *12*, 6781–6789. [CrossRef]

44. Wang, L.; Wu, F.; Xion, Y.; Fang, J. Origin and vertical variation of the bound fatty acids in core sediment of Lake Dianchi in Southwest China. *Env. Sci. Poll. Res.* **2013**, *20*, 2390–2397. [CrossRef] [PubMed]

45. Tall, L.; Cattaneo, A.; Cloutier, L.; Dray, S.; Legendre, P. Resource partitioning in a grazer guild feeding on a multilayer diatom mat. *J. North. Am. Benthol. Soc.* **2006**, *25*, 800–810. [CrossRef]

46. Finlay, J.C. Patterns and controls of lotic algal stable carbon isotope ratios. *Limnol. Oceanogr.* **2004**, *49*, 850–861. [CrossRef]

47. Hill, W.R.; Middleton, R.G. Changes in carbon stable isotope ratios during periphyton development. *Limnol. Oceanogr.* **2006**, *51*, 2360–2369. [CrossRef]

48. Huang, I.Y.; Lin, Y.S.; Chen, C.P.; Hsieh, H.L. Food web structure of a subtropical headwater stream. *Mar. Freshw. Res.* **2007**, *58*, 596–607. [CrossRef]

49. Lewis, W.M., Jr.; Hamilton, S.K.; Rodríguez, M.A.; Saunders, J.F., III; Lasi, M.A. Foodweb analysis of the Orinoco floodplain based on production estimates and stable isotope data. *J. North. Am. Benthol. Soc.* **2001**, *20*, 241–254. [CrossRef]

50. Crenier, C.; Arce-Funck, J.; Bec, A.; Billoir, E.; Perrière, F.; Leflaive, J.; Guérold, F.; Felten, V.; Danger, M. Minor food sources can play a major role in secondary production in detritus-based ecosystems. *Freshw. Biol.* **2017**, *62*, 1155–1167. [CrossRef]

51. Nakano, S.; Fausch, K.D.; Kitano, S. Flexible niche partitioning via a foraging mode shift: A proposed mechanism for coexistence in stream-dwelling charts. *J. Anim. Ecol.* **1999**, *68*, 1079–1092. [CrossRef]

52. Gladyshev, M.I.; Arts, M.T.; Sushchik, N.N. Preliminary estimates of the export of omega-3 highly unsaturated fatty acids (EPA+DHA) from aquatic to terrestrial ecosystems. In *Lipids in Aquatic Ecosystems*; Kainz, M., Brett, M.T., Arts, M.T., Eds.; Springer: New York, NY, USA, 2009; pp. 179–210.

53. Logan, J.M.; Jardine, T.D.; Miller, T.J.; Bunn, S.E.; Cunjak, R.A.; Lutcavage, M.E. Lipid corrections in carbon and nitrogen stable isotope analyses: Comparison of chemical extraction and modelling methods. *J. Anim. Ecol.* **2008**, *77*, 838–846. [CrossRef]

54. Hastings, N.; Agaba, M.; Tocher, D.R.; Leaver, M.J.; Dick, J.R.; Sargent, J.R.; Teale, A.J. A vertebrate fatty acid desaturase with $\Delta 5$ and $\Delta 6$ activities. *PNAS* **2001**, *98*, 14304–14309. [CrossRef] [PubMed]

55. Sushchik, N.N.; Gladyshev, M.I.; Moskvichova, A.V.; Makhutova, O.N.; Kalachova, G.S. Comparison of fatty acid composition in major lipid classes of the dominant benthic invertebrates of the Yenisei river. *Comp. Biochem. Physiol. B* **2003**, *134*, 111–122. [CrossRef]

56. Torres-Ruiz, M.; Wehr, J.D.; Perrone, A.A. Are net-spinning caddisflies what they eat? An investigation using controlled diets and fatty acids. *J. North. Am. Benthol. Soc.* **2010**, *29*, 803–813. [CrossRef]

57. Guo, F.; Kainz, M.J.; Sheldon, F.; Bunn, S.E. Effects of light and nutrients on periphyton and the fatty acid composition and somatic growth of invertebrate grazers in subtropical streams. *Oecologia* **2016**, *181*, 449–462. [CrossRef] [PubMed]

Permissions

The contributors of this book come from diverse backgrounds, making this book a truly international effort. This book will bring forth new frontiers with its revolutionizing research information and detailed analysis of the nascent developments around the world.

We would like to thank all the contributing authors for lending their expertise to make the book truly unique. They have played a crucial role in the development of this book. Without their invaluable contributions this book wouldn't have been possible. They have made vital efforts to compile up to date information on the varied aspects of this subject to make this book a valuable addition to the collection of many professionals and students.

This book was conceptualized with the vision of imparting up-to-date information and advanced data in this field. To ensure the same, a matchless editorial board was set up. Every individual on the board went through rigorous rounds of assessment to prove their worth. After which they invested a large part of their time researching and compiling the most relevant data for our readers.

The editorial board has been involved in producing this book since its inception. They have spent rigorous hours researching and exploring the diverse topics which have resulted in the successful publishing of this book. They have passed on their knowledge of decades through this book. To expedite this challenging task, the publisher supported the team at every step. A small team of assistant editors was also appointed to further simplify the editing procedure and attain best results for the readers.

Apart from the editorial board, the designing team has also invested a significant amount of their time in understanding the subject and creating the most relevant covers. They scrutinized every image to scout for the most suitable representation of the subject and create an appropriate cover for the book.

The publishing team has been an ardent support to the editorial, designing and production team. Their endless efforts to recruit the best for this project, has resulted in the accomplishment of this book. They are a veteran in the field of academics and their pool of knowledge is as vast as their experience in printing. Their expertise and guidance has proved useful at every step. Their uncompromising quality standards have made this book an exceptional effort. Their encouragement from time to time has been an inspiration for everyone.

The publisher and the editorial board hope that this book will prove to be a valuable piece of knowledge for researchers, students, practitioners and scholars across the globe.

List of Contributors

Jiapeng Chen
Key Laboratory of Marine Drugs, Chinese Ministry of Education, School of Medicine and Pharmacy, Ocean University of China, Qingdao 266003, China

Hongbing Liu
Key Laboratory of Marine Drugs, Chinese Ministry of Education, School of Medicine and Pharmacy, Ocean University of China, Qingdao 266003, China
Laboratory for Marine Drugs and Bioproducts, Pilot National Laboratory for Marine Science and Technology (Qingdao), Qingdao 266237, China

Mónica Estupiñán, M. Elisabete Bilbao, Iñaki Mendibil and Laura Alonso-Sáez
AZTI, Marine Research Division, Txatxarramendi Irla s/n, 48395 Sukarrieta, Spain

Igor Hernández, Eduardo Saitua and Jorge Ferrer
AZTI, Food Research Division, Astondo Bidea, Building 609, 48160 Derio, Spain

Naren Gajenthra Kumar
Department of Microbiology and Immunology, School of Medicine, Virginia Commonwealth University, Richmond, VA 23298, USA

Elvin T. Price, Benjamin Van Tassell, Donald F. Brophy and Daniel Contaifer
Department of Pharmacotherapy and Outcomes Sciences, School of Pharmacy, Virginia Commonwealth University, Richmond, VA 23298, USA

Parthasarathy Madurantakam
Department of General Practice, School of Dentistry, Virginia Commonwealth University, Richmond, VA 23298, USA

Salvatore Carbone
Department of Kinesiology & Health Sciences, College of Humanities & Sciences, Virginia Commonwealth University, Richmond, VA 23220, USA
VCU Pauley Heart Center, Department of Internal Medicine, Virginia Commonwealth University, Richmond, VA 23298, USA

Dayanjan S. Wijesinghe
Department of Pharmacotherapy and Outcomes Sciences, School of Pharmacy, Virginia Commonwealth University, Richmond, VA 23298, USA
da Vinci Center, Virginia Commonwealth University, Richmond, VA 23220, USA

Institute for Structural Biology, Drug Discovery and Development, Virginia Commonwealth University School of Pharmacy, Richmond, VA 23298, USA

Sydney Moyo
Department of Oceanography and Coastal Sciences, Louisiana State University, Baton Rouge, LA 70803, USA

Zuzana Bláhová, Martin Pšenička and Jan Mráz
South Bohemian Research Center of Aquaculture and Biodiversity of Hydrocenoses, Faculty of Fisheries and Protection of Waters, University of South Bohemia in České Budějovice, Zátiší 728/II, 389 25 Vodňany, Czech Republic

Thomas Nelson Harvey
Centre for Integrative Genetics (CIGENE), Department of Animal and Aquacultural Sciences, Faculty of Biosciences, Norwegian University of Life Sciences, 1430 Ås, Norway

Gianfranca Carta, Elisabetta Murru, Claudia Manca and Sebastiano Banni
Department of Biomedical Sciences, University of Cagliari, 09042 Monserrato, CA, Italy

Andrea Serra and Marcello Mele
Department of Agriculture, Food and Environment, University of Pisa, 56124 Pisa, Italy

Anzhelika A. Kolmakova
Institute of Biophysics of Federal Research Center "Krasnoyarsk Science Center" of Siberian Branch of Russian Academy of Sciences, Akademgorodok, 50/50, Krasnoyarsk 660036, Russia

Nadezhda N. Sushchik, Olesia N. Makhutova, Anastasia E. Rudchenko and Michail I. Gladyshev
Institute of Biophysics of Federal Research Center "Krasnoyarsk Science Center" of Siberian Branch of Russian Academy of Sciences, Akademgorodok, 50/50, Krasnoyarsk 660036, Russia
Siberian Federal University, Svobodny av., 79, Krasnoyarsk 660041, Russia

Larisa A. Glushchenko and Svetlana P. Shulepina
Siberian Federal University, Svobodny av., 79, Krasnoyarsk 660041, Russia

Chun-Kuang Shih, Ngan Thi Kim Nguyen, Amalina Shabrina and Te-Hsuan Tung
School of Nutrition and Health Sciences, Taipei Medical University, Taipei 11031, Taiwan

Yu-Tang Tung
Graduate Institute of Metabolism and Obesity Sciences, Taipei Medical University, Taipei 11031, Taiwan

I-Hsuan Lin
Research Center of Cancer Translational Medicine, Taipei Medical University, Taipei 11031, Taiwan

Shih-Yi Huang
School of Nutrition and Health Sciences, Taipei Medical University, Taipei 11031, Taiwan
Graduate Institute of Metabolism and Obesity Sciences, Taipei Medical University, Taipei 11031, Taiwan
Center for Reproductive Medicine & Sciences, Taipei Medical University Hospital, Taipei 11031, Taiwan

Khalid Abdullah Al-Ghanim, Fahad Al-Misned and Zubair Ahmed
Department of Zoology, College of Science, King Saud University, Riyadh 11451, Saudi Arabia

Shahid Mahboob
Department of Zoology, College of Science, King Saud University, Riyadh 11451, Saudi Arabia
Department of Zoology, Government College University, Faisalabad-38000, Pakistan

Salma Sultana, Tayyaba Sultan and Bilal Hussain
Department of Zoology, Government College University, Faisalabad-38000, Pakistan

Tehniat Shahid
House No. 423, Block M-1, Street No. 14, Lake city, Lahore 55150, Pakistan

Sami Taipale
Department of Biological and Environmental Science, Nanoscience center, University of Jyväskylä, 40014 Jyväskylä, Finland

Elina Peltomaa
Institute of Atmospheric and Earth System Research (INAR)/Forest Sciences, University of Helsinki, 00014 Helsinki, Finland
Helsinki Institute of Sustainability Science (HELSUS), University of Helsinki, 00014 Helsinki, Finland

Pauliina Salmi
Faculty of Information Technology, University of Jyväskylä, FI-40014 Jyväskylän, Finland

Klim A. Petrov
Institute for Biological Problems of Cryolithozone of Siberian Branch of the Russian Academy of Sciences, 41 Lenina av., Yakutsk 677000, Russia

Lyubov V. Dudareva
Siberian Institute of Plant Physiology and Biochemistry, Siberian Branch of Russian Academy of Sciences, 132 Lermontova str., Irkutsk 664033, Russia

Vasiliy V. Nokhsorov
North-Eastern Federal University, 48 Kulakovskogo str., Yakutsk 677000, Russia

Kirill N. Stoyanov
Siberian Federal University, 79 Svobodny pr., Krasnoyarsk 660041, Russia

Natalia V. Zhukova
National Scientific Center of Marine Biology, Far East Branch of the Russian Academy of Sciences, Vladivostok 690041, Russia
School of Biomedicine, Far Eastern Federal University, Vladivostok 690950, Russia

Yoshie Miura, Reina Suganuma and Naoyuki Miyata
Faculty of Bioresource Sciences, Akita Prefectural University, Kaidobata-Nishi 241-438, Shimoshinjo Nakano, Akita city, Akita Prefecture 010-0195, Japan

Megumu Fujibayashi
Faculty of Bioresource Sciences, Akita Prefectural University, Kaidobata-Nishi 241-438, Shimoshinjo Nakano, Akita city, Akita Prefecture 010-0195, Japan
Department of Urban and Environmental Engineering, Faculty of Engineering, Kyushu University, 744, Motooka, Nishiku, Fukuoka 819-0395, Japan

Shinji Takahashi
Technical Division, School of Engineering, Tohoku University, 6-6-06 Aoba, Aoba, Sendai, Miyagi 980-8579, Japan

Takashi Sakamaki
Department of Civil and Environmental Engineering, School of Engineering, Tohoku University, 6-6-06 Aoba, Aoba, Sendai, Miyagi 980-8579, Japan

Index

Printed in the USA
CPSIA information can be obtained
at www.ICGtesting.com
JSHW051407091023
49903JS00006B/315

9 781641 167697